Water Co-Management

Water Co-Management

Editors

Velma I. Grover
Adjunct Professor
Faculty of Environmental Studies
York University
Toronto, ON
Canada

Gail Krantzberg
Professor and Director
Centre for Engineered and Public Policy
School of Engineering Practice
McMaster University
Hamilton, ON
Canada

CRC Press
Taylor & Francis Group
Boca Raton London New York

CRC Press is an imprint of the
Taylor & Francis Group, an **informa** business
A SCIENCE PUBLISHERS BOOK

CRC Press
Taylor & Francis Group
6000 Broken Sound Parkway NW, Suite 300
Boca Raton, FL 33487-2742

Printed in the United States of America on acid-free paper

International Standard Book Number: 978-1-4665-7843-2 (Hardback)

Library of Congress Cataloging-in-Publication Data

Water co-management / editors, Velma I. Grover, Gail Krantzberg.
pages cm
Includes bibliographical references and index.
ISBN 978-1-4665-7843-2 (hardback)
1. Watershed management--International cooperation. 2. Water-supply--Management--International cooperation. 3. Natural resources--Co-management. I. Grover, Velma I.
TC409.W3674 2013
333.91--dc23

2012039601

Visit the Taylor & Francis Web site at
http://www.taylorandfrancis.com

Science Publishers Web site at
http://www.scipub.net

CRC Press Web site at
http://www.crcpress.com

Preface

Nothing truly valuable can be achieved
except by the unselfish cooperation of many individuals.
— Albert Einstein, 1940

Although, the term co-management is of recent origin, its use can be traced back to the late 1970s in the management of salmon under the Boldt Decision by the U.S. Treaty Tribes in Washington Statebutthecurrentpractice of power sharing in resource management can be traced back to the 1890s (Berkes 2009).

Co-management is a term so often used interchangeably with other terms such as stakeholders' engagement and public participation; however it is slightly different from both these terms, in that arrangements are made for managing the resources by sharing decision-making among stakeholders. The papers by Berkes (2009) and Carlsson and Berkes (2005) inspired us to look more closely at the concept of co-management and since our work is in the field of water, the result is this book on "water co-management". The aim of the book is to bring out the differences between water co-management and public participation in local water issues. Though co-management has been applied successfully in other sectors of natural resources management such as forestry, the focus in this book is on river basins, coastal areas, fisheries and groundwater co-management.

Implementation of co-management takes time and understanding among stakeholders but it results in more sustainable and long lasting solution to complex issues. Most of the examples in the book refer to co-management of water resources but in a move towards more holistic or integrated approach, the ideal condition will be to look at different sectors together. One of the forward looking chapters in the book also explores co-managing water and energy resources together (focus of discussion in the chapter is more on how to co-manage different sectors rather than co-management in strict sense of involving stakeholders). Since there will be conflict between old knowledge, cultural differences and existing way of doing things with new recommended co-management model, there is a need to be flexible and adapt with lessons being learnt within the process itself; thus the practice of adaptive co-management becomes crucial to make the planning process iterative based on ground reality and lessons being learnt along the way. This becomes all the more important as we move towards co-management since it is no longer top-down inflexible policy or plan being implemented, but rather a more flexible approach to management practices. Some authors have also explained the importance and use of adaptive co-management in this book.

The audience of the book includes everyone who, in one way or the other, is involved at the local, regional, national or international level in the governance and management of water resources and is searching for alternate water management methods and principles.The wealth of examples and tools discussed in the book in applying co-management in different geographical, ecological and cultural settings can also be used by academic community (both faculty and students) to further explore lessons learnt from water co-management in different areas and how to build on the success or failures of the examples provided.

Velma I. Grover
Gail Krantzberg
Hamilton, ON, Canada

Berkes, Fikret. 2009. Evolution of co-management: Role of knowledge generation, bridging organizations and social learning. Journal of Environmental Management 90 pp. 1692–1702.

Carlsson, L. and F. Berkes. 2005. Co-management: concepts and methodological implications. Journal of Environmental Management 75 pp. 65–76.

Contents

1

Co-Management
Principles and Complexities

Velma I. Grover[1,*] and *Gail Krantzberg*[2]

Introduction

Sustainability, sustainable management or integrated management of natural resources (also known as common-pool resources) have become buzz words in the growing literature around environmental management. Researchers now focus on complex and dynamic linkages between ecological systems, social, and economic aspects of society to achieve sustainability. A lot of researchers focus on case studies to show the existence of various management systems of natural resources while others focus on theoretical frameworks of management systems that support sustainable use of such resources. The concept of "co-management" has emerged from both types of research. This book focuses on co-management of one of such common-pool resources "water". An attempt has been made to combine research with action. The idea is that success and failures described in case studies with lessons learnt can be useful stepping stones for both policy makers and managers to apply to their own situation. This chapter briefly touches on some definitions, principles and complexities involved in co-management (with reference to adaptive co-management) followed by a road-map of the book.

[1]Professor, Adjunct, York University, Toronto, Canada.
Email: velmaigrover@yahoo.com
[2]Professor and Director of the Centre for Engineering and, Public Policy in the School of Engineering Practice, McMaster University, ETB 510, 1280 Main St. W, Hamilton, ON, L8S 4L7.
Email: Krantz@mcmaster.ca
*Corresponding author

Co-management—or sharing of power to manage—links community leaders with government agencies to enable users/stakeholders to fully engage in local decision making. Since the book is about water co-management, we can imply from the above definition that: as applied to water management issues, it means managing water collaboratively (to include those who have an interest in and a legal right to revitalize and protect precious resources like fisheries, and water levels in lakes or rivers).

Stakeholder engagement and co-management are similar words and are used interchangeably but some authors have made a clear distinction between these two terms. Stakeholder or multi-stakeholder arrangements usually have strong horizontal linkages among user groups at the same level of organization, as well as vertical linkages across levels of organization, between stakeholders, government and government agencies. By comparison, co-management has at least one strong vertical linkage involving the government and a user group and some formalized arrangement for power sharing. In other words co-management needs to have some kind of institutionalized arrangement for intensive user participation in decision-making (Berkes 2009).

Although, the term co-management is of recent origin, its use can be traced to the late 1970s in the management of salmon under the Boldt Decision by the US Treaty Tribes in Washington state, however the practice of power sharing in resource management can be traced back to 1890s (Berkes 2009).

Definitions of Co-Management

While the word co-management has been used over centuries, there is no one definition of co-management. Co-management has been defined differently by different authors over a period of time. Some of the definitions are:

> "Power-sharing in the exercise of resource management between a government agency and a community organization of stakeholders" (Pinkerton 1992).
>
> "At the heart of what we understand as 'co-management' ...[is] a process of collective understanding and action by which human communities and other social actors manage natural resources and ecosystems together, drawing from everyone's unique strengths, vantage points and capacities" (Borrini-Feyerabend el al. 2004).

Collaborative management, or co-management, has been defined as:

> "The sharing of power and responsibility between the government and local resource users" (Berkes et al. 1991).

Singleton (1998) defines co-management as:

> "The term given to governance systems that combine state control with local, decentralized decision making and accountability and which, ideally, combine the strengths and mitigate the weaknesses of each."
>
> "The sharing of responsibilities, rights and duties between the primary stakeholders, in particular, local communities and the nation state; a decentralized approach to decision making that involves the local users in the decision making process as equals with the nation-state" (TheWorld Bank 1999).

Different Aspects of Co-Management

Based on different definitions and explanations, there are a few common themes, aspects or facets of co-management that emerge. Six different aspects of co-management are described below (based on Berkes 2009):

- **Co-management as power sharing:** As can be seen in the definitions above, one of the core elements of co-management is power sharing. Coming to an equitable power sharing can be a challenge but co-management attempts to enable it.
- **Co-management as institution building:** Co-management often involves the working together of government and non-government bodies. In some cases when co-management is being implemented for the first time the institutions (and people) might not be ready for partnerships or do not have the capacity to interact. Evolution of co-management in a conducive environment (where favorable policy arrangements can help two-way feedback between government policy and local institutions) helps build the capacity of institutions to interact with each other as well.
- **Co-management as trust and social capital:** Co-management requires the working together of various stakeholders under certain power sharing arrangements. This not only requires institutional capacity building but building of trust between parties to negotiate and resolve conflicts to co-manage a resource. Co-management also helps learning by participatory approaches and provides a platform for single loop and double loop learning that can result in fundamental changes in behavior. Successive loops of learning-as-participation helps combine certain elements of adaptive management and co-management, with each cycle starting with observation and problem identification. With the help of either bridging organizations or other groups established under co-management arrangement, problem-solving networks organize themselves in identifying problems and opportunities leading

to solution formulations (and this becomes an iterative process with each cycle of problem identification and solution formulation providing lessons learnt).

- **Co-management as a process:** Co-management requires formal or institutionalized power sharing arrangement and thus is a process that involves institution building, development of trust and social capital. Arrangements are formed as a result of a process of negotiations and deliberations. Hence co-management is not an end point but a process to build relationships (which are constantly changing).
- **Co-management as problem solving:** In this case co-management is seen as task-oriented, concentrating mainly on the function rather than formal structure of the arrangement. Co-management as problem solving enables parties to transfer learning from one situation to another, and tackle increasingly more complex problems.
- **Co-management as governance:** Co-management as a governance tool essentially means involvement of various public and private stakeholders with multiple linkages across levels and domains with overlapping centers of authority. This type of institutional dynamics is also suitable for adaptive co-management and governance.

Other authors (Pahl-Wostl and Hare 2004) have added another aspect of co-management: **co-management as social learning.** As human dimension and governance issues take more important roles in management of natural resources, understanding collective decision-making and social learning becomes central. A social learning approach for resource management combines "content management as well as a social involvement process to achieve both technical and relational outcomes". The authors discuss strategies for urban water management in Switzerland and emphasize the role of social learning for management regimes that require changes in social practices, roles and responsibilities (Pahl-Wostl and Hare 2004). Social learning is important to understand transformation processes in human-technology-environment systems which again emphasizes linking scientific exchange and joint research among different disciplines

Co-management can be also looked at as web of networks.

WHY and WHEN Co-Management should be Considered

The need for co-management arises when (Borrini-Feyerabend et al. 2004):

- There are conflicts in sharing a common resource such as water;
- There is an increased complexity and uncertainty of ecosystem and natural resource management questions;

- There is an emerging interest in good governance principles and processes;
- Globalization and decentralization phenomena is on rise.

Co-management is not a solution for all problems. For example co-management can bring about some degree of power sharing but cannot fully eliminate the existing power structure (Calrson and Berkes 2005). However, as discussed by Calrson and Berkes, co-management can be useful in situations:

- when it comes to allocation of tasks (for example management systems involve operating at both small and large scale with both systems needing different skills and knowledge capacity. The co-management process brings all these skills together and tasks can be allocated to the right group);
- to facilitate exchange of resources between the different actors involved;
- in linking different types and levels of organization (unlike Max Weber's interpretation of bureaucracy where various layers of organization are linked with each other within a framework of hierarchy, co-management offers a platform where representatives of different types and levels of organizations coordinate their activities in relation to a resource system without hierarchy);
- in reduction of transaction costs; When organizations come together for co-management, costs for a few overlapping items (for example, maybe two organizations were collecting same or similar data—this duplication can be reduced/eliminated thus reducing the cost) can be carried out by one organization instead of all of them doing similar tasks. This will reduce the cost in general;
- in risk sharing (just like in agriculture where diversified crops are grown, just in case one crop fails there will still be a resource base for their living—in co-management governance/management system the dependence or reliance is not just on one actor but is shared among different actors and institutions);
- in establishing conflict resolution mechanisms (to establish a co-management system requires codification of rights and responsibilities of involved parties negotiation and a bargaining process is involved which helps establish relationship between involved parties, this also works as a means of reducing or resolving conflicts).

Benefits of Co-Management

Since neither local users nor the government can manage the complex natural resources, partnerships between different stakeholders and actors are important for natural resource management. Also since various natural resources (and problems) are cross scale and interlinked, they need to be solved concurrently. Co-management seems to be an answer for such governance and management challenges, though the arrangements and partnerships can vary for different resources and different locations (Calrson and Berkes (2005)). Some of the benefits of co-management include:

1. Decentralizes the decision making process (Pinkerton 1989);
2. Participatory democratic process (Pinkerton 1989);
3. Consensus building and less conflict (Pinkerton 1989);
4. Leads to trust building, inclusive decision-making and better data collection (Pinkerton 1989);
5. Enhancement of long term planning (Pinkerton 1989);
6. Allocation decisions (Pinkerton 1989);
7. Protection of resources from environmental damage (Pinkerton 1989);
8. Enforcement of regulations (Pinkerton 1989);
9. Bridging organizations and leadership (Berkes 2009): science and local knowledge can be brought together "by bridging organizations that provide an arena for knowledge co-production, trust building, sense making, learning, vertical and horizontal collaboration, and conflict resolution". Lessons learnt from all the successful case studies, show that bridging and knowledge co-production are two key characteristics of successful co-management systems and bridging organizations and leaderships are key factors that enable a co-management system to deal with knowledge issues.

Challenges in Applying Co-Management

As pointed out by Calrson and Berkes (2005), there are a number of complexities rarely accounted for in the conventional conceptualizations of co-management. Some of these complexities include (based on Calrson and Berkes 2005):

- *Complexities of the State*: The State is fragmented and is multifaceted consisting of various authorities and agencies responsible for management of a particular natural resource. Hence the State is not a unitary organization either in its structure or function. When we look at water management/governance a rich web of relations and agreements link different parts of the public sector to an equally heterogeneous set of private actors. So the State is not one entity but a complex set of public and private actors.
- *Complexities of the community*: Communities are not homogenous units, they have different interests based on their gender, socio-economic background and ethnicity. Communities are multi-dimensional, cross-

scale social political units that keep changing with highly unpredictable behavior. Communities also vary in their response to co-management, for example, in some cases they unite on a problem and respond with a single voice while in other cases there is a lack of cohesion in any response.

- *Complexities of the dynamic and iterative nature of the system*: Since there are many stakeholders involved in management of dynamic, complex natural resource, the process of governance (and co-management) needs readjustment (adaptive co-management).
- *Complexities of the conditions available to support the system*: "[C] ollaborative arrangements are highly dependent on to what extent parties recognize the legitimacy of one another. Thus, successful co-management depends on whether external circumstances are conducive for developing such systems. These exogenous factors involve some sense of security of resource tenure, their right to organize, availability of appropriate financial resources, facilitation support and so on, together constituting an important determinant extraneous to the characteristics of the parties or the natural resources they are supposed to manage."
- *Complexities of co-management as a governance system*: Most of the resource management regimes are embedded in some institutional context or other. Usually there are three layers of rules shaping any institutional arrangement: constitutional rules (mainly specify terms of governance, defining who possesses the decision-making right for use of a resource or share its benefits), collective choice rules (regulating how decisions are made) and operational rules (mainly considers daily activity level). These three rules form a hierarchy with constitutional rules at the top and operational rules at the bottom. "Even though co-management might be demonstrated in the day-to-day activities (guided by operational rules) of a community of users, these are performed under the umbrella of collective choice rules. Thus, constitutional rules set the framework for decision- making on the middle level where co-management is predominantly exercised. The difference between operational rights and collective choice rights is crucial. Ostrom and Schlager emphasize that 'it is the difference between exercising a right and participation in the definition to future right to be exercised (Ostrom and Schlager 1996)." So co-management systems can be seen as governance structures composed of a variety of actors related to each other by a number of relations that includes the State, local resource users, commercial users, public and private actors, non-governmental organizations.
- Complexities as a process of adaptive learning and problem solving, and

- Complexities of the ecosystem that provides the resources that are being managed: Eco-systems do not behave linearly, and are very uncertain and complex systems. The complexity impacts the governance models including co-management of natural resources. To incorporate uncertainties and complexities, adaptive management is/should be preferred over command and control style resource management as it offers more flexible options of governance.

In terms of the relationship to water—water co-management can be applied to water governance in general, to water governance along coastal regions, along rivers, for near shore zone governance and fisheries (among others).

Road Map for the Book

As discussed in the above section co-management recognizes that various natural resources are linked together and for a more effective management it is important to understand the linkages and manage the interlinked natural resources together. In the chapter following this Introduction **"Sustainable Development and the Water Energy Nexus"** Nasr explores linkages between water and energy and the need for co-management of the two systems. As explained by the author, both water and energy cycles are inter-linked. In the natural water cycle—water evaporates from water bodies and comes down in some form of precipitation making it a closed cycle. In earlier days water needs of humans were limited to drinking, some irrigation and navigation. However increased use of water for drinking, irrigation, industries and also to generate energy has created cracks in the natural water cycle. Humans now use energy to extract water from the natural cycle, which they use for more energy generation, as well as for agricultural and their own domestic freshwater needs. Most of this water is returned into the system but a large proportion of this water is in the form of untreated sewage, whose treatment requires yet more energy for treatment. Additional pollutants are emitted as part of human activity, and they also affect natural processes of evaporation, condensation, and precipitation. As a result, the link between the two water cycles, the natural and human, is now largely mediated by human energy consumption and production. By the middle of the 20th century, as human development accelerated, a "Water-Energy Nexus" had emerged. By the end of the 20th century, the water-energy nexus had come "full circle", as increased energy demands brought about the development of biofuels. In many cases, such biofuels are derived from agricultural products that are grown not for food, but for the exclusive purpose of energy generation. Energy is now competing for water with food. Human development is now intractably linked to this symbiotic relationship between water and energy. For this development to be sustainable, water and energy must be managed jointly (co-management of water and energy systems).

In the next chapter, **"Working Toward Co-Management of the Raccoon River Watershed in Iowa: The Role of Civil Society"** Flora and Delaney look at the emergence of a civil society organization, the Raccoon River Watershed Association (RRWA) and its role in co-management of the watershed. The authors examine how grass-roots can contribute to co-management to the degree a civil society organization can contribute to pluralistic management through collaboration with state and market institutions, sharing the benefits and responsibilities of watershed management. The chapter provides some interesting lessons learnt from the process.

Carr, Loucks and Bloschl in **"An Analysis of Public Participation in the Lake Ontario—St. Lawrence River Study"** have evaluated the impact of public involvement in a major review of the Lake Ontario and St. Lawrence River water level operating system (the LOSL Study). LOSL involved public meetings and stakeholder involvement to produce an operating policy for the system that was acceptable to all stakeholders impacted by the water levels and flows in that region. The authors applied criteria indicative of good participation processes, and that identify non-tangible or intermediary achievements and discuss their findings (such as, that the processes of participation were generally very effective and useful to both the study team and the public and point out many other achievements of the study beyond its original aims). As discussed by the authors, participant involvement in the development of the operating plans resulted in more innovative approaches and raised their legitimacy, particularly when the plans were perceived to be supported by sound science.

The next chapter **"Managing a Multi-objective Bi-national Study on Rationalizing the State of Upper Great Lakes Water Levels and Development of Improved Regulation Plans"**, by Moin describes an IJC study on lower lake levels in Upper Great Lakes basin. The first part of the chapter examines all the issues related to the conveyance of the St. Clair River and factors affecting Lake Michigan-Huron levels. The second component considers revisions to the regulation of the Lake Superior through the control works at Sault Ste. Marie. The author explains how shared vision planning, stakeholder engagement, and key elements of a co-management process, were implemented throughout the process.

The ethnographic essay, **"Towards Community based Management of Water Resources: A Critical Ethnography of Lake and Ground Water Conservation In Pushkar, India"** by Joseph examines whether the emergent and tentative steps being taken towards community-based catchment management could have wider implications for groundwater conservation.

The author examined two problems addressed by the Pushkar Project: lake pollution and lake sedimentation and presents an interesting discussion on how cultural and religious practices are leading to lake pollution. The author then discusses how decentralized initiatives, based on locally relevant religious concepts and legal precedents, could produce the cultural changes necessary for sustainable solutions and community-based management strategies.

The Great Lakes Water Quality Agreement has been the unifying principles for inter-jurisdictional shared water management for Canada and the United States for 40 yr. Beginning in 2009, both governments agreed to renegotiate a renewed agreement bringing it up to date with scientific advances and complex governance challenges. This is the first substantial amendment to the agreement since 1987 and represents a watershed point in the history of the Great Lakes regime. Fundamental to its success is the incorporation of programs and policies that address the pressures being experienced in the near shore zone of the Great Lakes. In **"Governance reform to secure resilience in the near-shore waters of the Great Lakes basin"**, Krantzberg discusses that effective interventions will require a new governance model that engages local participants, facilitated by governments, in co-management of this changing and dynamic environment.

Switching the focus to fisheries, Whillans et al. in their chapter **"Co-Generated Knowledge for Co-Management of a Mobile Resource: The Anishinabek/Ontario Fisheries Resource Centre"**, document a unique innovation in co-management of fisheries: the creation and operation, under formal agreement between two co-managing parties (The Union of Ontario Indians, representing the Anishinabek Nation, and the Ontario Government), of an independent institution (Anishinabek/Ontario Fisheries Resource Centre (A/OFRC)) for the generation and assembly of scientific and traditional ecological knowledge for the co-management of fish populations and habitat of shared interest. The focus in this case study is on the knowledge generation aspect of co-management. It is interesting to note that the independent mandate in this case is to uncover the truth about fishery problems without the taint of real or perceived bias that might cloak investigations undertaken by either of the signatory organizations. As the authors describe in the chapter that the A/OFRC offers the opportunity to either of the parent organizations, of requesting and receiving knowledge that is needed to make decisions. Existence of A/OFRC minimizes the co-optive peer pressure that both co-managing parties might feel, if they were directly in control of the knowledge generation. The balance is delicate—the role of ownership in knowledge generation in order to achieve trusted deep understanding weighed against the impartiality that is accorded to independent inquiry.

Socio-political institutions, such as rules of fishing created through the relationship between central governments and local level resource users, called co-management, have been created to manage freshwater fisheries in developing countries. Lawrence in **"Taking the "Co" out of "Co-Management": The De-Legitimization of Fishing Communities on Lake Victoria, E. Africa"** defines and discusses the importance of legitimacy and accountability for a successful fishery management institutions, followed by description of the concepts of community participation and co-management and how each inform the fishery management program on Lake Victoria. The author also describes the historical process and factors that led to creation of the fishery management program on Lake Victoria, including current theoretical notions of cooperation between central level government and local level resource users. Lawrence then uses data to determine (and discuss) how legitimacy of the "Beach Management Units" is affected through the current relationship with higher levels of government on Lake Victoria and how these relationships affect fisheries management.

As co-management has many aspects and can be applied to different fields, Boros et al. in, **"Barriers and Opportunities to Adaptive Co-Management of the Tisza River Basin in Hungry"** review the historical development of river management policies in the Tisza basin to show how the emergence of more frequent recurrences of extreme weather events (such as rising flood crest levels) and failure of the conventional system to deal with floods justifies a transition from a conventional to a more experimental river management regime. The authors discuss the role of adaptive co-management for water management and explore the water retention potentials on large areas within the floodplain. Since it is not possible to renew and adapt the current flood control strategy to the new threats only and exclusively by optimizing the water related technical measures, a broad consensus of stakeholders needs to be achieved if society wants to pursue experiments in alternative land use at landscape and regional scales.

As a delta country, the Netherlands has particularly immediate concerns regarding the increasing frequency of high and low water levels as a result of climate change. In the parts of the Netherlands which are well above sea level, water management is often practiced through co-management and many goals (such as water retention, nature, recreation, economic diversification, agriculture, and cultural history) in addition to water management are integrated. This also results in involvement of a lot of organizations and NGO's. The next chapter by de Boer, **"Co-Management of Water Resources in a High Density Environment as a Stimulus for Sustainable Regional Development"** examines planning for multi-functionality, increasing space for rivers and the efforts to connect natural areas that are at the heart of new initiatives in the Dutch rural areas to meet biodiversity and water quality and quantity goals from national and European levels, while contributing rather than restricting the economic viability of the region. The subject of co-management in this chapter is river valley renaturalization situated along the Regge River. The chapter concludes with a systematic inventory of boundary spanning strategies in co-management that were uncovered by the author through interviews with local project managers and stakeholders. Although contexts vary across times and places, these strategies provide some worthwhile insights for replication.

Shifting both the geographic region and co-management area to Chile and irrigation water co-management aspect, Leon and Fuster in, **"Irrigation Water Co-Management In The Limarí River Basin, Chile"** walk down the history of irrigation water management practices in Chile which went from public to private hands and ultimately the best solution that came out was co-management. The authors discuss various laws, policies and codes and management practices in Chile with pros and cons of different irrigation water management system and why co-management might work better than earlier regimes.

The next chapter shifts the focus from river (adaptive) co-management to (adaptive) coastal zone management. The complexity of managing water and coastal systems is influenced by two simultaneous drivers: need to involve various actors, and accommodate their sometimes conflicting demands; and the need to cope with uncertainty and change. Whereas the first challenge can be addressed by more interactive and participatory forms of management, such as co-management, the second challenge can be addressed by more adaptive forms of management. As discussed by Hermans, Cunningham and Slinger in, **"Adaptive Co-Management and learning: Developments in Coastal Zone Management in the Netherlands from 1985 to 2010"** adaptive management is the answer to deal with uncertain complex coastal problems. The authors first review key principles associated with adaptive management and learning, followed by a discussion of events in the case of coastal management in the Netherlands from 1985–2010. The authors used concepts from game theory and the notion of critical assumptions as key elements to connect decision-making and learning to explain the events. The chapter then compares the learning needs as identified in this reconstruction with the manifested learning efforts, followed by a conclusion with observations on the role of learning in the adaptive co-management of the Dutch coastline over the last 25 yr.

There are not too many international transboundary co-management arrangements for groundwater. As discussed by Walter in, **"The Roles of Knowledge in the Emergence Co-Management Initiative for Transboundary Groundwaters: The case of the Génévois Aquifer"**, the co-management of shared waters is the product of political struggles at different levels of governance; it claims that the co-management of transboundary groundwater is the functional response to known problems according to the preferences and possibilities of the powerful. Despite the consensus about the fact 'power matters', existing theories advance multiple and at times contradictory interpretations of the role of knowledge in the emergence of co-management schemes for transboundary groundwater. This chapter examines the different roles of knowledge in the political process that led to the co-management of the Génévois Aquifer shared between Switzerland and France. The case demonstrates that formal mechanisms can be used effectively to manage shared groundwater, which has been instrumental to the development of international guidelines for the co-management of transboundary aquifers at the global level.The author also reviews the process of social recognition of groundwater problems, their entry into the political agenda, the implementation of policies for their resolution, and the impacts of scientific knowledge in the different stages of the policy process. The analysis focuses on constitutive dimensions of decision-making scenarios: the original scientific assessments and local joint problem-framing that effectively reshape interests and challenge the legitimacy of established institutional frameworks.

The next chapter also focuses on groundwater co-management but moves the geographical focus from Europe to South America. Vilar in **"Guarani Aquifer Co-Management and the Brazilian Experience"** analyzes the existing institutional framework to promote the co-management of the Guarani Aquifer with a focus on the Brazilian case (since Brazil has more advanced water policies in comparison to the other countries; the biggest part of the aquifer is located in the Brazilian territory; and this country is its main user).This chapter begins with an analysis of the co-management as a transboundary-aquifer management strategy followed by characteristics of the Guarani Aquifer and the international efforts to promote its management; analysis of the Brazilian groundwater policy and the co-management perspectives.

In the last chapter, **"Institutional Dynamics of Khorezmian Water Management on the Farmer Level in Uzbekistan: From Ancient Practices to the Present Situation"**, Hirsch has given a historical record of the practices and experiences of various water management practices (including co-management arrangements) in North West Uzbekistan: from ancient times until the present time. She describes local water management (on the level of water users associations) in Uzbekistan with an aim to identify the dynamics between current practices and the practices set forth by tribal heads before the Soviet era in Uzbekistan and later under the Soviets in 1926. The author explains how vulnerable the current water management system is in regard to its ecological, social and technical dimensions and lessons learnt from application of co-management. The chapter has useful information that will lead others to consider co-management as an option for managing resources in their communities.

Conclusion

As can be seen from this introductory chapter, co-management has been defined in different ways by different authors, and can be applied to different aspects of water (near shore water, coastal zones, rivers, water management, irrigation and so on). Case studies in the book also support this multidimensional aspect of co-management and apply co-management in various contexts. Some of the lessons shared in the chapters can be used for replication in other places where co-management is being considered as an alternative for improved water management decision making.

Natural resources management now involves numerous stakeholders (and their conflicting interests) and adaptive forms of management (learning as part of the process of management and adjusting the management/governance plan as the process evolves). As discussed by Hermans et al. (in this book)—the first challenge can be addressed by more interactive and participatory forms of management, such as co-management, the second challenge can be addressed by more adaptive forms of management. Adaptive co-management is a combination that promotes multilevel institutional linkages, shared responsibility between different the stakeholders involved, a combination of knowledge sources, as well as learning (Olsson et al. 2004; Berkes et al. 2003). In other words, adaptive co-management:

- Combines aspects of adaptive management and co-management;
- Leads to integration of ecology, economics, society (governance);
- Involves the concepts of shared responsibilities for using and allocating resources among multiple parties.

Perhaps the answer to complex and dynamic natural resources management is adaptive co-management (discussed by several authors in this book).

References

Berkes, Fikret, Peter George and Richard Preston. 1991. "Co-Management: The Evolution of the Theory and Practice of Joint Administration of Living Resources." TASO Research Report, Second Series, No. 1. Paper presented at the Second Annual Meeting of the IASCP, 26–29 September, University of Manitoba, Winnipeg, Canada.

Berkes, F. 2009. Evolution of co-management: Role of knowledge generation, bridging organizations and social learning. Journal of Environmental Management 90: 1692–1702.

Borrini-Feyerabend, G., M. Pimbert, M.T. Farvar, A. Kothari and Y. Renard. 2004. Sharing Power.Learning by doing in co-management of natural resources throughout the world, IIED and IUCN/ CEESP/ CMWG, Cenesta, Tehran.

Carlsson, L. and F. Berkes. 2005. Co-management: concepts and methodological implications. Journal of Environmental Management 75: 65–76.

Ostrom, E. and E. Schlager. 1996. The formation of property rights. pp. 127–156. *In:* S. Hanna, C. Folke and K.-G.Mäler [eds.]. 1996. Rights to nature: ecological, economic, cultural, and political principles of institutions for the environment.Island Press, Washington, D.C., USA.

Olsson, P., C. Folke and F. Berkes. 2004. Adaptive co-management for building social-ecological resilience. Environmental Management 34: 75–90.

Pahl-Wostl, C. and M. Hare. 2004. Processes of Social Learning in Integrated Resources Management. Journal of Community & Applied Social Psychology. J. Community Appl. Soc. Psychol. 14: 193–206.

Pinkerton, E. (ed.). 1989. Co-operative Management of Local Fisheries.New directions for improved managementand community development, University of British Columbia Press, Vancouver, Canada.

Singleton, S. 1998. Constructing Cooperation: the Evolution of Institutions of Comanagement. University of Michigan Press, Ann Arbor. p. 7

The World Bank. 1999. Report from the International Workshop onCommunity-Based natural Resource Management (CBNRM),Washington, DC, 10–14 May1998.URL:http://www.worldbank.org/wbi/conatrem/

2

Sustainable Development and the Water Energy Nexus

George Nasr

Introduction

"Majesty, we found oil", said the American geologists, to which Ibn-Saud replied; *"I would rather you found us water"*. Ibn-Saud, the founder of Saudi Arabia understood this relationship more than most; while he realized the role of oil in the modern world, he knew that his arid kingdom could not live without water.

Ever since the dawn of civilization, water and energy have been deeply linked in human development. On Earth, freshwater is essential to ensure the basic needs of living organisms. Similarly, freshwater ensures the basic needs of civilization. In its early stages, civilization was built on the human ability to collect and transfer the water it needed to sustain large settlements, planting and irrigating crops, and raising livestock. At first, most of the energy needed to carry out those essential tasks was obtained by marshalling the labour of humans and animals, or by collecting such resources as wood. Over time, as civilization expanded, it grew to become a super-organism "feeding" on various resources to maintain its various functions. With this expansion came the need for more energy, which was often magnified further by the need to pump and convey more freshwater. As society's hunger for energy grew, it started exploiting other resources, such as hydropower and buried hydrocarbons stores.

As society's hunger for energy grew, so did its need for water. Once food surpluses were secured, the use of water was no longer limited to basic

Faculty of Engineering, Lebanese University, Beirut, Lebanon.

human needs in domestic supply and agriculture. Once used mainly on food production, water was into goods and services, particularly through energy-generation processes. Water is now increasingly used in energy generation, either directly through hydropower, or indirectly with thermal energy sources such as coal, oil, or nuclear, where it is used to power steam turbines and cool power plants. By the middle of the 20th century, humans had come to significantly alter the natural Hydrological Cycle, creating a parallel "Socio-Economic Water Cycle" (see Fig. 2.1).

Water and energy are both essential to this socio-economic water cycle. Humans use energy to extract water from the natural cycle, which they use for more energy generation, as well as for agricultural and their own domestic freshwater needs. Most of the water is returned into the system. However, a large proportion of this water is in the form of untreated sewage, whose treatment requires yet more energy. Additional pollutants are emitted as part of human activity, and they also affect natural processes of evaporation, condensation and precipitation.

As a result, the link between the two water cycles, the natural and the human, is now largely mediated by human energy consumption and production. By the middle of the 20th century, as human development accelerated, a "Water-Energy Nexus" had emerged. It is now an important parameter that powers the interactions between humans and the natural ecosystems they depend on.

By the end of the 20th century, the water-energy nexus had come "full circle", as increased energy demands brought about the development of biofuels. In many cases, such biofuels are derived from agricultural

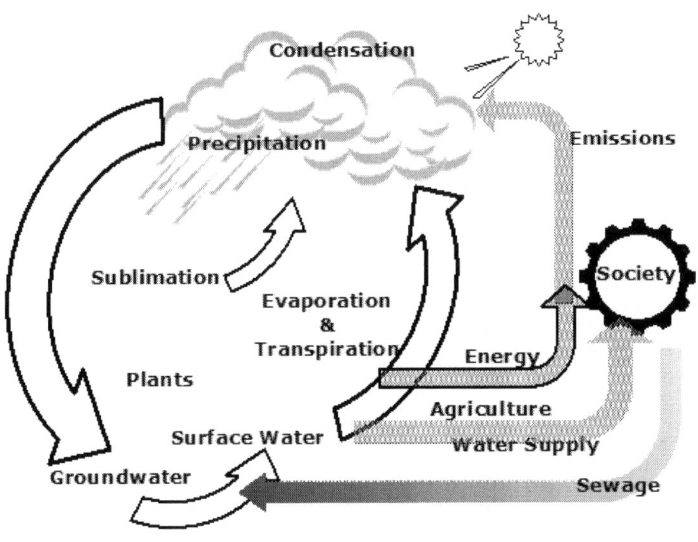

Figure 2.1 The Socio-Economic Water Cycle.

products that are grown not for food, but for the exclusive purpose of energy generation. Energy was now competing for water with food. Human development is now intractably linked to this symbiotic relationship between water and energy. For this development to be sustainable, water and energy must be managed jointly.

Water for Food: Securing Surpluses

Civilization was born from the need to manage water resources, and it expanded and thrived whenever cheap energy was available. In areas where rain is plentiful, farmers do not need complex and large organizational structures, since they can often rely on rainfall to feed their crops. However, in those areas where rainfall is less reliable, farming was still possible when humans organized themselves into larger grouping, "scaling up" together to adopt a more systematic approach to water supply, such as irrigation. At the smallest level, this required a degree of organization of necessary labour and resources allocation. Thanks to irrigation, settlements that could be supported on rainfall could grow to become large cities.

The technique of agriculture did develop independently in the American continent and the New Guinean plateaus, and it may have developed independently in China as well. By 5800 BC, hamlets of farmer communities that cover areas no larger than one hectare appear in those areas, surrounded by fields and simple irrigation works. However, it is in the "Fertile Crescent" of the Middle East that key staple foods were first farmed, and where the most productive beasts of burden were domesticated.

The Fertile Crescent is an arc that extended northward from the eastern Mediterranean coast, and followed the general outline of the Tigris Euphrates as it flowed downward towards the Persian Gulf. Development patterns differed across this vast expanse as people adapted differently to local climatic patterns. Along the coast, rainfall was reliable but the terrain was hilly, so farmers adapted by terracing the slopes and grew rain-fed crops. In the hinterland, rainfall was less reliable but land was more plentiful and more fertile, so farmers needed to get organized together to carry out large water infrastructure works.

This is the case of the north-eastern plains of Mesopotamia. The Sumerian inhabitants who first settled along the Tigris and Euphrates needed irrigation to ferry water and grow food on the fertile lands "between the Rivers".[1] In addition to basic staples such as wheat and barley, the inhabitants of the region farmed rye, oats, flax, alfalfa, plums, carrots, peas, lentil and beans (Diamond 1997). Sumerian settlements soon grew into

[1]They referred to this region as the land "between rivers" ("Beth-Nahrain"). The region's name is now derived from its Greek name, made up of their word for "middle" ("mesos", or "μεσοσ") and "river" ("potamos", or "ποταμοσ").

large cities, and the "land of Uruk"[2] became the first intensive agricultural society, growing high quality food such as wheat. By 5200 BC, Sumerian settlements grew to larger areas covering 10 hectares or more, and had become home to over 80 percent of the Sumerian population of the time (Adams 1983). Wherever agriculture could be organized in similar manner, the same pattern would emerge, as evidenced by early Meso-American civilizations.

At first, much of the energy necessary for creating and maintaining the complex irrigation networks was marshalled by human power. Any further expansion was limited by the availability of the energy sources necessary to "power up" technological innovations. In regions such as Mesoamerica, this and other factors created a set of "techno-environmental parameters" that limited "the possibility and probability" of expansion (Bailey and Llobera 1981).

In contrast, the Mesopotamian societies could overcome the basic limitations of "manpower" thanks to the more powerful "engines" available through animal domestication. Their environment permitted Sumerians easy access to a large variety of animals that could be domesticated, such as bovines and equines, sheep and goats (Diamond 1993; Tainter 1990). They were soon able to rely on the additional labour of cows and donkeys. These "beasts of burden" allowed them increased power to enhance the output from technological innovations such as the ploughshare to augment agricultural productivity, and the Archimedes wheel to pump water uphill and grow crops on terraced lands.[3]

Thanks to this diversity of domestic plants and energy, the food supply for humans was both larger and more stable. The surplus of food now allowed a decreasing number of farmers to feed a larger society, thus facilitating the emergence of specialized trades and cultured elites who pursued the arts and sciences. Civilisations would now be defined by their access to energy sources and their use of it, and those with access to the energy from wood or domesticated animals spreading farther and wider, to establish ever larger empires. On the downside, farmers became dependent on a climate remaining relatively constant. In the short to medium term, they could face limited climatic changes by deploying more energy, either pre-emptively or reactively. They could pre-empt changes by expanding energy to build structures for the storage and diversion of water. Then, when faced with events such as droughts , they could use animals to bring

[2]In ancient times, the region was often referred to "the land of Uruk", after the first major Sumerian city, or "Bilad Al-Iraq" in Arabic, which became the modern name of the country formed around Mesopotamia.

[3]The mythical "Hanging Gardens of Babylon" were likely build in this process, on an artificial mountain with stepped terraces, or "ziggurat".

water from farther distances, or dig wells to pull it from deeper locations. Either way, the infrastructure would remain inadequate to ensure longer term sustainability until the industrial revolution and the advent of the steam age.

Complex Societies: Emergence of the Water-Energy Nexus

Once food supplies were secure, there was a greater demand for non-food goods and services. Domesticated animals were also a source of some of the early manufactured goods, as in the case of sheep and goats in Mesopotamia, or the Llamas in southern America. This required more energy for processing and manufacturing. Some societies were able to be comparatively more efficient, as in Mesopotamia where bovines could be used for labour and food, in addition to being a source of manufactured goods. Thanks to this, from the complex societies of southern Mesopotamia and the Indus Valley, civilization could spread to the rest of the world.

Energy became an increasingly important factor for the spread of civilization. The "scale" necessary for this spread is largely defined by the amount of available energy, with various sources of energy "working together to make a large scale behaviour" of the complex societal system. As they undergo large scale expansions, or "behaviours", complex societies would require increased "coordination between more parts and/or larger amounts of energy". Because of the limits on early sources of energy, ancient empires needed to involve larger numbers of individuals and beasts of burden to maintain themselves and carry out their large-scale projects, and thus involved large hierarchies (Bar Yam 2002).

In such societies, water remained a limiting factor to expansion, as no society could conceivably expand in places where it would need too much energy to bring in the water it needed. By the end of the 18th century, new engines were developed that relied on mineable sources of energy. With the Newcomen steam engine, king coal had arrived, making possible the industrial revolution. Ever more complex societies were now to do even more with only the addition of energy.

Industrial Development: Energy for Water

The steam engine released unprecedented amounts of energy. More goods could be processed, and ferried over greater distances. The industrial revolution was largely based on the use of increased amounts of energy. They allowed larger scale operations through the use of new sources of energy, and far greater automation of tasks that amplified any necessary activity to the desired scale. After Newcomen, new designs came along that increased efficiencies and released ever larger amounts of energy.

Those new sources of energy made available ever newer sources of water. Hitherto inaccessible sources of water were reachable thanks to pumps powered by diesel and electricity, leading to the expansion of farming in places where it would have been impossible before, such as the deserts of Arizona and Saudi Arabia. However, the "quality" of this "new water" was not the same, either because it was non-renewable, or because of its high energy cost. Paradoxically, the availability of cheaper energy in the short term made water dearer in the long term. In their use of energy for water, humans were both outpacing the natural hydrologic cycle and lowering the "quality" of the stores they depended on.

As humans outpaced the natural hydrologic cycle, new sources of water often become a non-renewable resource, mostly because the low cost of energy allows humans to pump it out at rates that tend to outpace replenishment rates. As usage cycles started outpacing replenishment rates, farmers effectively became miners of "fossil" water. This happened first in desert countries where ancient groundwater stores allowed an expansion of irrigation, as in Saudi Arabia. There, looming shortages of water forced the phasing-out of a 30-yr wheat growing programme by 2008. Then, in countries where groundwater stores were initially renewable, excessive pumping caused similar problems. This is the case of Syria, where following a prolonged series of droughts that started in 2006, over pumping exacerbated aquifer depletion, leading to the evacuation of more than 50,000 families from some areas of the country. By 2010, two to three million of the country's 20 million inhabitants faced food insecurity (De Schutter 2010).

Even when new sources of water were discovered, water quality remained an issue. By the middle of the 20th century, cheaper sources of energy made possible even newer sources of water through the desalination of brackish or seawater. This remains the only realistic solution ever devised to create "new water", and arid areas such as the Middle East are now home to about 35 percent of the "water factories" in the world, generating 65 percent of total capacity. However, those developments do not "break" the nexus, but only serve to reinforce it in two crucial ways. First, because the technique remains energy intensive, it is only applicable in cases where other alternatives are not practical, such as domestic use. Second, the left-over brine generated can be a dangerous pollutant if not properly disposed of.

Industrial Expansion: Water for Energy

Yet, for all the progress made since then, humanity remains in the "steam age", as almost all energy-generation processes follow the same basic process. The process is based on using a source of heat to turn water into steam whose energy either pushes pistons, as in Newcomen's and Watt's early engines, or rotates turbines, as in later designs. This remains the case of new energy sources such as nuclear, geothermal, and even most solar plants. All those processes share a common characteristic; the "heat generating" element is "consumed" into an energy-generating process that re-circulates most of the water, and "loses" some of it.

As more energy is required, more water is therefore needed, and the expansion of human activity remains limited by the availability of water. By the end of the 20th century, energy production remained by far the largest user of water across the various sectors of human activity, particularly in developed countries. As an example, in Canada, not counting the recent demand for agricultural products for biofuels, production of energy by such means as thermal power uses up more than 60 percent of the water used, far above manufacturing (18.5 percent), municipal (9.5 percent), agriculture (8 percent), and mining (4 percent) (Brandes and Ferguson 2003). This adds up to significant amounts of water; in the United States, thermoelectric power plants use 136 billion gallons/day of freshwater, which is equivalent to an average of 25 gallons for each kilowatt-hour (kWh) of electricity produced, as shown in Table 2.1 (Younos et al. 2009).

Some water is still lost in these processes, even if much of this water is returned to be re-circulated into the hydrologic cycle. When water is abstracted and used in various processes, some of it is retained and thus "incorporated" in processes related to energy generation, such as scrubbing pollutants or cooling and cleaning machinery.

This water is thus "consumed" (see Table 2.2). This is often the case of mining and oil production. The production of fuels such as oil and gas requires the injection of water into wells, which is thus removed from the water cycle. This water is often too polluted to be discharged and is left behind in ponds, which effectively excludes it from the hydrologic cycle. Furthermore, other alternative energy sources, such as ethanol, biodiesel, and hydrogen, require large volumes of water that is essentially converted into another form. These processes are extremely water-thirsty; for each 1,000 KWh generated, water usage that ranges between 32,400 litres and

Table 2.1 Water Needs of Different Energy Generation Techniques (Younos et al. 2009; Hutson et al. 2004; US DOE 2006).

Fuel Source	Water Use (Liters/100 KWh)r
Hydroelectric	260
Geothermal	1,680
Solar Thermal	2,970–3,500
Thermoelectric (Coal, Oil, Gas)	14,200–24,800
Nuclear	31,000–74,900

Table 2.2 The Structure of Water Use in Canada (Brandes and Ferguson 2003).

Sector	Gross Use (MCM/Yr)	Recirculation (%)	Consumption (%)	Discharge (%)
Thermal	40,405	28.9	1.3	69.8
Manufacturing	12,995	53.5	4.3	42.2
Municipal	5,314	-	2.2	97.8
Agriculture	3,991	-	77.4	22.6
Mining	1,715	69.8	3.0	39.2

375,900 litres for fuel ethanol, 180,900 litres and 969,000 litres for "First Generation"[4] biodiesel. This is a far larger amount than the 500–2,100 litres for coal, or even 15,500–31,200 litres for petroleum/fuel oil (Younos et al. 2009). An increasing amount of water is now abstracted for these processes, so much so that the need to meet industry's thirst for water is now competing with the necessity to safeguard the supply of humans and the ecosystems they depend on.

As human water consumption increases, industrial water consumption will increasingly compete with more non-industrial water uses. In the latter part of the 20th century, the increase in global water use did not merely parallel human population growth, it exceeded it. While the world population increased by a factor of 2.7, freshwater withdrawal and consumption increased by at least twice that amount. The yearly amounts of withdrawal and consumption of freshwater have increased by factors of 5.9 and 5.6, respectively, in the period from 1900 to 2000. The amount of yearly freshwater withdrawal is projected to total at least 5,235 km^3 in the year 2025. Many consider this a critical amount, since it would then represent more than 50 percent of the total amount of practically available freshwater.

As agricultural and municipal water demand increases, so will their demand for the energy to convey and supply it. The necessary increase in energy will thus bring about additional demands for water to supply the power plants, as well as the mining and oil production operations that sustain them. In the long run, however, there does not need to be a conflict between the need for water and energy. This is because humans are a unique species in the way they meet their needs; their material existence is limited by "ideas" not "things" (Romer 1993). People do not need wood, oil or coal per se, but rather a way to heat their homes or cook food, and so it is conceivable that alternative sources of energy can be developed that are less water-thirsty.

[4]First Generation Biodiesel are made from products that would have otherwise been used for food, such as sugar, starch, and vegetable oil.

The Water Energy Nexus

As humanity enters the 21st century, it finds itself in a situation where water and energy are durably linked in a "Water-Energy Nexus" (Fig. 2.2). In addition to sustaining human populations and their food producing activities, water is now used to make energy directly through bio-fuels, or indirectly through the extraction and processing of hydrocarbons, and the generation of electric power.

Energy is increasingly vital to pump water from the groundwater, and to ferry it across long distances, to population centres where humans cluster in ever increasing numbers. By 2008, with a majority of the world's population living in urban centres, ever more energy was needed for water; in addition to extraction, it was also needed for treating the increasing amounts of sewage. However, in part because of the high-energy costs, extraction gets a priority over treatment, and water quality generally decreases. One billion of the world's people now live in urban slums that tend to be overcrowded and polluted, with limited access to clean water and sanitation, if any.

Water now lubricates a complex society and its various industrial activities, while energy is now vital to bring irrigation to fields and freshwater to cities, and to sustain a growing industrial base. For the purposes of sustainable development, energy management is now intractably linked to water management.

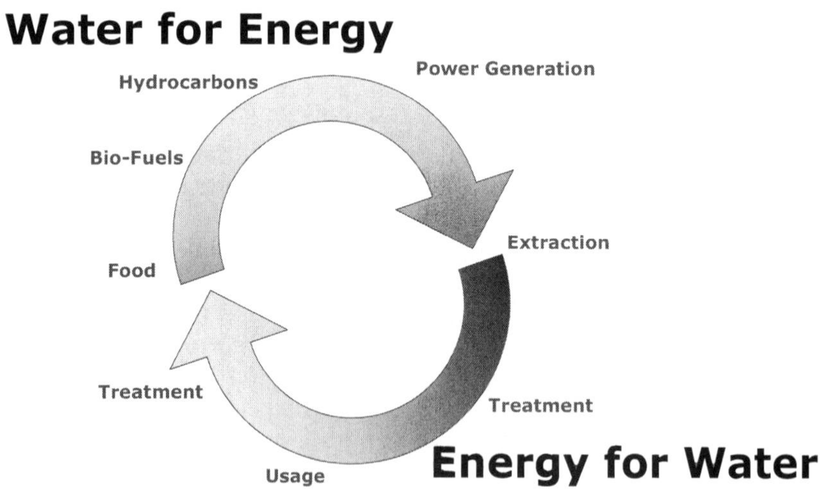

Figure 2.2 The Water-Energy Nexus.

The Water-Energy Nexus and Sustainable Development

On current patterns of development, the demand on water supply may continue growing further as human civilization continues to increase in size and complexity. However, while demand on freshwater resources has dramatically grown over the course of the entire 20th century, the picture is more nuanced in its last 25 yr. The rate of growth of consumption of freshwater resources appears to be slowing down, possibly thanks to conservation methods and new technologies. Population growth has also slowed down. There appears to be a marked slowdown in the rates of increase of population and water withdrawal that was apparently started in the 1970's. This follows a pattern whereby, as nations start a period of growth, they increase demands on natural resources. Then, as they grow wealthier, they appear to become more efficient users of natural resources. However, even if population growth rates is slowing down, total population has grown to unprecedented amounts over the course of the 20th century; while it took the entire human history to reach 1 billion people by 1800, then the population doubled by the 1930's, and went on to exceed 6 billion by the onset of the 21st century.

As a result, total demand for water and energy is likely to continue growing in the short to medium term, challenging water managers to secure adequate supplies. Traditionally, managers focused on securing the amounts of water needed, by establish structures for retaining, abstracting, and conveying this precious resource, and for evacuating waste and storm water. However, while this "hard path" (Wolff and Gleick 2002) to water did bring about substantial public health benefits, the limitations of such a "hard path" engineering approach had become evident by the end of the 20th century, in a context of rising urbanization and spreading "metro-cities" and increasingly threatened ecosystems.

Throughout human history, there are many examples of the limitations of the "hard path" for water. Early societies often reached the limits of all efforts to expand irrigation and agriculture, as in the case of Mesopotamia. Their use of intensive irrigation caused slow but steady soil degradations due to salination and the resulting changes in geochemical balances. As a result, they kept expanding and intensifying irrigation, further accelerating the process of degradation. Around the year 3000 BC, the salt content of the soil had increased and the land degraded to such an extent that wheat could no longer be grown in the dominions of the powerful city-states of Akkad and Sumer. Farmers responded by gradually replacing more salt tolerant barley crops, and archaeological evidence shows no more wheat cultivation by the year 1700 BC. Yet, continuing intensive irrigation caused further salination until barley yields started decreasing as well. Faced with diminishing yields, the people moved away to other city-states where salination was less of a problem. The kingdoms of Sumer and Akkad vanished, precipitated in their doom by the continuous shifting of the river beds of the Tigris and Euphrates. Once the leading centre for world agriculture, the region is today so unproductive that "applying the term Fertile Crescent today to the area of Syria and Iraq [...] would be a cruel Joke" (Diamond 2004).

There are now fewer places people could "move away" to. In any case, there are now many more humans who would need to "move away". By the 21st century, it had become obvious that more comprehensive developmental approaches were needed that look beyond simply securing supplies of water and energy. Beyond the need to improve efficiencies in resource usage across sectors of activity, such paradigms are rooted in the principle that the needs of the present should not compromise those of future generations. In such a "Sustainable Development" paradigm (see Fig. 2.3) the emphasis is on balancing the immediate necessities of "People", the need to generate "Profits" in the short to medium term, and the limitations of the ecosystems of our "Planet".

Sustainable development therefore requires that human society be viewed as a complex system subject to limitations imposed by the current state of technology and societal organization, all operating within an "Earth System". As humans strive to reach a balance among these various interactions, they need to optimize their use of water and energy, both part of the same nexus.

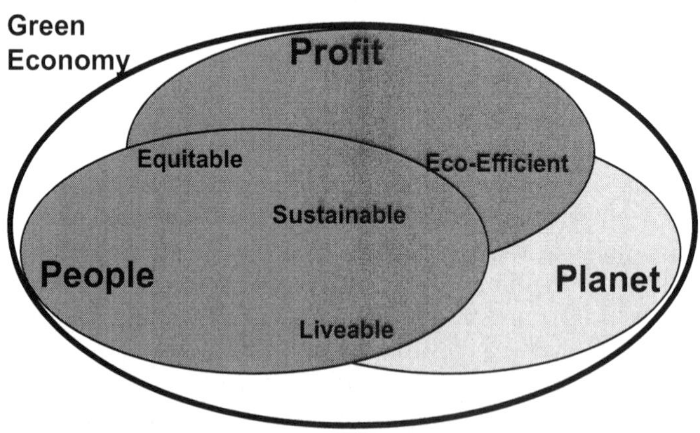

Figure 2.3 Sustainable Development: People, Planet, and Profits (UNEP 2007).

The Water-Energy Nexus and Climate Change

The concept of sustainable development was formulated because humanity faces climatic change. While climatic changes are common throughout Earth's life, they are less common in human history, and even less so in the past 200 years, when human populations undertook their dramatic expansion. Anatomically modern humans appear to have originated in Africa about 200,000 years ago, and to have reached "full behavioural modernity" about 50,000 yr ago. However, for all their abilities, they were eking out a precarious existence as hunter-gatherers until the onset on the Holocene, about 12,000 yr ago. It was soon after the onset of this relatively warmer climatic period that human civilizations appear, their fortunes ebbing and flowing in parallel to climate variations (Fagan 2008).

Coincidence may not equal causation. However, as humanity grew to unprecedented numbers over the past century, our civilization also grew to unprecedented levels of complexity. This makes our civilisation ever more vulnerable to "Black Swan", or unforeseeable adverse events (Taleb 2007). It is in this context that the growing need for water and energy is manifesting itself. This change is dramatically altering the fundamentals of resources management. In the past, water managers could safely assume that key hydrological parameters remained constant over time. Nowadays, such "hydrologic stationarity" is no longer a valid assumption; any resource management plan needs to take these dire consequences into account (Bates et al. 2008).

An extreme case is that of Lake Chad, located at the intersection of Chad, Niger, Nigeria and Cameroon, in Central Africa. Starting in the 1960's, the monsoon rains that are the Lake's primary source of water had been decreasing significantly and the average local temperatures started rising. However, this was the time when the region was in the midst of an expansion of irrigation, as planners were racing to develop their countries in an effort to emulate European development patterns of the previous century. Under the combined effect of the prolonged drought and the increasing demands of irrigation, the amount of water coming into the lake decreased by as much as 75 percent. The lake started to shrink and will soon disappear under the continuing dual pressures of climate change and population stress. In 1963, the lake used to span across the borders of Niger, Nigeria, Chad and Cameroon. By 2004, it was reduced to 1/20th of its former size; a small patch of water and marshlands that straddle the border between Chad and Cameroon (Coe and Foley 2001).

Development in the 21st century cannot be based on past assumptions. Existing management practices need to be robust enough to address both the impacts of climate change on both water needs and the effect of societal development on energy demand. This requires a more dynamic interaction between science, which focuses on understanding the various mechanisms of the Earth's dynamic systems and their interactions, and policy making, whose role is to decide on the level of acceptable risks and agree on optimal courses of action.

Implications for the Future: Complexity and Fragility

The past development of human societal systems has occurred in a context of climatic stability. Yet the fact remains that the Earth is a dynamic system made up of various interdependent and interconnected systems that vary on various timescales. Current societies are increasingly networked, with various interactions moving through interconnected nodes. While this allows for greater efficiency and communication among various members and components of society, it also makes it easier for adverse impacts to be transmitted across the network. The result is societal "fragility", which makes our modern civilization extremely vulnerable to changes in prevailing conditions (Taleb 2011).

These challenges can best be addressed through new approaches. This is because the central importance of Water-Energy in our modern society means that any managerial solutions to the challenges cannot be "uni-modal", but needs to address the various aspects of the challenge. Resources management needs to address the need to both use energy and water more efficiently, and manage the Energy-Water nexus optimally. Two aspects must be considered; technical, and policy-related.

From a technical perspective, energy management needs to focus on two aspects. On one hand, managers need to minimize the energy expended for water extraction, generation, conveyance, and treatment. On the other hand, they need to maximize the returns of using water for energy generation. This requires that managers need to consider both the returns on energy and/or water used, and the net energy and/or water that results from the processes considered, once all inputs have been accounted for. From a technical perspective, energy and water can no longer be managed separately, but jointly.

From a policy perspective, the 21st century adds unprecedented challenges. It adds "systemic uncertainty" to these technical challenges. This uncertainty then requires a continuous evaluation of the processes considered to make the necessary adjustment, which is best achieved through a form of "Adaptive Management", or "Adaptive Resource Management". Decision-making needs to follow an iterative process that can adapt to the uncertainty over time by constantly accruing information and adjusting decisions in a structured, systematic manner. In this manner, decision makers would not be taking managerial decisions, they would be learning about the system and the changes it is undergoing (Holling 1978). To do so, decision makers need to consider both parameters of society's "engine"; energy and water.

For both policy making and technical management to meet these challenges, they need to incorporate a new "horizon" and based on more "learning". This requires a long-term outlook that extends beyond regular policy horizons such as the electoral cycle. The main challenge has two aspects; a management-policy aspect, and a knowledge-policy aspect.

First, it is necessary to find the balance between the need for long-term management outlook the necessity to achieve and show short-term outcomes. Benchmarks should therefore be developed that allow to measure the ongoing progress, and thus permit policy makers to "show results" within their own decision cycles.

Second, it is necessary to cope with the fact that empiricism requires that most knowledge be essentially provisional. This means that a balance must be found between the need to act on current knowledge, and the importance of revising or updating this knowledge as a new understanding develops.

These challenges are even greater in the context of the 21st century. Over the course of their history, humans have dramatically altered the Earth's Hydrologic Cycle by adding a new Socio-Economic Water Cycle. Inspite of this, managers of the past have often operated on assumptions of a relatively constant environment. This had dreadful results when the environment changed suddenly, as evidenced by the ruins of many defunct civilizations. The current climatic change is only the latest such variation. While it may not be unprecedented in the history of Earth, it remains unprecedented in the history of a society that has grown to unprecedented levels of complexity and interdependence. Since climate change creates conditions where the likelihood of unforeseen "Black Swan" events are greater, societal fragility is magnified. In this modern context, the assumptions of the past are even less valid. Society can no longer rely on assumptions of cheap energy, "hydraulic stationarity", or climatic stability.

Having reached unprecedented levels of complexity, humanity needs a comprehensive approach to manage the increasingly complex Water-Energy Nexus it has created. Even in the past century's stable environment, managers had struggled to jointly manage water and energy. Nowadays, there is an ever greater need for an open management process in which the development of strategies is constantly balanced with scientific understanding and empiricism, and seeks to engage various stakeholders across various temporal and spatial scales. These systems need to "co-manage" water and energy, where energy sources are the "fuel" that feeds society's engine and water is the "lubricant" that is constantly recycled.

References

Bailey, A.M. and J.R. Llobera. 1981. The Asiatic Mode of Production: Science and Politics, Routlege, London, UK. p. 225.
Bar Yam, Y. 2002. Complexity Rising: From Human Beings to Human Civilization, a Complexity Profile. Encyclopaedia of Life Support Systems (EOLSS), Oxford, UK.
Bates, B.C., Z.W. Kundzewicz, S. Wu and J.P. Palutikof (Eds.). 2008. Climate Change and Water, IPPC Technical Paper VI, Intergovernmental Panel on Climate Change, IPPC Secretariat, Geneva, 200 pp.
Brandes, O.M. and K. Ferguson. 2003. Flushing the Future? Examining Urban Water Use in Canada, POLIS Project on Ecological Governance, University of Victoria, Victoria, BC.
Coe, M. and J. Foley. 2001. Human and natural impacts on the water resources of the Lake. Chad basin, Journal of Geophysical Research, Vol. 104, No. D4, pp. 3349–3356, Feb. 27, 2001.
De Schutter, O. 2010. Preliminary Remarks on: Mission to Syria from 29 August to 7 September 2010. Office of the Special Rapporteur on the Right to Food, Special Procedures Of The Human Rights Council (HRC), Office Of The United Nations High Commissioner For Human Rights (UNHCR), United Nations.
Diamond, J. 1997. Guns, Germs and Steel: The Fates of Human Societies, W.W. Norton, NY, NY.
Diamond, J. 2004. Collapse: How Societies Choose to Fail or Succeed, Viking—Penguin Group, London, UK.
Fagan, B.M. 2008. The Great Warming: Climate Change and the Rise and Fall of Civilizations. Bloomsbury Publishing, New York, USA.
Holling, C.S. 1978. Adaptive Environmental Assessment and Management, Wiley, NY, NY.
Hutson, S.S., N.L. Barber, J.F. Kenny, K.S. Linsey, D.S. Lumia and M.M. Maupin. 2004. Estimated Use of Water in the United States in 2000, Circular 1268, U.S. Geological Survey. Available on the Internet at (as of Sept. 25, 2011): http://pubs.usgs.gov/circ/2004/circ1268/
Romer, P. 1993. Ideas and Things, The Economist, September 11, 1993.
Tainter, J.A. 1990. The Collapse of Complex Societies. Cambridge University Press, Cambridge, UK.
Taleb, N.N. 2007. The Black Swan: The Impact of the Highly Improbable. 2nd edn. Random House, New York, NY.
Taleb, N.N. 2011. Anti-Fragility (Manuscript), Random House, New York NY.
UNEP. 2007. Life Cycle Management: A Business Guide to Sustainability, United Nations Environment Programme (UNEP).
USDOE. 2006. Report to Congress on the Interdependency of Energy and Water. United States Department of Energy (USDOE), Washington, DC.
Wolf, G.H. and P.H. Gleick. 2002. The Soft Path for Water. *In:* P.H. Gleick, W.C.G. Burns, E.L. Chalecki, M. Cohen, K.K. Cushing, A.S. Mann, R. Reyes, G.H. Wolff and A.K. Wong [eds.]. The World's Water 2002–2003: The Biennial Report on Freshwater Resources, Island Press, Washington, DC.
Younos, T., R. Hill and H. Poole. 2009. Water Dependency of Energy Production and Power Generation Systems. Virginia Water Resources Research Center (VWRRC), VWRRC Special Report No. SR46-2009.

3

Working Toward Co-Management of the Raccoon River Watershed in Iowa
The Role of Civil Society

Cornelia Butler Flora[1,*] and *Michael Delaney*[2]

Introduction

Non-point pollution is increasing, as row crops genetically modified to be input responsive has increased the use of fertilizers and pesticides in the Corn Belt. With corn and soybean prices at record highs, land previously used as conservation reserves and buffers has been moved to row crop production. As states claim the need for fiscal austerity, enforcement of already weak regulations has declined. In such cases, watershed health cannot be left solely to the governments—local, state, and federal—although all should be reminded of their obligation to protect the environment to insure the well-being of the people. In this case study, we look at the emergence of a civil society organization, the Raccoon River Watershed Association (RRWA) and its role in co-management of the watershed. It

[1]Charles F. Curtiss Distinguished Professor of Agriculture and Life Sciences and Sociology Iowa State University, 315 East Hall, Iowa State University, Ames, IA50011-1070.
Email: Cflora@iastate.edu
[2]Professor, Des Moines Area Community College, President, Raccoon River Watershed Association, 1112 45th St.Des Moines, Iowa50311.
Email: mikedelaney1@mac.com
*Corresponding author

addresses the question of how grass-roots organizing can contribute to co-management of a watershed. It also addresses the degree to which a civil society organization can contribute to pluralistic management through collaboration with state and market institutions, sharing the benefits and responsibilities of watershed management. The complexity of the issues and the hesitancy of the state actors make this a complex and dynamic process. The Community Capitals Framework provides a mechanism to analyze the process of co-management and its effectiveness.

The Setting

Located in Central Iowa in the Corn Belt of the United States, the Raccoon River flows for much of its length as three streams (see Figs. 3.1 and 3.2).

The North Raccoon River is, by far, the longest of the three, at 315 km. It rises north of Marathon in northeastern Buena Vista County and initially flows southwardly into Sac County, where it turns southeastward for the remainder of its course through Calhoun, Carroll, Greene, and Dallas counties. It passes the towns of SacCity, Jefferson, Perry and Adel.

The Middle Raccoon River, 148 km long, rises in northwestern Carroll County and flows generally southeastwardly through Guthrie and Dallas

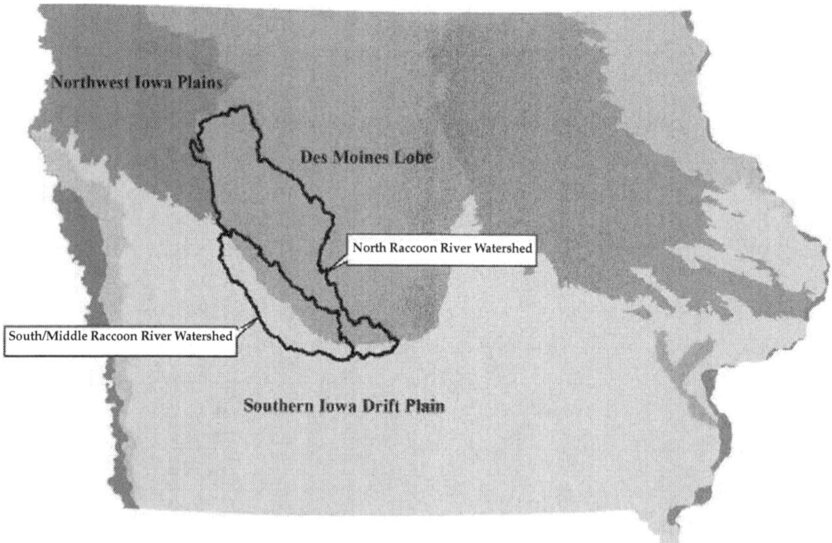

Figure 3.1 Location within Iowa and land forms.

Source: Raccoon River Masterplan 2011

Color image of this figure appears in the color plate section at the end of the book.

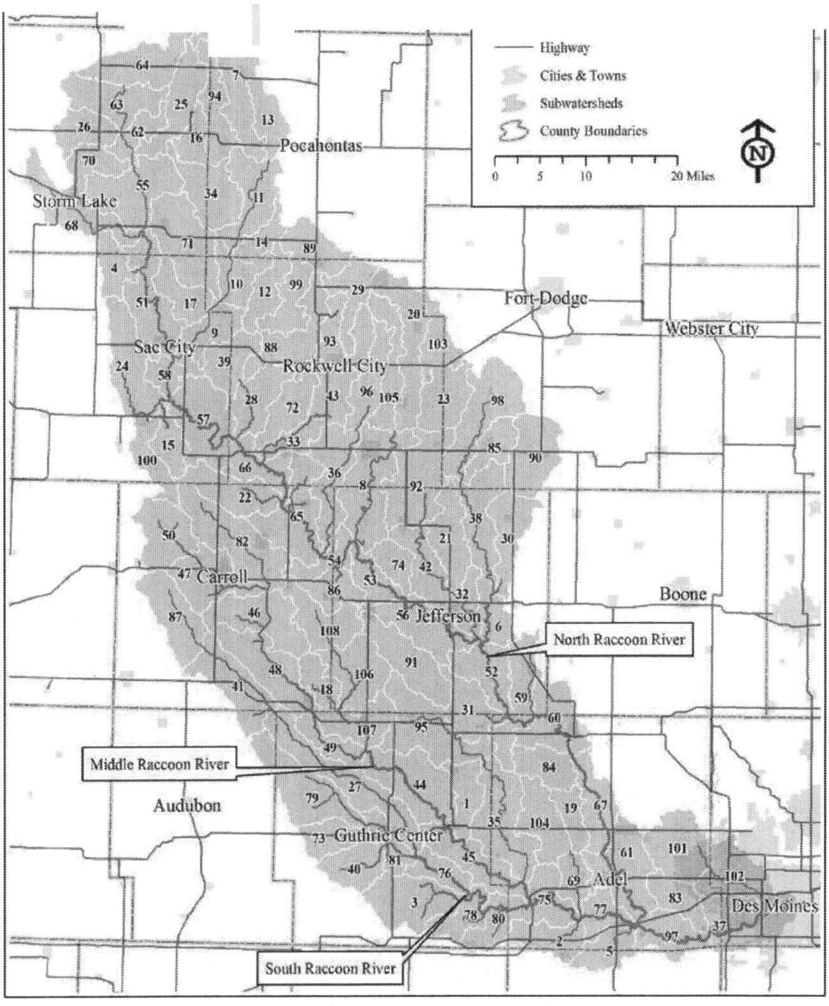

Figure 3.2 Raccoon River Watershed.

Source: Raccoon River Masterplan 2011

Color image of this figure appears in the color plate section at the end of the book.

counties, past Carroll, Coon Rapids, Springbrook State Park, Panora, Linden and Redfield. The three dams located along its course are the Lake Panorama dam and the Lennon Mill dam at Panora, and the Redfield dam at Redfield. Even though the Lake Panorama dam, completed in 1970, was built for recreational purposes only, it has provided some additional benefits by controlling flooding along the Middle Raccoon River. The Middle Raccoon River flows into the South Raccoon River about 1.6 km south of Redfield.

The South Raccoon River, about 116 km long, rises in northeastern Audubon County and flows generally southeastwardly through Guthrie and Dallas counties, past the town of Guthrie Center. South of Redfield, after the Middle Raccoon River flows into the South Raccoon River, a flow gage operated jointly by the United States Army Corps of Engineers (Rock Island District), the United States Geological Survey (Iowa District), while the Iowa Department of Transportation provides data about potential flooding threats.

The north and south forks join in Dallas County just west of Van Meter, and the Raccoon River flows generally eastwardly into Polk County past Walnut Woods State Park and West Des Moines. It joins the Des Moines River just south of downtown Des Moines, the state capital.

The Raccoon River watershed contains some of the richest farmland with the best rainfall regime in the United States. As a result, most of the land is in corn and soybeans, row crops that are constantly re-engineered to be more input responsive. Each new generation of "stacked" seed means more inputs are applied to attain maximum yields. The result is a great deal of agricultural runoff of soil and nutrients. One side of the river is part of the Des Moines Lobe, very flat soil carved out by glacier action. On the other is an older drift plain, a steeper land formation with extremely productive soils. In addition, the watershed is home to an increasing number of hog confinement operations, with the potential of spills from the manure lagoons contaminating the river. As part of the Mississippi River basin, the Raccoon River is also the major source of drinking water for the communities along its path, including the capital city of Iowa, Des Moines. At Des Moines, the Raccoon River joins the Des Moines River, which runs to the Mississippi River at the southeast border of Iowa and Illinois, flowing to the Gulf of Mexico.

The loss of biodiversity, soil organic matter, and topsoil is hastened by the continuing incursion of industrial agriculture, which is justified by the myth of "feeding the world", reinforced by policies which are to "support the family farm" but primarily supporting input manufacturers and dealers and output aggregators, particularly grain companies and meat packers. A number of input manufacturers have headquarters in the state, including John Deere tractors Pioneer Seed and DuPont Crop Protection (merged, as one is bred to be supported by the other). Confined animal feeding operations (CAFOs) and effluent from meat packing plants also contribute to nitrogen load and fecal bacteria count undermining water quality (Centner 2010; Donham et al. 2007; Environmental Integrity Group 2004).

Besides the pollution from agriculture and industry, the Raccoon River and its tributaries have filled with garbage—unused washing machines and other major appliances, general junk that one would have to pay the landfill to take, even old cars. That garbage is not only unsightly; it is dangerous to those who use the river and its banks for sport and recreation. These river users are the most eager to maintain water quality as well a clean riverbed and bank.

Community Capitals Framework

The Community Capitals Framework (CCF) (Flora and Flora 2008) was developed in response to the need of community developers to maintain a holistic approach to their work (see Fig. 3.3). A number of scholars point to the multiple meanings of natural resources and the need to acknowledge all of them (Cortese 2003). The CCF has been used by scholars in different parts of the world when addressing economic development and natural resource issues, particularly in conjunction with livelihood strategies (Cepeda Gomez et al. 2008; Emery and Flora 2006; Flora 2011, 2004, 2008, 1998; Flora and Kroma 1998; Gutierrez-Montes 2005). The framework addresses seven capitals, which all contribute—or detract from—the goals of a healthy ecosystem, social inclusion and economic security. Natural, cultural, human, social, political, financial and built capitals overlap and can at times be fungible. However, the theory posits that ignoring some capitals and privileging others eventually leads all to decline (Flora and Flora 2008).

In natural resource management, natural capital is clearly a major variable, and indicators of its changing quality are often measured by groups that wish to maintain or improve it. In the Raccoon River watershed, the most salient aspects are water quality, water quantity (particularly related to the tilling of fields and rapid water runoff exacerbating flooding),

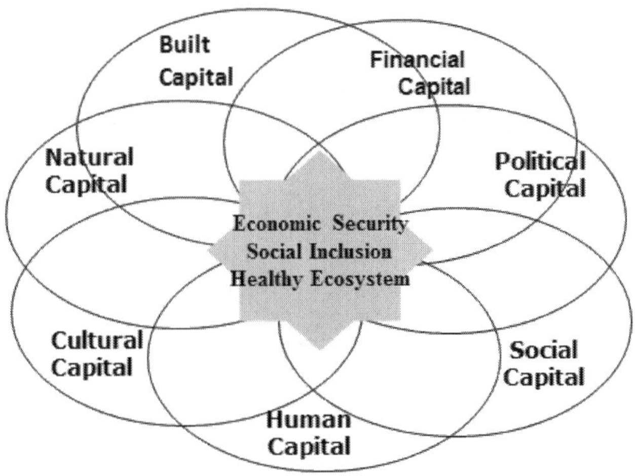

Figure 3.3 The Community Capitals Framework.

biodiversity, with a focus on birds, flora of all sorts, fish, mussels, insects, particularly dragon flies and damselflies, and landscape.

Cultural capital is the way different groups see the world—how the seen is connected to the unseen. Building on Bourdieu (1974), which recognizes the cultural hegemony of those with control of multiple resources, cultural capital includes cosmology, ways of knowing and of being, and what is thought possible to change. Given the hegemony of industrial agriculture in the state, it is particularly difficult for many in the Raccoon Watershed to imagine doing agriculture and managing land differently. Cultural hegemony of industrial agriculture is very strong, so that anything that might interfere with maximizing agricultural profits is "anti-progress". History—place based collective memories—are an important part of cultural capital and are effective in mobilizing action, as knowing what once was gives promise to what could be (Gasteyer ad Flora 2000; Flora and Kroma 1998).

Human capital is made up of the characteristics and potentials of each individual in the watershed. It includes individual levels of education, skills, health and self-efficacy. Two major elements of human capital are involved in the efforts to improve the watershed—education to make adults and particularly children aware of the natural system in which they are embedded, and human health, as the quality of the water impacts the health of urban and rural residents. Of particular concern for those who like to raft, float, and canoe in the river are the diseases that are carried by polluted water. There is particular concern for young children coming in contact with the water in any way.

Social capital at a watershed level involves interactions that can occur with a degree of frequency and comfort. It includes mutual trust, reciprocity, groups, a sense of a shared future and a collective identity. Social capital has two dimensions—bonding, which is within the group and bridging, which is with outsiders. Social movements must build both, particularly in the case of natural resource management (NRM) where an identity with the place is critical for involvement and occurs through the building of social relationships around that place (O'Neil 2004; Kline 2005).

Political capital is the ability to turn norms and values into standards, with rules and regulations to insure that they are adhered to and enforced. Those standards, with their defining rules and regulations, determine the allocation of resources, including the resources necessary to enforce the rules. This definition of political capital is based on the French convention theory (Boltanski and Trèvenot 2006; Trèvenot 1997, 2001). One of the major issues faced in Iowa in the second decade of the 21st century is enforcement of the existing rules and the tendency to give exemptions to the rules to well-connected firms.

Financial capital is forms of currency used to increase capacity of any of the capitals. It is often privileged because it is easy to measure. There is a tendency to put other capitals in terms of financial capital. But it is difficult to adequately access damage to a watershed, even when the polluter is identified, generally as point pollution. When fines are assessed, they seldom reflect the full damage to the waterway and its wildlife. Financial capital can incentivize a variety of behaviors that can have positive or negative impacts on a watershed. Payments for environmental services and fines for polluting can influence land use. Raising a "green" product can increase its value in the marketplace. Taxes and tax exemptions often motivate either environmental exploitation or environmental conservation.

Built capital includes roads, bridges, machinery of all kinds, technology, including information technology, terraces and grass waterways, school kitchens where local food can be prepared, windmills and ethanol plants. Each of these has an influence on how land is used in a watershed. These are very connected to financial capital. For example, in Iowa high rates of cost sharing and rapid depreciation inspired many land managers to convert land from pasture to much more erodible row crops, despite the intended function of stopping erosion. Build capital holds a great deal of promise to increase water quality. For example, Schimmelpfennig and Ebel (2011) found that the adoption of precision agriculture, which encompasses a suite of farm-level information technologies, can improve the efficiency of input use and reduce environmental harm from the over-application of inputs such as fertilizers and pesticides. Yet very few producers have adopted Geographic Positioning System (GPS) maps or variable-rate input application technologies.

Capitals and Watershed Management

Mobilizing Social Capital

Many of the early attempts at organization in the Raccoon River Watershed were reactive, trying to stop financial capital from negatively impacting natural capital. The Raccoon River watershed has long been contested terrain between the forces of industrial crop and animal agriculture and proponents of environmental integrity. In the 1960s, local residents, led by the Krause family, organized against the Corps of Engineers constructing a dam on the Raccoon River. Supporters of the project, including many of their neighbors, viewed them as "stopping progress". They persisted to resist dam construction, and formed the Citizens United to Save the Valley, a group of local residents seeking to preserve the integrity of the Raccoon River watershed. More recently, groups in the watershed have mobilized against the expansion of confined animal feeding operations

(CAFOs), again confronting powerful financial interests with extensive political connections.

Growth Machine Solutions to Environmental Problems

Harvey Moloch (1976) developed the concept "Growth Machine" to explain development patterns in cities, where the push to increase exchange value of property and generate profits leads to constant expansion. Scholars of ecological modernization build on this concept by stressing how problems of natural capital deterioration are "fixed" by ever-greater investments in technology. The Growth Machine is generally supported by the local construction industry, banks, newspaper, and politicians, who see any large construction effort as creating jobs—increasing revenue for all the entities involved.

The need to be in compliance with the Clean Water Act and the threat of sanctions (political capital) motivated market actors to get involved to improve water quality in the Raccoon River. In the mid-1980s, scientists at the Des Moines Water Works noticed a substantial increase in nitrates in the Raccoon River, their water source for the entire city. To dilute the pollutants, the Water Works built a new intake in the Des Moines River to supplement the existing intakes from the Raccoon River and other sources. The increasing level of nitrogen in the drinking water of Des Moines and the 400,000 people served by the Des Moines Water Works spurred activity to reduce the level of pollution. Instead of mobilizing social capital and human capital, as was done with the drinking supply for New York City through pay for whole farm planning in the watersheds that supply water to the city (Pfeffer and Wagenet 2010), Des Moines chose the high tech solution. Because of the high rate of nitrates in the city water that came from the Raccoon River, in 1991 the Des Moines Water Works built the world's largest nitrate removal plant (built capital) at a cost of nearly US$ 4-million. In 2000, the system was in operation at a record of 106 dy at a cost of more than US$ 300,000. Due to the impaired nature of the river and unwillingness of Iowa authorities to require anything seen as increasing farmers' costs, new infrastructure to remove agricultural pollutants is again required.

Even with the huge investment in the nitrate filtration plant (which dumped the extractive nitrates back into the Des Moines River for downstream cities to deal with), nitrogen content in the Raccoon River continued to increase. There were high levels of nitrogen in the water administered by the municipality, and the major water source for the city was the Raccoon River. In 1994 fertilizer companies and the Soybean Association organized farmers in the Raccoon River watershed into Agriculture's Clean Water Alliance (ACWA) to respond to the water quality problems of the city of Des Moines and the threat of new rules and regulations that would limit their use of inputs, particularly nitrogen. The Raccoon River Project of ACWA depended on State and Federal funding and the Northwest Area Foundation. The goal is to educate the farmers in the watershed to better manage the application of nitrogen to their crops of corn and soybeans. That particular effort of working directly with farmers came to end when the funding ended, but the ACWA continues with a regional presence around the Raccoon River to reduce the nutrient loss—specifically nitrates from farm fields—and keep the nutrients from entering the Des Moines and Raccoon Rivers and their tributaries. They work with a variety of input dealers to provide scientific monitoring.

The 2011 membership is primarily agricultural chemical applicators and purveyors, including cooperatives and private applicators http://www.acwa-rrws.org/members.html. The organization has sellers Dow AgroSciences and Mosaic as associate members. Dow AgroSciences, a major input seller of industrial agricultural inputs, according to its webpage, "is a wholly owned subsidiary of The Dow Company, began in 1989 as DowElanco, a joint venture between the plant sciences businesses of The Dow Chemical Company and Eli Lilly and Company. In 1997, DowElanco was renamed Dow AgroSciences when Dow acquired 100 percent ownership of the business from Lilly" http://www.dowagro.com/about/. Mosaic, according to its webpage, "is the world's leading producer and marketer of concentrated phosphate and potash, two of the primary nutrients required to grow the food the world needs. Their business engages in every phase of crop nutrition development, from the mining of resources to the production of crop nutrients, feed and industrial products for customers around the globe" http://www.mosaicco.com/about.html.

The Soybean Association currently provides the seven staff of the ACWA, providing management, monitoring, communications and finance and administration. ACWA claims a wide range of other partners, including the USDA National Laboratory for Agriculture and the Environment of the Agricultural Research Service, Des Moines Water Works, Lake Panorama Association and the US Geological Survey. They reach out to input providers, increasing their knowledge in order to enhance water quality as they convince land managers of the need to apply only as many inputs as are needed. There are some inherent contractions in a program that is aimed at *reducing* input sales and implemented by those whose profits depend on the volume of those sales.

Most of those interested in maintaining the status quo of industrial agriculture and seek projects to put on bandages and avoid regulation are driven by industrial agriculture's concerns for financial capital (Cochrane 1993; Gould et al. 1996; Schnaiberg and Gould 1994; Magdoff et al. 2000). Urban financial interests are represented in the Des Moines Water Works concern with the costs of removing nitrogen from the city's drinking water.

Together, these groups seek linear projects to protect short term financial capital and the natural capital it depends upon.

Cultural Capital as a base for NRM

RRWA was founded on cultural capital: the value of natural capital and the need to protect it. In 2005, Michael Delaney, a professor of sociology at Des Moines Community College and owner since 1988 of 30 acres on each side of the Raccoon River which he had converted to native habitat, brought together a group of friends and neighbors to discuss the deterioration of the river. They decided to do something to stop that degradation of the natural capital that they collectively so highly valued.

They convened a public meeting to address the question of the degradation of the river. As a result, they formed a civil society organization, the North Raccoon River Watershed Association. The founding members included paddlers, a descendant of the first Europeans in the river valley, two property owners with land near the river, a neighbor from Des Moines, and another professor from DMACC. They decided to invite the people in the area to join in their mission to preserve and enhance the river.

They began with regular meetings and minimal dues. They created a webpage, www.northraccoon.org, wrote a constitution and gained legal status as an organization. A critical collective decision was to NOT seek or accept grants from governments, corporations, or foundations, but to be self-supporting through donations from their members and volunteer efforts. This critical decision was made based on observations of other watershed associations' dependence on government agencies and industrial agriculture for funding, which then set the priorities of the associations. They understood that the cause of pollution of the Northern Raccoon River was the product of laws and practices that would not be easy to change. They did not want to lose their potential power for change with obligations to donors from either the public or private sectors.

This decision to maintain their independence was extremely important. RRWA was pressured by industrial agriculture actors, fertilizer companies and the Soybean Association to join their efforts through the ACWA to support industrial agriculture and thus help protect them from potential legislation that might limit their actions and application of inputs on the land. The RRWA leadership was told that another watershed association was not necessary since the Clean Water Alliance had already been formed. The Soy Bean Association asked them to sign on to a large federally funded grant. The RRWA turned them down because they did not want to be seen as co-opted by agribusiness.

In their first year as a formal organization 2005, the RRWA launched Operation Massive Assessment, in which they conducted water quality testing and linked it to secondary data indicating the condition of the watershed. They also interviewed key personnel in many public and private agencies.

They decided that a first action step was to launch an effort to clean up the river and its banks. The RRWA approached the city of Perry to clean up the access road, which had become a *de facto* dump. RRWA asked permission to remove the trash to make the river more appealing to anglers and paddlers. In collaboration with the city of Perry, they took out three huge trucks full of garbage from the river. RRWA calculated that no one could oppose cleaning up the river and eliminating trash. The effort was a victory in public relations, gaining a lot of recognition for the RRWA. Seventy members joined the RRWA during their first year.

At the end of the first year, in collaboration with Cornelia Butler Flora of Iowa State University and Director of the North Central Regional Center for Rural Development, the RRWA analyzed the resources of the association and its members, using the community capitals framework. They found their strengths were in natural, cultural, human and social capital. But they discovered deficiencies in political and financial capital.

The RRWA responded to the analysis using their strong capitals to strengthen their weak capitals. They then set out to include individuals with political clout who shared their basic values and members with financial interest in the health of the watershed. A young commercial art student from DMACC designed a logo, which the RRWA used on cards, tee shirts, and sweat shirts. The slogan on the sweatshirts read, "Fix it and they will come back." This slogan, a variation from that of Field of Dreams,[1] reinforced the idea that the river could indeed recuperate and that life could return to the river. With renewed flora and fauna, people would again use it for recreation and to enjoy the natural capital that the river provides.

RRWA expanded the board of directors, diversifying the talents and influence present on the board. They increased their membership by offering educational meetings. They invited local people who were experts in canoeing, the art of fishing, measuring water quality, and knowledge of wild flowers and animals.

At the same time that they expanded their membership base, the RRWA affiliated with the Environmental Council of Iowa and regularly attended meetings of other environmental groups interested in changing environmental policies. They began their own efforts to enforce the rules to implement the Federal Clean Water Act. They also directed their efforts to influence state legislation and local policy, including elimination garbage on

[1]Field of Dreams was a popular movie made in 1989. It depicted a farmer that build a baseball diamond in his corn field, based on a voice that tells him, "Built it and they will come." The games on the field are played by legendary baseball greats, long dead. And it becomes an enormous tourist attraction, bringing families to reminisce about those simpler, more pristine times—and saving the family farm.

the banks of the Raccoon River and the control of the use of ATVs (All Terrain Vehicles). They joined with other groups in Raccoon watershed, expanding beyond the northern watershed to the middle and southern branches of the river in order to more effectively fight for the entire Raccoon River.

In the RRWA's second year, 2006, a group of land developers proposed to put in a dam on the southern part of the watershed in order to form a lake that was calculated to generate 4,000 lots. The intention was to construct elegant and expensive homes around the shore of the lake. The people of the area organized and asked the help of the RRWA to oppose the dam. Utilizing all their political and civil society connections, the RRWA helped the landowners stop the construction of the dam and the enrichment of a few Iowans. With this success, the RRWA recruited more members.

By the 2006, there were 280 members. The RRWA purchased liability insurance. In part because of the existence of the human and social capital of RRWA, the Department of Natural Resources decided to bring Project Aware to the Raccoon River. Through this alliance, they helped mobilized 400 volunteers to collect 27 tons of trash from the North and middle Raccoon River. Once again, membership increased.

In 2006, the Manning family, local agricultural suppliers, proposed to establish a CAFO in the watershed. The Manning's neighbors, who would be impacted by the CAFO, asked the RRWA to help in their fight against the CAFO. The RRWA testified against the CAFO to the Iowa Commission for Environmental Protection. The Commission decided against the CAFO. But the Manning family threatened to sue the state of Iowa for that decision, and the state gave in and the CAFO was constructed.

In 2011, the RRWA is in their sixth year. They have nearly 500 members and an executive committee of 13 diverse people and US$ 6,000 in their bank account. They use their money to support student research on the watershed. As part of a flexible advocacy coalition, they work where they can to influence public policy and stop private misuse of land in the watershed.

Mobilizing their Capitals

The RRWA uses a variety of mechanisms to transform the watershed. They gather information, educate the public, strengthen the grassroots organization, form networks, grow the organization and win political power.

Their most important sources of power are information and networks. The computer and the Internet are essential for the success of the RRWA. The organization never used ordinary mail to communicate. From the beginning, the RRWA has utilized e-mail and their web site to share information. They can inform and mobilize a good number of motivated and informed people from one day to the next using e-mail. They can connect individuals with their personal capital (financial, social, political, cultural, or human) with others, in seconds, to create more social capital and achieve their goals.

The RRWA has an important presence in the Iowa legislature. Elected officials know the association. Members of the organization work with legislative staffers, and some of their members are members of the state legislature. They know all the legislators in the watershed and work with candidates sympathetic to good management of the watershed who have the potential to win against candidates whose positions are damaging the health of the watershed. Even with the shift of power from Democrats to Republicans in state government in 2011, the RWWA maintains key relationships with policy makers.

Networks

Given that the RRWA doesn't seek funding from foundations or government agencies, they are not seen as competition for other environmental organizations. There is not a vertical structure that can devour the capital resources of the RRWA. They aid the efforts of organizations which share their vision of a healthy watershed. Almost all of the members also belong to other environmental local and national environmental organizations.

Members of the RRWA are also on the the boards of directors of other environmental organization and agencies that are important in Iowa, including the Iowa Environmental Council, the Izaak Walton League, Iowa Natural Heritage Foundation, County Conservations, Sierra Club, Audubon, Nature Conservancy, Central Iowa Paddlers, Iowa Whitewater Coalitions, Concerned Citizens of Iowa, Thousand Friends of Iowa, Pheasants Forever, Ducks Unlimited, Iowa Farm Bureau, Iowa Department of Land Stewardship, Iowa Department of Natural Resources, as well as others.

A recent effort is to influence the power elite in Iowa. A few of the members are part of that power elite. "Old money" has a certain kind of influence that cannot be found in the legislature or in other environmental organizations. The RRWA, in private conversations, asks, "Why has the Iowa elite permitted corporations from outside Iowa to destroy Iowa?" Part of the answer is that the old elite no longer live in Iowa, but in Arizona, Florida, or California, thus have little interest in what happens in the state. Further, they have little influence against corporate power. It is possible that with the vertical organization of the Iowa agricultural economy, closely linking corn and soybeans to biofuel production heavily dependent on Federal subsidies, outside forces drive the Iowa economy (Perrin and Sesmero 2007), RRWA is gradually learning how to not let them also drive environmental policy in the state.

Some members are discouraged by the inability of the RRWA to make immediate impacts on water quality. While the RRWA has had some important victories, the Raccoon River continues to be degraded. The RRWA hoped to quickly change the policies at the state level that are against the public interest. But the power of the industrial agriculture companies, which profit from the pollution, seems at times invincible. For example, Cargill is expanding its CAFOs in Iowa. Cargill is the largest agribusiness company in the U.S. Their revenue in 2007 was more than US$ 119,000,000,000. How is it possible to combat their influence? It seems at times that transnational corporations are in control of the Iowa environment (Perrow 2007).

With this realization, the RRWA has expanded its agenda to national policy, with attention to the 2012 Farm Bill and the Environmental Protection Agency in order to improve the conditions of the river. In order to be more effective at the national level, the RRWA is working with national organizations that are very influential in Washington, D.C., such as Water Keepers, the Clean Water Network, the Sierra Club (national and Iowa) and the local Izaak Walton league.

The growth and influence of the RRWA is due in part to its social capital. The members have a really good time in RRWA activities. The RRWA has many events on the Raccoon River, with contests, demonstrations, music, comedy, and education. Each year they sponsor a conference on "Life on the Raccoon River". They give scholarships to university students to study the quality of water and life in the Raccoon River.

The goal of the RRWA is to follow their path of building consensus among Iowa residents about the current and future conditions of the watershed and the standards and rules around watershed quality that lead to a sustainable future. The constant goal that goes beyond a single project is being constantly engaged to make sure those standards and rules are enforced at the counties, the state of Iowa and at the national level. They seek a variety of opportunities to negotiate with those who are mandated with rule enforcement in order to achieve the desired future conditions of the Raccoon River watershed.

In May of 2009 we made another evaluation of the capitals of the RRWA, noting the relative increase in political and financial capital. The most important lesson learned was the necessity to balance multiple capitals in the evolution of a pro-watershed association.

Lessons Learned

The RRWA works hard to maintain flexibility and effectiveness in the light of intense pressure to maintain the row crops and CAFOs. As active participants in the process of governance of the watershed, the RRWA has a high degree of centrality among environmental actors in Iowa. By focusing on a particular watershed, particularly a very strategic one, they can complement the efforts of other groups.

The RRWA was part of a coalition of environmental and some farm organizations in Iowa working together over a 10 yr period to craft and pass Iowa's Water and Legacy Amendment on the Iowa ballot in November of 2010. The Amendment created a mechanism to "protect Iowa's valuable soils, preserve our farming heritage, restore wetlands, and protect against future flood damage" http://www.iowaswaterandlandlegacy.org/home.aspx. The Amendment created the Natural Resources and Outdoor Recreation Trust Fund. The amendment did not include a tax increase. Trust Fund revenue will come from allocating 3/8 of one percent from sales tax revenue the next time the Iowa Legislature approves a sales tax increase. The amendment passed with 63 percent of the vote—winning passage in 79 of 99 counties and all five of Iowa's congressional districts. This amendment created the trust fund to ensure Iowa's natural resources are protected for future generations if the sales tax is increased in the future. By focusing on non-structural solutions to flooding, it will not be used for levies or creating new areas for the Growth Machine to expand (see discussion in Freudenberg et al. 2009).

Its passage, despite initial bi-partisan support, occurred in an election where the Tea Party, not known for their environmental enthusiasm, organized very hard to get out the vote of those opposed to same-sex marriages to oust judges who supported their legality. In addition the Iowa Farm Bureau mobilized against it at the last minute, calling the amendment a "tax increase".

Environmental policy is threatened by fiscal constraints and neo-liberal policies that glory in the absence of the state as a regulatory presence. Projects for environmental protection are vulnerable to political and economic pressures. RRWA is an example of the many grassroots local environmental groups, drawn together by the multiple values of the ecosystem and the capitals it enhances, to provide consistent pressure and counteraction for environmental quality. Yet, as evidence is gathered from other settings, these coalitions are fragile despite efforts to maintain independence and effectiveness (Novellino and Dressler 2010).

The issues around the river are important for urban and rural residents, as they find commonality in defending it because of its cultural, social, and human capital implications. Struggling for co-management means building strong and flexible networks with a wide variety of market, state and civil society actors.

References

Boltanski, L. and L. Thévenot. 2006. On Justification. The Economies of Worth. Princeton University Press, Princeton.

Bourdieu, P. Cultural reproduction and social reproduction. pp. 71–84. *In:* R. Brown [ed.]. 1974. Knowledge, Education and Social Change. Taylor & Francis, London.

Centner, T.J. 2010. Limitations on the Confinement of Food Animals in the United States. Journal of Agricultural and Environmental Ethics 23: 469–486.

Cepeda Gómez, C.I. Gutiérrez-Montes, A. Imbach, F. Alpizar and N. Windevoxhel. 2008. Tiburón ballena y bienestar comunitario en Holbox, Quintana Roo, México. Recursos Naturales y Ambiente 55: 109–117.

Cochrane, W.W. 1993. The Development of American Agriculture: A Historical Analysis. Minneapolis, MN. University of Minnesota Press.

Cortese, C.F. 2003. Conflicting uses of the river: Anticipated threats to the resource. Society and Natural Resources. 16: 1–18.

Davenport, M.A. and D.H. Anderson. 2004. Getting from sense of place to place-based management: An interpretive investigation of place meanings and perceptions of landscape change. Society and Natural Resources. 18: 625–641.

Donham, K.J., S. Wing, D. Osterberg, J.L. Flora, C. Hodne, K.M. Thu and P.S. Thorne. 2007. Community health and socioeconomic issues surrounding concentrated animal feeding operations. Environmental Health Perspectives 115(2): 317–320.

Emery, M. and C. Flora. 2006. Spiraling-up: Mapping community transformation with the community capitals framework. Journal of the Community Development Society 37: 19–35.

Environmental Integrity Project 2004. Threatening Iowa's Future. Environmental Integrity Project, Washington, DC. http://www.environmentalintegrity.org/pdf/publications/Report_Threatening_lowa_Future.pdf (accessed March 17, 2012).

Flora, C.B. Sustainable production and consumption patterns: Community capitals. pp. 115–122. *In:* G.B. Softing, G. Benneh, K. Hindar, L. Walloe and A. Wijkman [eds.]. 1998. The Brundtland Commission's Report—Ten Years. Scandinavian University Press, Oslo.

Flora, C.B. The Creation of Social Incentives for the Conservation of Biodiversity. pp. 135-144. *In:* R.E. Rhoades and J. Stallings [eds.]. 2001. Integrated Conservation and Development in Tropical America: Experiences and Lessons in Linking Communities, Projects, and Policies. SANREM and CARE,Athens, GA.

Flora, C.B. 2004. Social aspects of small water systems. Journal of Contemporary Water Research and Education 128: 6–12.

Flora, C.B. 2008. Social capital and community problem solving: Combining local and scientific knowledge to fight invasive species. Learning Communities: International Journal of Learning in Social Contexts (Australia), & Kritis: Journal of Interdisciplinary Development Studies(Indonesia). Biosecurity Bilingual Monograph 41–54.

Flora, C. B. Mobilizing community capitals to support biodiversity. pp. 355–364. *In:* J.L.Pujol [ed.]. 2011. The Importance of Biological Interactions in the Study of Biodiversity.InTech http://www.intechopen.com/articles/show/title/mobilizing-community-capitals-to-support-biodiversity

Flora, C.B. and M. Kroma. Gender and sustainability in agriculture and rural development. pp. 103–116. *In:* G.E. D'Souza and T.G. Gebermedhin [eds.]. 1998.Sustainability in Agricultural and Rural Development. Ashgate Publishing Company,Brookfield, VT.

Flora, C.B. and J.L. Flora. 2008. Rural Communities: Legacy and Change. 3rd edn. Westview Press, Boulder, Co.

Freudenberg, W.R., R. Gramling, S. Laska and K.T. Erikson. 2009. Catastrophe in the Making: The Engineering of Katrina and the Disasters of Tomorrow. Island Press, Washington, DC.

Gasteyer, S. and C.B. Flora. 2000. Measuring ppm with tennis shoes: Science and locally meaningful indicators of environmental quality. Society and Natural Resources13: 589–597.

Gould, K.A., A. Schnaiberg and A.S. Weinberg. 1996. Local environmental struggles: citizen activism in the treadmill of production. Cambridge University Press. Cambridge and New York.

Gutierrez-Montes, I.A. 2005. Healthy Communities Equals Healthy Ecosystems? Evolution (and Breakdown) of a Participatory Ecological Research Project Towards a Community Natural Resource Management Process, San Miguel Chimalapa (Mexico). Iowa State University PhD dissertation. Ames, IA.

Iowa Department of Natural Resources. 2011. Raccoon River Water Quality Master Plan. http://dl.dropbox.com/u/6461952/Binder%20RRWQMP%20Final.pdf

Kline, J.D. 2005. Public demand for preserving local open space.Society and Natural Resources. 19: 645–659.

Magdoff, F., J. Foster and Buttel. 2000. Hungry for Profit: The Agribusiness Threat to Farmers, Food, and the Environment. Monthly Review Press, New York.

Molotch, H. 1976. The city as growth machine: Toward a political economy of place. American Journal of Sociology 82: 309–332.

Novellino, D. and W.H. Dressler. 2010. The role of "hybrid" NGOs in the conservation and development of Palawan Island, The Philippines. Society & Natural Resources 23: 165–180

O'Neill, K.M. 2004. Can watershed management unite town and country? Society and Natural Resources 18: 241–253.

Perrin, R.K. and J.P. Sesmero. 2007. How much ethanol, ultimately? Cornhusker Economics. October 17. http://agecon.unl.edu/c/document_library/get_file?folderId=2369822&name=DLFE-18179.pdf.

Perrow, C. 2007. The Next Catastrophe: Reducing Our Vulnerabilities to Natural, Industrial, and Terrorist Disasters. Princeton University Press. Princeton, N.J.

Pfeffer, M. and L. Wagenet. Communities of Interest and the Negotiation of Watershed Management. pp. 109–120. *In:* L.W. Morton and S. Brown [eds.]. 2010. The Citizen Effect: Pathways to Better Water Quality. The Penn State University Press, Pennsylvania.

Schimmelpfennig, D. and R. Ebel. 2011. On the Doorstep of the Information Age: Recent Adoption of Precision Agriculture, EIB-80. Economic Research Service. U.S. Dept. of Agriculture.

Schlosberg, D. and R. Sara. 2008. Ecological modernization, American style. Environmental Politics 17: 254–275.

Schnailberg, A. and K.A. Gould. Treadmill predispositions and social responses: Population, consumption, and technological change. pp. 68–91. *In:* A. Schnailberg [ed.]. 1994. Environment and Society :The Enduring Conflict. St Martin's Press, New York.

Trèvenot, L. 1997. Un gouvernement par les normes: Pratiques et politiques des formats d'information, Cognition et Information en Sociétè 8: 205–241.

Trèvenot, L. Constituer l'environnement en chose publique. Une comparaison franco-américaine. *In:* J.-P. Blais, C. Gillio and J. Ion [eds.]. 2001. Cadre de Vie, Environnement et Dynamiques Associatives, Paris, PUCA.

4

An Analysis of Public Participation in the Lake Ontario—St. Lawrence River Study

Gemma Carr[1,a,*] *Daniel P. Loucks*[2] and *Günter Blöschl*[1,b]

Introduction

The Great Lakes (Fig. 4.1) are the largest freshwater bodies in North America. They are located on the boundary between Canada and the United States. Managing the water levels and flows associated with the five Great Lakes and connecting rivers, including the St. Lawrence River that flows from the most downstream lake to the Atlantic Ocean, is the responsibility of a bi-national organization established to manage all border waters between Canada and the United States. This organization, called the International Joint Commission (IJC) operates under a Boundary Waters Treaty established in 1909. The IJC serves at the pleasure of the governments of the two countries, and its effectiveness depends in part on what the governments want with respect to the management of the transboundary

[1]Center for Water Resource Systems, Vienna University of Technology, Karlsplatz 13/222, A-1040, Vienna, Austria.
[a]Email: carr@waterresources.at
[b]Email: bloeschl@hydro.tuwien.ac.at
[2]Cornell University, Ithaca, NY 14853, USA.
Email: DPL3@cornell.edu
*Corresponding author

waters. When requested to do so by the two federal governments, the IJC has the authority to resolve disputes over the use of water resources that cross the international boundary. Most of its efforts for the Great Lakes have been devoted to carrying out studies requested by the governments and advising the governments about problems.

Both countries are stakeholders in the Great Lakes Basin. Especially the Canadian provinces and the US states bordering the lakes have an interest in their levels, flows, and water quality. The economy of that region depends in part on the freighters that transport products into and out from the region to other parts of the world. The flows between some of the Great Lakes and to the St. Lawrence River provide hydropower benefits to both countries (Eberhardt et al. 1996). In addition there are benefits derived from water supply, recreational facilities and ecosystem services. Those living on the shores of the Lakes and River benefit from their waterfronts, but also must weather the waves, wind and erosion that characterize shorelines during storms.

Hence in a narrow sense the Great Lakes system is co-managed. Co-management—a situation in which two or more distinct social entities share management functions, entitlements and responsibilities—takes place

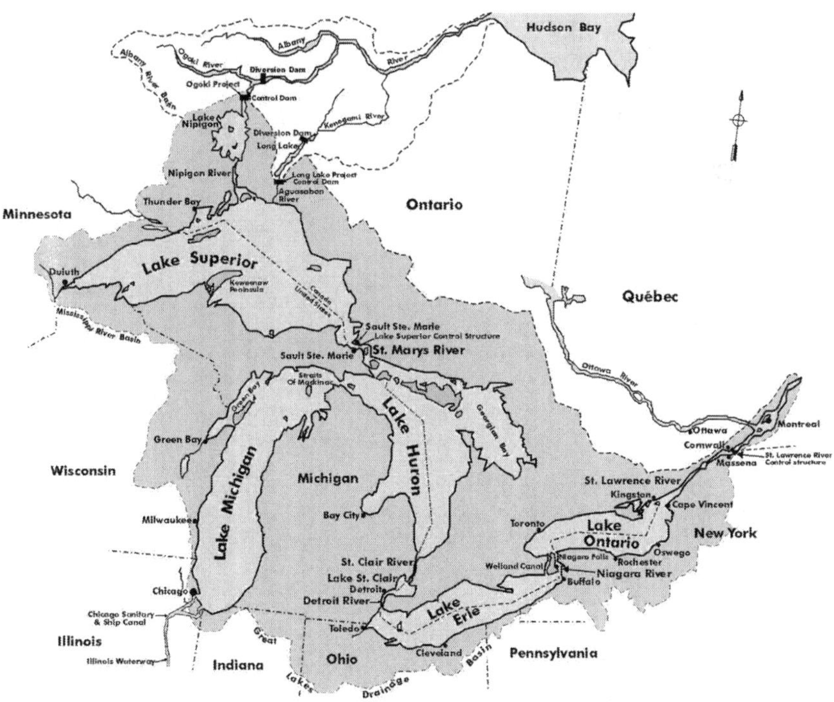

Figure 4.1 The Great Lakes Basin and St. Lawrence River (from GLIN 2011).

through the IJC. It serves as a means whereby both countries can negotiate differences of management goals as they arise. But in a broader sense co-management means managing water together as stakeholders who have an interest and legal right to manage as well as those who are interested but may not have any legal rights. Co-management can be a highly dynamic, evolving, adaptive and forward looking process.

Background to the Lake Ontario—St. Lawrence River Study

The International Joint Commission

The IJC has six members, called commissioners, three appointed from each country by the heads of the federal governments. They are political appointees. They are served by a small staff of experts. The authors of the 1909 Boundary Waters Treaty saw the Commission not as separate national delegations, but as a single body seeking common solutions in the joint interests of the two countries. All members are expected to act independently of national concerns, and few IJC decisions have split along national lines.

The IJC has three responsibilities for the Great Lakes under the original treaty (IJC 1998). The first is the limited authority to approve applications for the use, obstruction or diversion of boundary waters on either side of the border that would affect the natural level or flow on either side. Under this authority, it is the IJC that determines how the control works on the St. Lawrence River will be operated to control releases of water from Lake Ontario. Ten individuals (five each from the US and Canada) form the IJC International St. Lawrence River Board of Control. Their task is to ensure that the quantity of water released from Lake Ontario on a weekly basis conforms to the current regulation plan (Clinton Edmonds and Associates 2002).

The current operating policy or plan, 1958-D, has been in effect since October 1963. It was designed for the hydrologic conditions experienced from 1860 to 1954. It has not performed well under the extreme high and low water supply conditions experienced since that time (Werick 2011). As a result, the Board of Control has on occasion deviated from the Plan, as authorized under the existing Orders of Approval. However, over time the Board of Control has increasingly deviated from the Plan to better meet changing needs and interests not considered when the plan was created.

The second responsibility of the IJC is to conduct studies of specific problems as requested by the governments. Funding comes from the two governments. The implementation of the recommendations resulting from IJC studies is at the discretion of the two governments. A number of such studies have been undertaken in the history of the IJC. This chapter is about one of them, the Lake Ontario St. Lawrence River Study (LOSL) which ran from 2000 to 2005.

The third responsibility of the IJC is to arbitrate specific disputes that may arise between the two governments in relation to boundary waters. The governments may refer any matters of difference to the Commission for a final decision. This procedure requires the approval of both governments and has never been invoked.

The LOSL Study

In April 1999, the International Joint Commission informed the governments that it was becoming increasingly urgent to review the regulation of Lake Ontario levels and outflows in view of dissatisfaction on the part of some interests, in light of environmental concerns and in response to potential climate change conditions. Thus on December 11, 2000, the Commission issued a directive to the International Lake Ontario—St. Lawrence River Study Board, which it had appointed, to:

i) review the current regulation of levels and flows in the Lake Ontario—St. Lawrence River system, taking into account the impact of regulation on affected interests;
ii) develop an improved understanding of the system among all concerned; and
iii) provide all the relevant technical and other information needed for the review.

The Study was to assess the current operating policy of the LOSL system (Fig. 4.2) and to suggest improved policies especially taking into account new goals or objectives not considered when the current policy was defined. Yet the Study Board recognized from the beginning, that it would be unlikely that a policy could be identified which would satisfy all interests.

The subsequent five-year, US $ 20 million Study was conducted with funding provided equally by the U.S. and Canadian governments and through participation of government agencies, individuals and non-governmental organizations in both countries. The tangible outcome was the creation of three potential operating plans (A+, B+ and D+) which were presented to the IJC. The IJC retains decision-making authority over the selection of the final plan to be presented to the national governments for approval. In 2006 (after the Study had ended), the IJC asked several of the Study experts to further develop Plan D+ in an attempt to get it to more closely align with the recalibrated needs of the stakeholders. The result was Plan 2007. The IJC announced that it aimed to implement the new plan by the end of 2008, but would carefully consider public opinion in its decision (IJC 2008). Plan 2007 went out for public consultation and comment in the

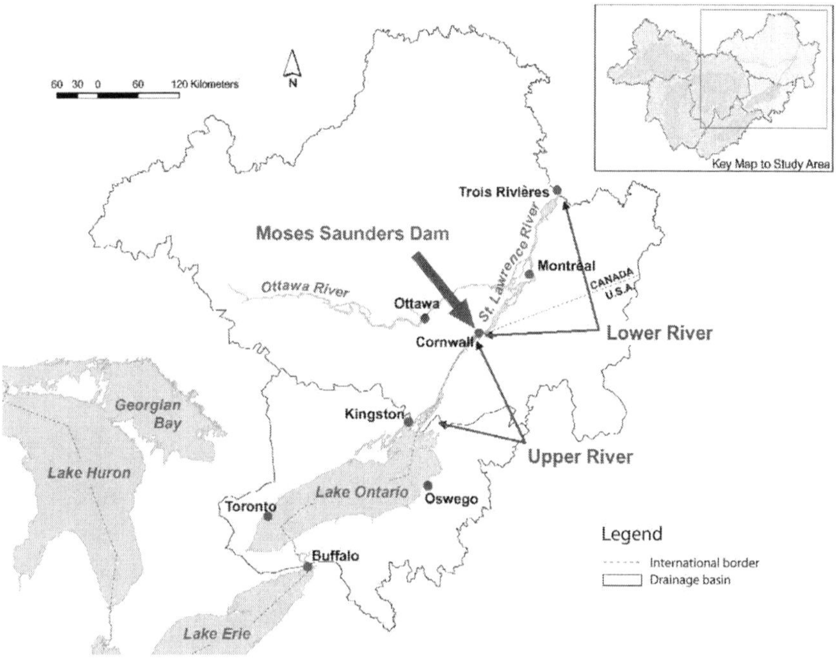

Figure 4.2 The Lake Ontario—St. Lawrence River System of interest in this study (from LOSL 2006). Lake Ontario water levels and St. Lawrence River flows are controlled by the releases from the Moses Saunders Dam.

summer of 2008 and was later dropped from consideration due to broad stakeholder objections. The IJC continues to move forwards towards refining and implementing an updated plan.

The Study Board created to oversee and carry out the five-year study consisted of seven representatives from each country, chosen for their expertise in some aspect of the study, or as representatives of key institutions, such as New York State and the province of Quebec. From their first meeting the Study Board recognized the importance of public or stakeholder buy in to any operating policy they might recommend to the IJC. The Commission required the Study Board to form a public interest advisory board. Hence it was important to devise ways of involving the public. To aid in that effort a Public Interest Advisory Group (PIAG) of influential citizens from the communities bordering Lake Ontario and the St. Lawrence River was formed. They had their own budget, which amounted to approximately US$ 2 million (10 percent of the study budget) (LOSL semi-annual progress reports 2000–2005) and reported to the Study Board as well as to the IJC. The co-chairs of the PIAG served on the Study Board. Their job was to keep both the Board and the public informed as to what planning or technical studies were taking place and any public issues or concerns that needed attention or that might impact the policies being developed. Except for the Study Board Co-chairs, the members of the Study Board and PIAG served without compensation. For day-to-day operational requirements and decisions, a public interest "Outreach Committee" was formed. This committee developed public involvement strategies, which were then approved by the PIAG, and ensured that Study Board information reached the PIAG in a timely manner. The sub-committee was led by the two Study co-chairs; and comprised of the PIAG co-chairs; the two study managers and the public information specialists hired to provide assistance to the PIAG. This group provided most of the intellectual input into the development and execution of the public involvement program, because it coordinated the needs of open public engagement with the requirements of the formal "shared vision planning process". Figure 4.3 illustrates the organizational structure of the LOSL Study.

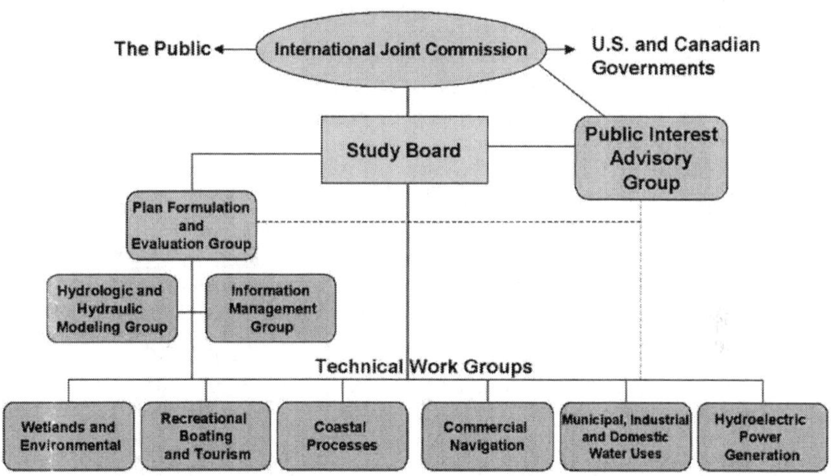

Figure 4.3 Organizational structure of the LOSL Study (from LOSL 2006).

Basin Interest Groups

In an attempt to better structure the planning process, multiple interests were lumped together into interest groups, defined by the uses to which they put the Lake and River. These primary stakeholder groups included:

- Power producers, NY Power Authority, Ontario Hydro, industrial, residential and commercial energy users who benefit from electrical power generation.
- Commercial shippers, Seaway Authority, various Port Authorities, seamen's unions, producers and consumers of bulk goods who benefit from commercial shipping (Fig. 4.4).

Figure 4.4 Freight being transported on the Great Lakes and St. Lawrence River (from http://www.boatnerd.com).

- Shoreline property owners, local communities who are concerned about shoreline maintenance and development.
- Boaters, marinas, local communities who benefit from recreational boating activities.
- Municipal Water Suppliers, populations in communities who get their public water supplies from the Lake or River.
- Environmentalists, anglers, trappers, hunters, hikers, bird watchers, tourists who enjoy and benefit from a diverse healthy natural environment.

The Role of Public Involvement

The Study was initiated in part because the current Plan was not satisfying some of the various interest groups—especially those that were not formally acknowledged in the Treaty of 1909, particularly the environment and the recreational boating industry. A further impetus for the study was dissatisfaction of property owners on the south shore of Lake Ontario. They had suffered in recent years from floods and erosion due to high water levels combined with storms on the lake, and thought the Board of Control should have been able to regulate the system to mitigate or prevent those damages. The study was also initiated because it was not obvious from the beginning just what policy of lake level and river flow regulation would best, or even better, satisfy every stakeholder interest group. The Study Board as well as the IJC knew that there would be conflicts among various interest groups in the basin. Hence stakeholder participation was viewed as being absolutely essential to guide the work toward defining preferred policies and hence for the successful conclusion of the Study.

Stakeholder and public involvement, or participation, has gained increasing momentum as a component of environmental management over recent decades. Strategies have shifted from informing and educating people on the "right" strategy, often determined by experts, towards co-production of management strategies and systems through collaborative work between "experts" and resource users. Co-produced strategies are considered to better reflect the realities of resource use and be more suitable and acceptable to resource users (Pahl-Wostl et al. 2007). The LOSL Study used a model of co-production to shape the public involvement process.

The 20 member, bi-national PIAG was responsible for providing public involvement guidance, consultation and assistance to the Study Board, and to periodically report to the IJC on its activities, findings and recommendations. They were strongly supported by the Outreach Committee, which provided much of the strategic advice to PIAG for their approval. The PIAG raised public awareness through disseminating Study progress and findings through information meetings, newsletters and other media. They also served as a conduit for public input into the Study through holding public meetings and workshops and conducting surveys. The PIAG, in consultation with the Study Board, also worked with grass-roots organizations and interests throughout the Study area and conducted public participation activities at strategic points in the Study to:

- identify and use local expertise and information;
- consult with the public on critical or potentially controversial Study findings before related Study components were approved by the Study Board;
- disseminate plain language information to enhance public understanding of the causes and problems related to fluctuating water levels and of the consequences of proposed solutions;
- identify and consider priorities and preferences of the public as alternatives were defined; and
- consult with the public on Study findings and recommendations prior to their adoption by the Study Board.

During the Study the PIAG and the Study Board gave several hundred presentations to the public. The aims of the public meetings and information sessions evolved through the duration of the Study. Earlier, public involvement focused on information provision and raising awareness. This shifted to information exchange whereby comments from the public surrounding their concerns over water levels were actively sought through surveys, questionnaires, and finally consultative public meetings. These meetings were relatively informal in that anyone could ask questions and give their opinions to the Study Board.

The remainder of this chapter explores the role of public involvement within a co-management setting as it took place in the LOSL Study. It attempts to evaluate the processes by which public and stakeholder participation took place, and to identify some of the outcomes and non-tangible achievements to date.

Methods Used to Evaluate Stakeholder and Public Involvement in the LOSL Study

Evaluation is essential to provide insight into how a program or approach is functioning or has functioned, and to identify strengths, weaknesses and potential improvements (Beierle 1998; Chess and Purcell 1999; Muro and Jeffrey 2006). It also forms part of a learning cycle (Blackstock et al. 2007). To evaluate public involvement in the LOSL Study an evaluation framework was devised that focused on the processes by which participation took place and the outcomes that emerged.

Evaluation Framework

Three main types of evaluation can be found in the participation literature (see Carr et al. 2012). (a) Process based evaluation focuses on how participation has taken place (Conley and Moote 2003) or the quality of the process (Beierle and Konisky 2000). (b) Intermediary outcome based evaluation identifies outcomes such as agreements over plans or proposals (Burgess and Chilvers 2006) or non-tangible outputs such as innovation in decision making, relationship building and empathy towards alternative values and ideals (Connick and Innes 2003). Intermediary outcomes do not relate to a direct change in resource management at the point in time at which they are evaluated, but they are likely to be essential to achieve resource management improvements. (c) Resource management outcomes are considered to be longer term responses. They are always evaluated with regards to specific interests such as the implementation of an agreement, or a measurable improvement in ecological health (Beierle and Konisky 2000).

Because an updated operating plan (i.e., a resource management outcome) had not been implemented at the time of our evaluation we chose to evaluate the Study based on processes and intermediary outcomes. A set of criteria for each of these evaluation types was selected from the literature based on suitability to the LOSL Study and the available resources (Table 4.1). Process criteria were derived from studies that have identified desirable characteristics of stakeholder participation through case study

Table 1. Framework and criteria used to evaluate the LOSL Study.

Process based evaluation	
Criteria	**Data sets for evaluation (performance indicators)**
Access Information and meetings are accessible to participants (Klinke 2009; Mostert et al. 2007; Walker et al. 2006).	Participant perspectives on their capacity to understand and get hold of information. Participant perspectives on the timing and location of public meetings based on public meeting transcripts and individual interviews with participants.
Cost effectiveness The costs of implementing the programme are balanced by the importance of the issue being addressed (Beierle 1998; Rowe and Frewer 2000).	Participant perspectives on cost effectiveness based on public meeting transcripts.
Deadlines and milestones There is a detailed agenda with deadlines and promise of investment money once agreements are reached (Jiggins et al. 2007).	Assessment of agenda, deadlines and milestones reported in study documents.
Facilitation Facilitation is impartial (Jiggins et al. 2007; Moote et al. 1997; Rowe and Frewer 2000). The process focuses on shared values rather than entrenched interests (Beierele 1998; Jiggins et al. 2007).	Assessment of facilitators' impartiality based on public meeting transcripts. Assessment of degree to which process focuses on shared positions and interests based on public meeting transcripts and meeting minutes.
Knowledge inclusion A variety of knowledge is included to help make informed decisions (Beierle 2002; Hedelin 2007; Reed 2008).	Assessment of participant input to selection of study performance indicators based on study board meeting minutes and public meeting transcripts.
Legitimate decision making Decision making is clearly displayed, based on evidence rather than political motivations and participants inputs have a genuine impact (Rowe and Frewer 2000; Webler et al. 2001).	Assessment of explanation of decision making process and impact of participant inputs based on public meeting transcripts and material distributed to public.

Table 1. contd....

Table 1. contd....

Process based evaluation	
Criteria	**Data sets for evaluation (performance indicators)**
Representation Participants represent a broad and cross-cutting section of interest groups (Blackstock et al. Chilvers 2009; Hedelin 2007; Mostert et al. 2007; Rowe and Frewer 2000).	Classification of participants at public meetings according to interest group (based on meeting transcripts) and number of interest groups represented at each meeting.
Intermediary outcome based evaluation	
Agreements are reached A final agreement on suitable strategy is indentified and broadly supported by all participants (Leach et al. 2002).	Assessment of recommendations made by Study to IJC in 2006.
Innovation Strategies are developed which are more creative and context specific (Connick and Innes 201; Newig and Fritsch 2009).	Assessment of impact of participant involvement in contributing management ideas and refining management options.
Interaction and network development The process leads to greater interaction between different interest groups and awareness to others activities, needs and values (Leach et al. 2002).	Assessment of participant awareness to needs and values of other interest groups based on public meeting transcripts and individual interviews with participants. Evidence of network development between two or more interest groups based on public meeting transcripts.
Institutional change Institutional functions, roles or structures are modified to reflect participant ideas, values or requirements.	Assessment of participants' comments that identify institution change requirements.
Shared knowledges and information Data, information and knowledge is generated that is accepted and trusted by all participants (Steyaert et al. 2007; Jiggins et al. 2007).	Participant trust in data and information generated by the Study based on public meeting transcripts.

analysis and empirical research into participant perspectives. Good process characteristics are those which produce a legitimate or fair process, and a process that is effective and well run (Rowe and Frewer 2000; Webler 1995; Webler 1999). Intermediary outcome criteria are based on work which has associated participation with a range of actions or non-tangible outcomes (Connick and Innes 2003). They relate to benefits such as increasing the connectivity between different stakeholders and government networks which may raise trust and the willingness of participants to invest in joint work (social capital) (Pretty 2003). They also relate to achievements, which may perhaps fall outside the original objectives or scope of work. These may include changes to strengthen or modify existing institutions, or the development of new organizations. A decision that is based on shared information created by many stakeholders and is accepted and trusted may reduce dissatisfaction over the final decision (Bentrup 2001). These outcomes might be essential to allow new strategies to be implemented in a quick and efficient manner.

Resources and Data Sets Available for Evaluation

The LOSL Study offers a substantial collection of material documenting the public involvement activities that took place. All resources used to conduct this evaluation (except for material collected by the authors during individual interviews with persons involved) are currently published on the internet (http://www.losl.org; http://www.ijc.org/en/activities/losl/index.php).

The availability of resources reflects the attention paid to documentation and reporting during the Study. This reflects the recognition by the IJC and the Study Board that transparency and access to information throughout the process are essential for co-management. Reporting, documenting and ensuring public access to all information were of high priority. This leads to a substantial and unique data set with which processes and outcomes can be evaluated. Data sets used in this research are:

- Transcripts from 14 of the 25 public meetings organized by the PIAG and Study Board in 2004 and 2005 (held to gain feedback on ongoing development of operating plans (2004) and to identify public opinion on the three plans put together by the Study (2005)). Incomplete transcripts and those in French have not been included in the analysis.
- Transcripts from 8 of the 10 public hearings organized by the IJC in 2008 (these formal hearings followed information sessions and were held as part of the IJCs consultation process for Plan 2007). Transcripts in French have not been included.
- Minutes from Study Board meetings held between 2000 and 2006.
- Half yearly and annual reports by PIAG and the Study Board.

- Semi-structured interviews with a member of PIAG and a professional member of the Study Board conducted by one of the co-authors (GC) in September 2010.
- First-hand experience from a Study Board member who co-authors this chapter (DPL).

Data Analysis

Data analysis was structured by the evaluation framework (Table 4.1). Standard methods for analyzing interview transcripts were employed and the material was read and phrases, paragraphs or dialogues were grouped according to dominant key themes such as institutional issues, process factors, data and analysis concerns, facilitation (Kitchin and Tate 2000). The material was further subdivided and recombined according to its relevance to each of the evaluation criteria. Similarities and trends in the experiences and opinions reported, as well as diversity and controversy then became apparent, and the achievement status of each of the criteria could be assessed.

Public Involvement in the LOSL Study

Over the five-year study period, hundreds of people and dozens of organizations participated directly in the Study. The volunteers of the Public Interest Advisory Group were central to the success of the undertaking, contributing significantly and uniquely to the work of the Study Board. PIAG members were fully integrated into the Study Team, providing advice, feedback and input during all phases of the Study process. This included representatives from First Nations whose issues are complex and knowledge is great, but not always in written form. The final PIAG report shows that between April 2004 and November 2005, 139 presentations were given to an audience of approximately 5850.

Our evaluation findings have been grouped according to whether they relate to the process by which participation took place, or to the outcomes that emerged from the process.

Process-based Evaluation

Access. A legitimate and effective process should ensure that all interested or affected individuals have access to resources and opportunities to take part. The PIAG used a range of technology to reach interested parties and to attempt to capture their involvement. Group members coordinated the Study's communications process, which included publication of the Ripple

Effects newsletter, creation of the website, stakeholder meetings, workshops, a speaker's bureau, roundtable meetings and public meetings. The Group published a glossary of terms and led the creation of Study banners and brochures. The PIAG were supported by two IJC communications assistants (one from US and one from Canada) for arrangements such as booking venues, newspaper and radio advertising and sending out invitations to mailing lists developed by PIAG and the Study Board throughout the duration of the study.

Regarding access to information, participant feedback suggests that some of the powerpoint presentations and graphs exhibited during the presentation were received very well at the meetings where they were shown and helped people to understand the complexity of the Lake Ontario-St. Lawrence River system in a very direct way. The Study Group (the Study Board and the PIAG) seemed to be receptive to advice and willing to make improvements. Only one comment was made during the 2004 public meetings (Olcott, New York, September 17, 2004) suggesting that the level of detail in the presentations was too great, and that simplification would help the audience to understand the system better. The Study Board and the PIAG adjusted their presentations before the 2005 public meetings and no comments were documented that suggests further confusion.

The PIAG recognized that one of their major challenges was ensuring that potentially affected stakeholders were aware of the Study and received information on meetings and publications (comment by PIAG member during public meeting, Trois-Rivières, Québec, September 17, 2004). During the course of the Study the PIAG compiled a mailing list and encouraged those on their list to sign up friends and colleagues who would be interested. Considering the scale of the Study and the number of people potentially affected by changing lake level regimes, there were very few comments made during the 2005 public meetings from individuals who felt their participation had not been adequately sought by the Study Group earlier in the process.

Regarding access to meetings, the PIAG created a public meeting plan. Their plan ensured that a wide variety of interest groups would have a local meeting. The timings of the meetings were carefully considered, taking place in summer when recreational boaters and seasonal property owners would be using the lake and available to attend.

Cost-effective. The extent to which participants viewed the process as good value for money may reflect how effective it was. There are several aspects to cost effectiveness. The importance of sound science on which to base decisions was highlighted during the two semi-structured interviews. One comment was also made by a participant in a 2005 public meeting:

> "I'm very happy that we're spending $ 20 million to come up with data. I think decisions that are of this magnitude ought to be based on data and not on interests, although we all have interests and I certainly as a boater and recreational user have interest in the river. So I'm glad we spent the money."
>
> (Massena, New York, June 22, 2005)

A second aspect of cost-effectiveness relates to the potential economic benefits that can be gained from implementing a new plan compared to the current plan. These gains effectively offset the costs of running the study. Interest groups who were set to benefit economically from a new plan were more inclined to view cost-effectiveness in these terms.

A third concern tended to be voiced by those interest groups who had more to lose than to gain from any new plan (shoreline property owners whose land could be put at greater risk of erosion from changes leading to higher lake levels). This group was concerned that the cost of the study would influence decision making leading to rejection of the existing plan solely because money had been spent formulating a new plan. A comment made by an IJC commissioner at one of the last Study Board Meetings suggests such a bias could influence decision making:

> "What do you think congress will say if we say the best thing to do is nothing after taking five years and spending $ 20 million?"
>
> (Study Board Meeting, Washington DC, December 5, 2005)

Deadlines and milestones. Clear deadlines and rewards, such as funding to assist with implementation, have been shown to help focus a participation process and encourage cooperation between participants (Jiggins et al. 2007). The mandate of the study was to review the existing criteria for regulation of the LOSL River level and flows, and to provide options and recommendations to the IJC in five years (IJC Plan of Study 1999). The IJC retained decision-making authority on selecting and implementing an option and did not set a deadline by which a decision would be made. The context in which the Study was operating was a significant factor. The existing 1909 Boundary Waters Treaty protects stakeholders' interests and therefore any decision that may jeopardize interests would need national government support to either accept the risks or agree that mitigation measures are in place or will be implemented that adequately protect interests that may be at increased risk due to changes in regulation. The IJC are therefore challenged in their decision making capacity by the institutional and political systems operating within each country.

Comments from the 2008 IJC public hearings suggest some stakeholders were frustrated with the delay in decision making. The extent to which the slow decision making leads stakeholders to give up on the process cannot be identified, but over 1000 comments were submitted to the IJCs consultation on Plan 2007 (http://www.ijc.org/en/activities/losl/comments_order_plan.php) which suggests that interest in the issue and willingness to contribute to the process remained, at least until 2008.

Facilitation. Impartial and unbiased facilitation has been identified as essential to encourage discussion and ensure everyone who wants to be heard is given opportunity to speak (Jiggins et al. 2007; Moote et al. 1997; Walker et al. 2006). The range of separate and diverse interests being met in the Study led to particular challenges in unbiased facilitation, especially towards the end of study when the options that would be put forward to the IJC were being debated in Study Board, PIAG and public meetings. Most members of all groups seemed to hold some affiliation or personal interest and identified their own preference towards either the status quo or one of the possible plans. Despite this, personal bias from PIAG or Study Board facilitators could rarely be detected in the meeting transcripts. However, the titles given to the developed plans were perceived by some participants to show bias:

> "My concern is, when you have this amount of data floating around, it doesn't take much for some small group of people to name it. And I think names become labels, which become very dangerous. For someone to claim that Plan D is a blended benefits plan which to a whole bunch of people who hadn't looked at it would say, blended benefits, sounds pretty good. I think that's doing a disservice to my $ 20 million."
>
> (Massena, New York, June 22, 2005)

> Participant: "Everything, every piece of material that I have seen come out from the Study Board including a letter to the editor in the Watertown Times from Mr. Stakhiv [US Study Chair] has been able to point out to you how you have been slanting the material.
>
> Study Board Member: We're not, we're not biasing any of the plans. We developed three plans. The mere fact that we developed an environmental plan, that you have available for consideration, and we're merely, we're sending three plans forward for the IJC to consider. The Study Board doesn't have any particular—
>
> Participant: Sir, The simple, the titles, the titles alone show a slant."
>
> (Alexandria Bay, June 23, 2005)

The second part of the facilitation criteria is concerned with whether the process attempts to build up shared values between the participants, rather than more deeply entrenched personal interests. The meeting minutes, reports and transcripts all suggest that from the very beginning of the study, the narrative was that compromise was needed as it would not be possible to devise a management plan to satisfy all interest groups at all times. During the 2004 public meetings the PIAG attempted to hold joint question and answer sessions, linking meetings at two different locations by telephone. The aim, as described by a member of PIAG, was to raise awareness to the need for compromise:

> "And what we heard last night … was, the people in Hamilton last night, or near Hamilton, were saying, well, we want the water levels down, and people in Massena were saying, well, if you drop them too much, this is what it's going to do to us. And so we want the people in the two areas to hear each other, hear the problems, and therefore hopefully understand when it comes to the time of making decisions, that it's going to be a tricky thing to try to balance and figure out how we can serve everybody around the system without hurting anybody disproportionately sort of thing. But we think, and the PIAG had big discussions about this, and we really think that the different areas have to hear each other to know what they are concerned about, so that you understand when the decisions are to be made, that there's going to have to be a lot of give and take, and okay—well, that's the word. Give and take. At certain times of the year, for different people, for different purposes."
>
> (Alexandria Bay, New York, June 23, 2005)

Knowledge inclusion. The PIAGs principal objective was to ensure that Study results consider the interest and "natural knowledge" of the public (LOSL 2006). Of primary interest to the Study Board was the relationship between what can be controlled or managed, i.e., lake levels and river flows, and indicator values representing the various interests. Technical working groups were established and paid by the Study Board to define these relationships. Individual members of the PIAG acted as liaisons to the various technical working groups of the Study. The PIAG helped to focus discussions in a practical way, giving the Board real world implications for decisions. PIAG members suggested metrics in the Coastal, Environment and Recreational Boating technical work groups and played an integral role in providing input from the public into the Study's Performance Indicators (quantitative indicators such as amounts of hydropower produced, freight tonnage shipped, days in boating season with levels above a specified threshold level, amount of shoreline erosion, etc.) (see Loucks 2006). Public meetings held during the summer of 2003, the Study newsletter and the website were all used to collect public feedback on performance indicators suggested by the technical work groups. A wide range of comments were made, particularly regarding environmental indicators. The appropriate technical work group then responded to each suggestion (PIAG Year 2–3 Report, Appendix H).

The 2004 public meetings identified many more suggestions and concerns with the performance indicators used. At this stage the technical work groups tended to defend their approaches, probably because they felt few changes could have been made to the analysis at this late stage in the Study. This is illustrated by a comment from a member of the Plan Formulation and Evaluation group:

> "The performance indicators are in pretty good shape, and they're pretty good performance indicators. That's not to say that we won't listen, but I think we have a year left on the study, approximately, and really the focus now should be on plan formulation and evaluation. I think if you took a good look at these performance indicators, they're a very good, robust set."
>
> (Alexandria Bay, New York, August 19, 2004)

During another meeting in 2004 a shore line property owner tried to suggest that a performance indicator which considers the taxes paid by shoreline dwellers would better address riparian's concerns (Oswego, New York, September 2, 2004). A member of the Coastal Processes technical work group responded that this was not being considered because the group felt confident that their other performance indicators were accurately capturing the effects of lake levels on riparians. During the 2005 public meetings a couple of participants with shoreline property interests returned to this issue and emphasized that they felt the lack of consideration for issues such as property value and tax revenues had not been adequately addressed and had led to inaccurate evaluation of lake level impacts on riparians. During the 2008 IJC hearings, four elected officials and three individuals from shoreline communities with an interest in keeping Plan 1958-D with deviations, argued that the omission of property values and tax revenues invalidated the economic evaluations performed by the Study.

Legitimate decision-making. The Study documents all show that throughout the process the Study Board and the PIAG clearly explained that the final decision on a plan would be taken by the IJC. No concerns appear to have been voiced by participants regarding the decision making arrangements. However, there was concern among some participants about whether their voices and opinions would truly be considered. At many of the meetings, elected officials spoke to express what many in the audience strongly supported. A number of resolutions were passed by municipalities, mostly in New York State, confirming the views expressed (observations of DPL during the LOSL Study 2000–2005).

Some of the public who lived along the River expressed concerns about the short timeline for making comments on various plans and hence felt that they were not being fairly dealt with and their opinions were not truly valued. Once it was explained that the IJC would hold hearings on the candidate plans recommended by the Study Board, many of those individuals seemed to be satisfied that their voices would be heard.

During one public meeting (Olcott, New York, September 17, 2004), a participant spoke of how 10 yr previously close to 400 participants had taken part in a public meeting. He related the turnout of 42 participants as being due to apathy among property owners resulting from their distrust and concern that nothing will change. Perhaps in an attempt to encourage participation, members of the PIAG regularly spoke at the public meetings about the importance of stakeholders voicing their opinions. They emphasized that the process would ensure that everyone's opinions and views would be documented and therefore heard by the Study group and the IJC.

During the 2008 public hearings held by the IJC, some interest groups made threats of legal action against either the IJC or the national governments if their preferred plan was not selected. The IJC, being an international treaty organization, explained that they are protected from litigation. These comments suggest that participants felt that, at this stage, legal systems were the only way their voices would truly be incorporated. It is important to note that Plan 2007 was a modified version of one of the plans developed by the Study and it was not produced in direct collaboration with the public. Plan 2007 was fairly unanimously rejected by all interest groups when it went to consultation in 2008 (IJC public hearing transcripts, June to Sept 2008). Several comments were made that criticised the lack of public participation in the development of Plan 2007. Some participants also felt their inputs into the Study had been ignored because the IJC had chosen to develop an alternative plan, rather than select one of those developed by the Study. It is perhaps possible that the lack of public involvement in creating Plan 2007 contributed to it being dropped by the IJC.

Representation. Broad representation of all interested and affected parties is considered to be central to a legitimate process as agreements reached by an unrepresentative group of stakeholders can be said to result from an undemocratic process and dismissed by critics of the agreement (Mostert et al. 2007; Rowe and Frewer 2000). Broad representation also ensures that a full understanding of the interactions and perspectives can be achieved (Hedelin 2007).

The Study was set up to ensure that the Study Board and the PIAG included members from all the interest groups operating on the Basin. The PIAG attempted to ensure that a representative group of stakeholders attended the public meetings through arranging meetings at places where they knew interest would be high. They distributed material, organized presentations and workshops and advertised the Study throughout its duration in an attempt to identify and capture the interest of as many people as possible. Specific meetings were held with First Nations communities throughout the Study.

Table 4.2 shows the number of interests represented by speakers (not including the Study Board or PIAG) at five public meetings held in 2005 for which full transcripts were available. This shows that all meetings heard from speakers from more than one interest group. Although, the meetings at Massena and Alexandria Bay were heavily weighted towards boating and North Rose was very heavily weighted towards riparian interests. No meetings heard from representatives from all interest groups which suggest that full representation within each meeting was rarely achieved.

Table 4.2 Interests represented at some of the 2005 public meetings.

Interest Group	Massena, June 22, 2005	Alexandria Bay, June 23, 2005	Oswego, July 14, 2005	North Rose, July 20, 2005	Greece, July 21, 2005	Total
Recreational boating	7	5		1	1	14
Environment	1	3	2		9	15
Riparian (shoreline property)			3	16	14	33
Commercial boating		2				2
Recreational boating and environment	1					1
Recreational boating and riparian					1	1
Unknown	2	5	1	2	5	15

Intermediary-outcome based Evaluation

Agreements are reached. Co-production of an operating plan should lead to a plan that is both technically feasible and acceptable to all because it has been produced by a combination of technical and stakeholder inputs. Creation of a plan to which all could agree in principle would naturally be an important outcome. The LOSL Study created multiple plans, from which three, based on stakeholder, Study Board and external scientific review, were presented to the IJC as management options. The public meeting comments showed that everyone could align themselves to one of the new plans or the status quo (Plan 1958-D with deviations). This achievement may be overlooked as a success but should perhaps be given considerable credit as it shows the Study outcomes did reflect the interests of the people it involved.

One of the most difficult issues for the Study Board was the environment. Many interests benefit from the environment, and whatever set of indicators used to show changes in the environment resulting from any policy, it was never clear how significant that change was in relation to changes in other interests. Furthermore the complex environmental quality model developed and used to derive values for environmental indices was an exercise in dealing with uncertainty (see Werick 2011). At one point in the study the US IJC Co-chair commissioner asked that the Study Board, with the help of the environmental technical working group, to quantify environmental improvement in terms of dollars, as other interests were. This was resisted by the environmental technical work group and Study Board.

Innovation. More creative and advanced management plans emerged as a result of public input. The plans put together by the Study during the final year were refined according to public input during the 2005 public meetings. The plans A, B, and D were enhanced and relabelled A+, B+ and D+ (Study Board Meeting Minutes, Aug 24 and 25, 2005).

A strong wish was expressed by the public in several locations that the performance of the plans be monitored, with a review, for example, every five years, to assess the results. This supported the Study Board's intention that adaptive management, by which adjustments are made to the plan to reflect changing environmental and socio-economic circumstances, would be included in any new plan.

Interaction and network development. Raised awareness to both the interests of other stakeholders and environmental concerns does seem to have been achieved by the Study. This is shown by comparing comments made in 2008 to those made in 2005. Twenty-eight participants representing either themselves or specific groups or industries made statements that specifically acknowledged the need to consider other interest groups during the 2008 public hearings. During the 2005 public meetings only one comment was made that showed awareness and consideration of other interests. Similarly, during the 2008 public hearings, 23 participants emphasized that their own interests should be given priority, but also stressed an interest and commitment to environmental improvements. This can be compared to only three comments from the 2005 public meetings that acknowledged environmental concerns, while arguing that their interests were of greater importance.

At the end of the study it became clear that there was broad public support for the plan that best improved the environment, yet depending on where the public lived, there were misgivings about variable flows that favoured the environment but detracted from boating and shore line stability. What some of the public learned was that tradeoffs were sometimes necessary. It became obvious that no single plan of all those considered resulted in satisfying all public interests. In meetings on the south shore of Lake Ontario (see Fig. 4.2), there was large support for the status quo, that is Plan 1958-D with deviations, because all of the candidate plans appeared to raise Lake Ontario levels. Concerns regarding shoreline erosion and flooding were noted at meetings in towns and cities were people lived near the shore. Even those who wanted much lower highs or higher lows also said that they wanted a more natural lake/river regime. When the debate was framed in terms of a natural or environmental plan versus any other kind of plan, residents from the River favoured environmental plans.

A valuable outcome emerged from the relationship between the PIAG and the International St. Lawrence River Board of Control. Public comments made throughout the Study highlighted frustration and sometimes anger with the Board of Control. The PIAG addressed this by formulating a sub-committee to advise the Board of Control on their communications strategy (LOSL Semi-annual Report 8, March–Sept 2004). Several joint meetings between the Study Board, the PIAG and the Board of Control took place over the course of the Study that are likely to have improved the Board's communication approach.

The PIAG itself could be considered to be a valuable human resource to the IJC. The members have strong networks throughout the basin and developed extensive knowledge and understanding of the system and its complexities. However, at the closure of the Study there was, according to one member of the PIAG, no follow up communications from the IJC which may jeopardize the value of this network:

> "That's one of my big complaints about the process. Is that they should have, for the study board members and the PIAG, they should have, even just twice a year, sent us an email saying this is the status, this is what we're doing. ... Because I think that, if the IJC approached me again to be on something, I'm not sure that I would."
>
> (Semi-structured interview with member of PIAG, September, 2010)

Little evidence is available in the research resources to identify whether network development between two or more interest groups took place during or as a result of the Study. The data set is unable to capture the informal discussions and networking held at the start and end of the meetings that are likely to be an important area for interaction.

Institutional change. Prior to the Study, some members of the PIAG were tough and active critics of the Board of Control's operations. Some urged that the Board of Control be restructured to better represent the full range of interests in the system. During the Study an institutional report was commissioned which made numerous recommendations for institutional improvements (Clinton Edmonds and Associates 2002). Institutional issues were fairly regularly raised at public meetings, particularly concerning the Board of Control and the procedure for obtaining permits for installing shore line protection from the US Department for Environmental Conservation (DEC). During one of the 2004 public meetings an active participant spoke about the importance of institutional review and reform:

> "For instance, when there were the TWGs, the technical working groups, put together, and they were put together with some excellent people, and they did excellent work. And they've come up with some recommendations they're giving to the [Study] Board to be evaluated and incorporated into a plan. But there should have been at the same time another group that was looking at the management structure and the overall procedures in how decision processes were going to take—how long they would take. Everyone knows that's been a common criticism. It takes too long to get a decision made. But that should have been going on parallel and feeding in at the same time, so that when we got to the end of the study we would have these independent recommendations coming in as well."
>
> (Alexandria Bay, New York, August 19, 2004)

Towards the end of the Study, an institutional workshop was held and recommendations were derived which were integrated into the final report (Report on the Institutional Issues Workshop, November 30–December 1, 2004). The Study board recommended that the IJC act on the findings and emphasized that their implementation would be independent of any new plan and could be acted on immediately. Changes to the structure and number of interest groups represented in the Board of Control are said to continue to be in progress (interview with member of the Study Board, Sept. 2010).

Shared knowledge and information. The strategy of co-production and natural knowledge inclusion used by the Study should create data and information which has been generated by all, and is therefore accepted and trusted. Participant trust in data can be evaluated by examining the comments made during the 2005 public meetings and the IJC 2008 hearings.

During the 2005 public meetings, the recreational boating sector and some of the shoreline property owners expressed concerns regarding potential flaws in the performance indicators, data collection and data analysis. Many of the concerns had been voiced during the study process but some are likely to have emerged as participants gained more understanding of the approaches used.

One of the reasons for exploring options for a revised management strategy was to reduce the need for deviations from the operating plan determined by the Board of Control. The role of deviations was brought up by several people on the South Shore of Lake Ontario during the 2005 meetings. There was concern that a fully automated (i.e., no human decision making) management system would be "irresponsible", as some circumstances such as flooding risks may require deviations to reduce the impacts. A couple of comments were made that questioned the models used to produce the plans, and voiced concerns that the plans would not perform as intended, therefore requiring human intervention. The Study Board generally agreed with the need for deviations, through some members argued that allowing deviations from a prescribed plan would inhibit the plan from achieving what it is designed to achieve, such as wetland restoration and boating economic gains (Study Board Meeting Minutes, Dec 5, 2005). Deviations would also make any assessment of the effectiveness of any plan on achieving its stated objectives much more difficult.

During the IJC 2008 hearings, at least eight individuals or group representatives who favoured keeping Plan 1958-DD used the critical peer review of the Study (completed in 2006) to support their arguments that a change in management plan would be based on scientifically flawed data and analysis.

Discussion of the Processes and Outcomes and Lessons Learnt

One of the most exciting aspects of the LOSL Study is its attempt to co-produce a management plan for a highly complex system at a large scale with many different interest groups. Scientific and lay experts interacted throughout the process in the development of performance indicators and refinement of management plans that seems to have led to more creativity and innovation in the plans submitted to the IJC. It also seems to have led to plans being developed which are recognized to be more legitimate. Plan 2007 was not viewed positively, perhaps because it had been created without direct public input and was viewed as less legitimate. Our evaluation suggests that legitimate plans emerge from legitimate processes that need to be constantly maintained through access, transparency and impartial facilitation.

Co-production, as a model for developing resource management plans has been used on smaller scales, for example for designing urban river restoration (Petts 2006). The challenges described by Petts (2006) included getting people to trust in the system. An important lesson seems to emerge that stakeholder opinions on performance indicators need to be satisfactorily addressed. The coastal processes technical work group appears to have neither included the suggestion that property taxes form part of the performance indicators nor explained, to the satisfaction of some of the public, the performance indicators they did chose to work with. This seems to have led some of the riparians to mistrust the Study's findings.

Sound science is important for making unbiased decisions but the scientific review of the Study found weaknesses in the science that damaged trust in the Study's outputs, and may have left people questioning the cost-effectiveness of the Study. This provided ammunition for interest groups who felt they would not benefit from a change in the operating plan and wanted the IJC to reject the options put together by the Study. Trust in the models and systems developed by the Study is also vital if the plan is to operate without deviations. These observations suggest that finding ways to develop and maintain trust in scientific work is essential to reduce the potential for decision making based on personal agendas rather than facts.

The Study adopted a narrative that everyone would have to compromise. It is unknown whether a greater focus on identifying and building shared values among the interest groups would have led to a different outcome. Many property owners concerned about erosion also described having an interest in the environment, as did recreational and commercial boaters. Treating the environment as another interest group placed the burden on those representing it to stake a claim on water resources, rather than forcing other users to accommodate the requirements of environmental protection and conservation. An alternative approach would be to view the desired environmental state as a constraint on all other interest groups, not as another interest group participant. Developing a shared value of environmental quality would provide a constraint and tradeoffs could still be made among all remaining interests.

Public meetings and hearings have received a fair amount of criticism in the literature. Chess and Purcell (1999) talk about public meetings as being used for a "decide, announce, defend," strategy. This can happen when organizations make decisions prior to the meeting and use it solely as a forum to announce and defend their decisions. Public meetings are also associated with divisiveness, rather than consensus building (Chess and Purcell 1999; Collins et al. 2007; Duram and Brown 1999) and have also been called non-deliberative (Beierle 1998; Innes and Booher 2004). The Study documents clearly show that no plan was agreed prior to the public meetings. The evidence compiled in this evaluation also suggests that the process was deliberative. Members of the Study Board and the Technical Work Groups recognized that public participants had a great deal to offer and seemed to embrace the opportunity to discuss their work with lay experts. Perhaps the Study Board recognized that the production of a management plan may be only one of their achievements. The development of valuable institutional recommendations emerged from a willingness to discuss the broad variety of issues affecting interest groups.

In contrast to the deliberative style of the LOSL public meetings, the IJC hearings gave participants the opportunity to voice their opinions and to argue their cases, they did not give space for mutual exchange, understanding and learning (Klinke 2009). The findings from our evaluation suggest that the processes (i.e., fair access, representation, unbiased facilitation etc.) by which participation takes place are of much greater importance than the mechanism used to conduct participation (i.e., public meetings, small group workshops etc.). This finding supports that from other studies (Chess and Purcell 1999; Webler 1999).

The PIAG contributed hundreds of hours of volunteer time and were a significant, unique and invaluable part of the Study. PIAG members were appointed on the basis of their expertise and ability to reach out to local interest groups and this appears to have worked well. Based on our evaluation, the Study achieved good access and broad representation overall, though the space for learning about other interests may have been limited because most public meetings tended to be dominated by only a few interest groups. The PIAG attempted to address this though the telephone link up among different groups at different locations during some of the public meetings.

The time dedicated by the PIAG was much more than initially expected. The organizers of future studies should encourage prospective members to be prepared to devote time when joining an advisory group. Adequate administrative support to these key volunteers appears to be essential. Work also suggests that individuals have different expectations from participation processes (Webler et al. 2001). These may be the expectation that their input will directly feed into decision making, that the process will lead to better decision making or that a fairer and more equitable process will emerge (Webler 2001). Identifying and addressing the values and expectations of the PIAG group should perhaps form an important initial stage to reduce dissatisfaction or "burn out" later in the process. Simple follow up emails to inform PIAG members of developments and progress was noted to be lacking. To the volunteers, this suggests to them that their input has not been valued. It may reduce their willingness to take part in future volunteer advisory work, effectively leading to the loss of valuable human resources who are highly educated and experienced in resource management. This suggests that follow-up communications should form part of any program.

Conclusions

This work has benefited from an excellent publicly available data set and offers one interpretation and evaluation of the LOSL Study. The evaluation has shown some considerable strengths to the process which should be adopted by other participation programs. These included good access to information and meetings, commitment to involving all potentially affected communities and interest groups resulting in broad representation, impartial facilitation and inclusion of a wide variety of knowledge. These aspects emerged from strong institutional commitment to public involvement from the IJC which was heavily backed by the Study Board, but perhaps most importantly, through the inclusion of a dynamic, dedicated and well supported Public Interest Advisory Group. The difficulty will be to extrapolate these lessons learnt to other studies and public involvement programs where financial resources may not be so readily available and commitment to co-production of a resource management solution may not be so strong.

The evaluation has shown that co-production of a management strategy could be considered to be a long process and sufficient time, human capital, (and financial resources) need to be allocated. Our work suggests that good processes are essential to develop stakeholder trust, which is crucial for co-production. Inadequate processes, such as narrow representation or exclusion of participant knowledge, seem to lead to indicators of stakeholder dissatisfaction, such as distrust in information and threats of legal action.

The Study produced three management plans, but was unable to identify a consensus option within the time available. However, the Study produced many other outputs that extended beyond its original objectives. These non-tangible outcomes that are produced by stakeholder involvement programs should be given greater recognition when planning and evaluating participation. For the LOSL Study, some intermediary outcomes are achievements within themselves, such as institutional changes that benefit the current operating system. Others help to create an environment where an updated management plan may be implemented more willingly, such as an increased understanding of others views and positions.

Acknowledgements

The authors would like to acknowledge financial support provided by the Austrian Science Funds (FWF) as part of the Vienna Doctoral Programme on Water Resource Systems (DK-plus W1219-N22). They would also like to express their sincere gratitude to Elaine Kennedy and Anthony Eberhardt for sharing their experiences of the Study, as well as Eugene Stakhiv, Russ Trowbridge and Janey Trowbridge for providing valuable reviews of the chapter.

References

Beierle, T.C. 1998. Public Participation in Environmental Decisions: An Evaluation Framework Using Social Goals. Discussion Paper 99-06. Resources for the Future, Washington DC, USA.

Beierle, T.C and D.M. Konisky. 2000. Values, Conflict, and Trust in Participatory Environmental Planning. Journal of Policy Analysis and Management 19(4): 587–602.

Bentrup, G. 2001. Evaluation of a Collaborative Model: A Case Study Analysis of Watershed Planning in the Intermountain West. Environmental Management 27(5): 739–748.

Blackstock, K.L., G.J. Kelly and B.L. Horsey. 2007. Developing and applying a framework to evaluate participatory research for sustainability. Ecological Economics 60: 726–742.

Burgess, J. and J.Chilvers. 2006. Upping the *ante*: a conceptual framework for designing and evaluating participatory technology assessments. Science and Public Policy 33(10): 713–28.

Carr, G., G. Bloeschl and D.P. Loucks. 2012. Evaluating participation in water resource management: A review. Water Resources Research 48, W11401.

Chess, C. and K. Purcell. 1999. Public participation and the environment: do we know what works? Environmental Science and Technology 33(16): 2685–2692

Chilvers, J. Deliberative and Participatory Approaches in Environmental Geography. *In*: N. Castree, D. Demeritt, D. Liverman and B. Rhoads [eds.]. 2009. A Companion to Environmental Geography, Wiley-Blackwell, Oxford, UK.

Clinton Edmonds and Associates Limited. 2002. Lake Ontario and St. Lawrence River—Changes in the Institutional Structure and Their Impact on Water Levels, 1950–2001. Report to International Lake Ontario—St. Lawrence River Study Board University of Ottawa, Institute of the Environment.

Collins, K., C. Blackmore, D. Morris and D. Watson. 2007. A systemic approach to managing multiple perspectives and stakeholding in water catchments: some findings from three UK case studies. Environmental Science and Policy 10: 564–574.

Conley, A. and M.A. Moote. 2003. Evaluating collaborative natural resource management. Society and Natural Resources 16: 371–386.

Connick, S. and J.E. Innes. 2003. Outcomes of Collaborative Water Policy Making: Applying Complexity Thinking to Evaluation. Journal of Environmental Planning and Management 46(2): 177–197.

Duram, L.A. and K.G. Brown. 1999. Assessing public participation in US watershed planning initiatives. Society and Natural Resources 12: 455–467.

Eberhardt, A.J., J.C. Rassam and L. Duckstein. Operation and management of the Lake Ontario—St. Lawrence River System: Present and future options. pp. 217–232. *In*: J. Ganoulis, L. Duckstein, P. Literarthy and I. Bogardi [eds.]. 1996. Transboundary Water Resources Management: Institutional and Engineering Approaches. NATO ASI Series. Springer-Verlag, Berlin.

GLIN. 2011. Great Lakes-St Lawrence River System. Great Lakes Information Network. Map produced by US Army Corps of Engineers, Detroit District, Detroit, Michigan. http://gis.glin.net/maps/.

Hedelin, B. 2007. Criteria for assessment of sustainable water management. Environmental Management 39: 151–163.

http://www.boatnerd.com. 2011. http://www.boatnerd.com/news/newsthumbs/images-07-3/smchl72607bh.jpg . Accessed August, 2011.

IJC. 1909. Boundary Waters Treaty.

IJC. 1998. The International Joint Commission and the Boundary Waters Treaty of 1909. September, 1998. Washington/Ottawa.

IJC. 1999. Plan of Study for Criteria Review in the Orders of Approval for the Regulation of Lake Ontario—St. Lawrence River Levels and Flows. Prepared for the International Joint

Commission By The St. Lawrence River-Lake Ontario Plan of Study Team. September, 1999. Washington/Ottawa.

IJC. 2008. Your Guide to the IJCs Proposed new Order of Approval and Plan. 2007. International Joint Commission. Washington/Ottawa.

Innes, J.E. and D.E. Booher. 2004. Reframing public participation: strategies for the 21st century. Planning Theory and Practice 5(4): 419–436.

Jiggins, J., E. van Slobbe and N. Röling. 2007. The organisation of social learning in response to perceptions of crisis in the water sector of The Netherlands. Environmental Science and Policy 10: 526–536.

Kitchin, R. and N.J. Tate. 2000. Conducting Research in Human Geography: Theory, Methodology and Practice. Pearson Education Ltd., Harlow, England.

Klinke, A. 2009. Deliberatie transnationalism—Transnational governance, public participation and expert deliberation. Forest Policy and Economics 11: 348–356.

Leach, W.D., N.W. Pelkey and P.A. Sabatier. 2002. Stakeholder Partnerships as Collaborative Policymaking: Evaluation Criteria Applied to Watershed Management in California and Washington. Journal of Policy Analysis and Management 21(4): 645–670.

LOSL Study Board. 2006. Options for Managing Lake Ontario and St. Lawrence River Water Levels and Flows. Final Report by the International Lake Ontario—St. Lawrence River Study Board to the International Joint Commission, March, 2006.

Loucks, D.P. 2006. Modeling and managing the interactions between hydrology, ecology and economics. Journal of Hydrology 328: 408–416.

Moote, M.A., M.P. McClaran and D.K. Chickering. 1997. Theory in practice: Applying participatory democracy theory to public land planning. Environmental Management, 21(6): 877–889.

Mostert, E., C. Pahl-Wostl, Y. Rees, B.Seare, D. Tàbara and J. Tippett. 2007. Social learning in European river-basin management: barriers and fostering mechanisms from 10 river basins. Ecology and Society 12(1): 19 [online].

Muro, M. and P. Jeffrey. 2006. A review of participatory test and evaluation approaches for water management. Water Science and Technology: Water Supply 6(5): 1–8.

Newig, J. and O. Fritsch. More Input—Better Output: Does citizen involvement improve environmental governance? *In*: I. Blühdorn [ed.]. 2009. In Search of Legitimacy. Policy Making in Europe and the Challenge of Complexity. Opladen, Farmington Hills.

Pahl-Wostl, C., M. Craps, A. Dewulf, E. Mostert, D. Tabara and T. Taillieu. 2007. Social learning and water resources management. Ecology and Society 12(2): 5[online].

Pretty, J.N. 2003. Social capital and the collective management of resources. Science 302: 1912–1914.

Petts, J. 2006. Managing public engagement to optimize learning: Reflections from urban river restoration. Human Ecology Review 13(2): 172–181.

Reed, M.S. 2008. Stakeholder participation for environmental management: A literature review. Biological Conservation 141: 2417–2431.

Rowe, G. and L.J. Frewer. 2000. Public Participation Methods: A Framework for Evaluation. Science, Technology, and Human Values 25(1): 3–29.

Steyaert, P., M. Barzman, J. Billaud, H. Brives, B. Hubert, G. Ollivier and B. Roche. 2007. The role of knowledge and research in facilitating social learning among stakeholders in natural resources management in the French Atlantic coastal wetlands. Environmental Science and Policy 10: 537–550.

Walker, G.B., S.L. Senecah and S.E. Daniels. 2006. From the forest to the river: Citizens' views of stakeholder engagement. Human Ecology Review 13(2): 193–202.

Webler, T. "Right" discourse in citizen participation: an evaluation yardstick. pp. 35–86. *In*: O. Renn, T. Webler and P. Wiedemann [eds.]. 1995. Fairness and Competence in Citizen Participation—Evaluating models for environmental discourse. Kluwer Academic Publishers, Dordrecht, Boston, London.

Webler, T. 1999. The craft and theory of public participation: a dialectical process. Journal of Risk Research 2(1): 55–71.

Webler, T., S. Tuler and R. Krueger. 2001. What is a good public participation process? Five perspectives from the public. Environmental Management 27(3): 435–450.

Werick, W. The Lake Ontario—St. Lawrence Rive Study. pp. 71–90. *In:* E. Bourget [ed.]. 2011. Converging Waters: Integrating Collaborative Modeling with Participatory Processes to Make Water Resources Decisions. Maass-White Series, The Institute for Water Resources, U.S. Army Corps of Engineers, United States.

5

Managing a Multi-objective, Bi-national Study on Rationalizing the State of Upper Great Lakes Water Levels and the Development of Improved Regulation Plans

Syed M.A. Moin

Introduction

The International Joint Commission (IJC) was founded in 1909 under the Boundary Waters Treaty to prevent and resolve potential disputes regarding many of the lakes and rivers along the border between the two countries. With the agreement of the US and Canadian governments, the Study was funded at a cost of Can $ 17.2 M shared equally between the US and Canada. Following the completion of an earlier Great Lakes comprehensive study in 1993, the IJC followed up on its recommendation to revisit regulation initially for Lake Ontario in 2000 and followed up with the International Upper Great Lakes Study (IUGLS) in 2007. Across the Upper Great Lakes

International Upper Great Lakes Study, 234 Laurier Avenue West, 22nd Floor, Ottawa, ON K1P 6K6.
Email: moins@ottawa.ijc.org

basin since 1997, low lake level conditions continued to be a concern for commercial shippers, property owners and other interests. This was particularly the case in the Georgian Bay region, which features a rocky archipelago of thousands of islands, many of which are inaccessible by boat when levels are extremely low. Lake Michigan-Huron is located downstream of Lake Superior and along with Lake Erie and their connecting channels constitute the Upper Great Lakes. This prompted the IJC to add to the Study specifications to address this issue as part of Lake Superior Regulation. These two phases of the Study are presented in two major sections.

The upper Great Lakes basin, the focus area of the Study, covers an area of about 686,000 km^2 Coordinating Committee on Great Lakes Basic Hydraulic and Hydrologic Data (CCGLHHD 1977). Figure 5.1 presents the study area comprising lakes Superior, Michigan-Huron, St. Clair and Erie. Figure 5.2 shows the general water surface profile of the Great Lakes system, including the St. Lawrence River. A unique feature is that about one-third of the upper basin area consists of the water surfaces of the upper Great Lakes and their connecting channels (see Table 5.1).

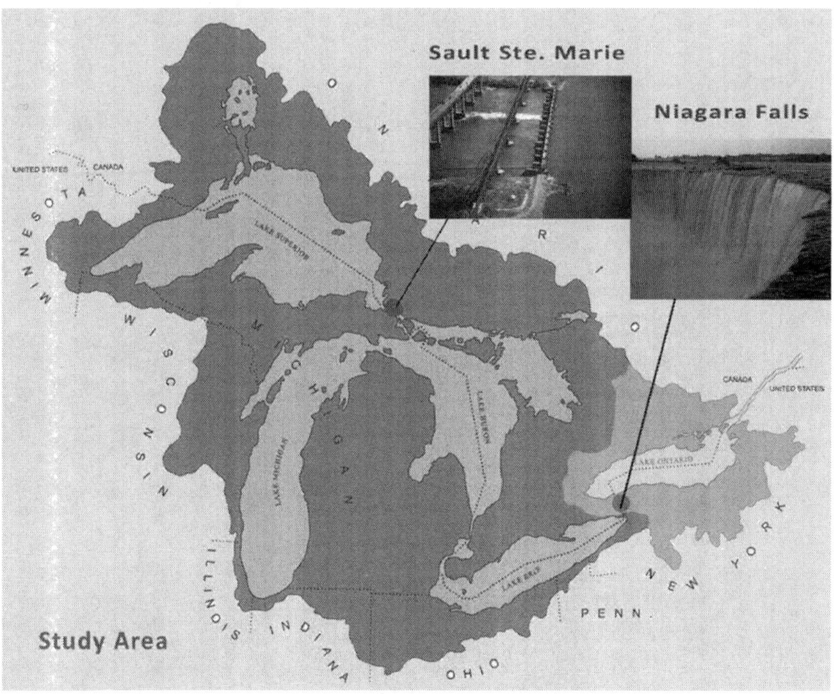

Figure 5.1 International Upper Great Lakes Study Area.

(Source: IUGLS 2009, 2012).

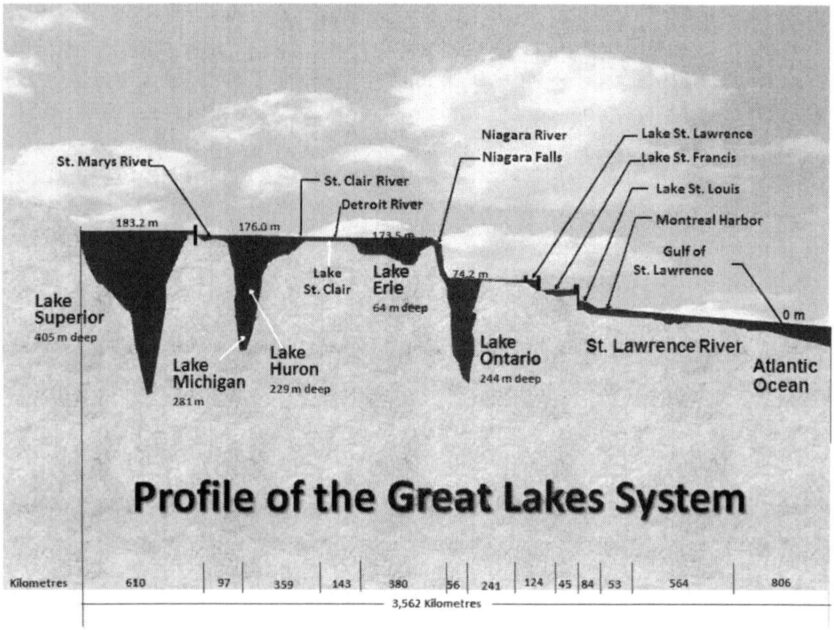

Figure 5.2 Water Surface Profile of the Great Lakes System (Source: IUGLS 2009).

(Source: modified from Great Lakes Commission and U.S. Army Corps of Engineers 1999). Note: Water surface elevations are at chart datum on IGLD 1985.

Table 5.1 Dimensions of the Great Lakes Basins (Source: CCGLHHD 1977).

Dimensions of the Great Lakes				
	Surface Area		Volume*	Depth*
	Water	Land		
	km^2	km^2	km^3	metres
Lake Superior	82,100	127,700	12,100	405
St. Marys River	230	2,600	-	-
Lake Michigan	57,800	118,000	4,920	281
Lake Huron	59,600	131,300	3,540	229
St. Clair River	55	3,300	-	-
Lake St. Clair	1,110	12,430	-	-
River	100	2,230	-	-
Lake Erie	25,700	58,800	484	64
Niagara River	60	3,370	-	-
Lake Ontario	18,960	60,600	1,640	244
St. Lawrence River *(to Cornwall/Massena)*	610	7,190	-	-

*Measured when the lake's water level is at chart datum on IGLD (1985). Note: No value provided for Lake St. Clair.

The Great Lakes basin is highly dynamic, characterized by changes in lake levels as a result of both natural and human factors operating on time scales from hours to decades to centuries (International Great Lakes Levels Board 1973; Nicholas 2003).

Three types of water level fluctuations occur on the Great Lakes: short-term; seasonal; and long-term:

- Short-period fluctuations (lasting from less than an hour to several days) can occur when sustained high winds blow over a lake producing a wind set-up or storm surge on the downwind shore of the lake. This results in lower water levels at the opposite shore of the lake. Such large events are almost always followed by seiches (oscillations) that can disturb water levels for two to three days.
- Seasonal fluctuations of the Great Lakes levels generally correspond to the basin's annual hydrologic cycle. The cycle is characterized by higher NBS during the spring and early summer, and lower NBS during the remainder of the year. Each Great Lake loses water through evaporation from its surface. The relative importance of evaporation varies from one lake to another, depending primarily upon the area of the lake surface as compared to the area of the watershed draining to the lake. Summer evaporation over the lakes is much less important than in colder months. The presence of ice cover on the lakes will reduce water losses through evaporation. Conversely, the absence of ice cover on the lakes in the fall and winter will increase the volume of water lost from evaporation. Much of the seasonal decline the lakes experience each fall and early winter is due to the increase in evaporation from their surfaces when cool, dry air passes over the relatively warm water of the lakes.
- Long-term fluctuations in the levels of the Great Lakes are the result of a number of years of above or below average precipitation or evaporation. Their magnitude and duration are irregular. Figures 5.2–5.5 illustrates the monthly mean lake levels of the upper Great Lakes from 1918 through 2008. Prior to 1918, there were insufficient water level data and gauge stations to determine the lake-wide average monthly mean lake levels accurately and consistently. Table 5.2 lists the long-term average and range of water level and outflow fluctuation for the upper Great Lakes for the period 1918–2008.

Part 1—St. Clair River Phase

Concerned with the state of the water levels, the Georgian Bay Foundation commissioned a study (Baird and Associates 2005) to explain the factors responsible for the low water levels. Commonly known as the Baird Report,

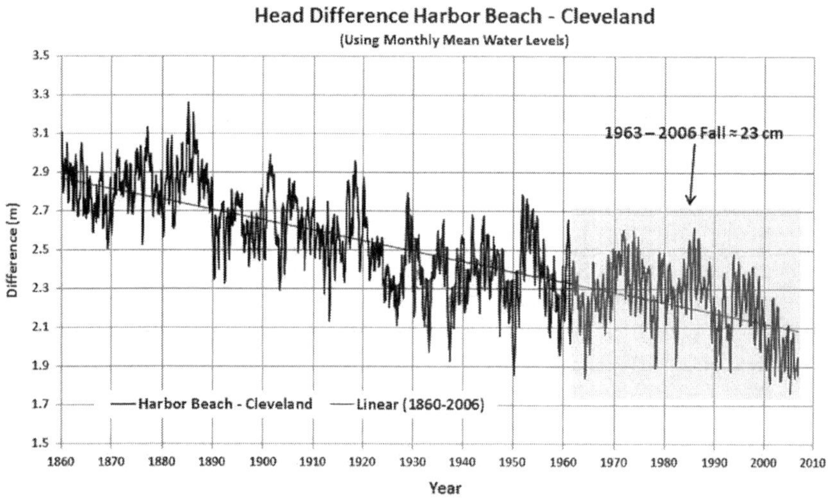

Figure 5.3 Showing the Lake-to-lake Head Drop from 1962–2006.
(Source: IUGLS 2009).

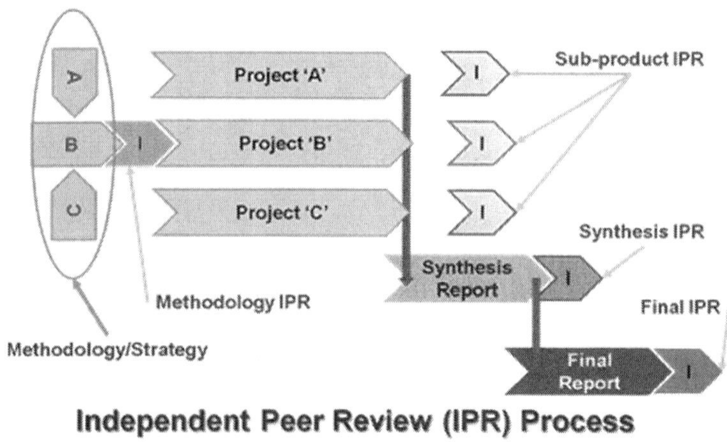

Figure 5.4 Schematic of the Independent Peer Review Process.

this study concluded that "the primary cause of the drop in Michigan-Huron lake levels is due to river bed erosion." It identified a number of possible human-induced causes for this erosion. In response to the concerns raised by the findings of the Baird Report, the International Joint Commission, a bi-national US and Canada body established in 1909 to prevent and resolve water management issues, established a team to consider ways to resolve the questions surrounding possible human-induced and natural changes to the St. Clair River. The Study funded at Can$ 17.2 million consists of two major components; the first part described in this chapter is to examine

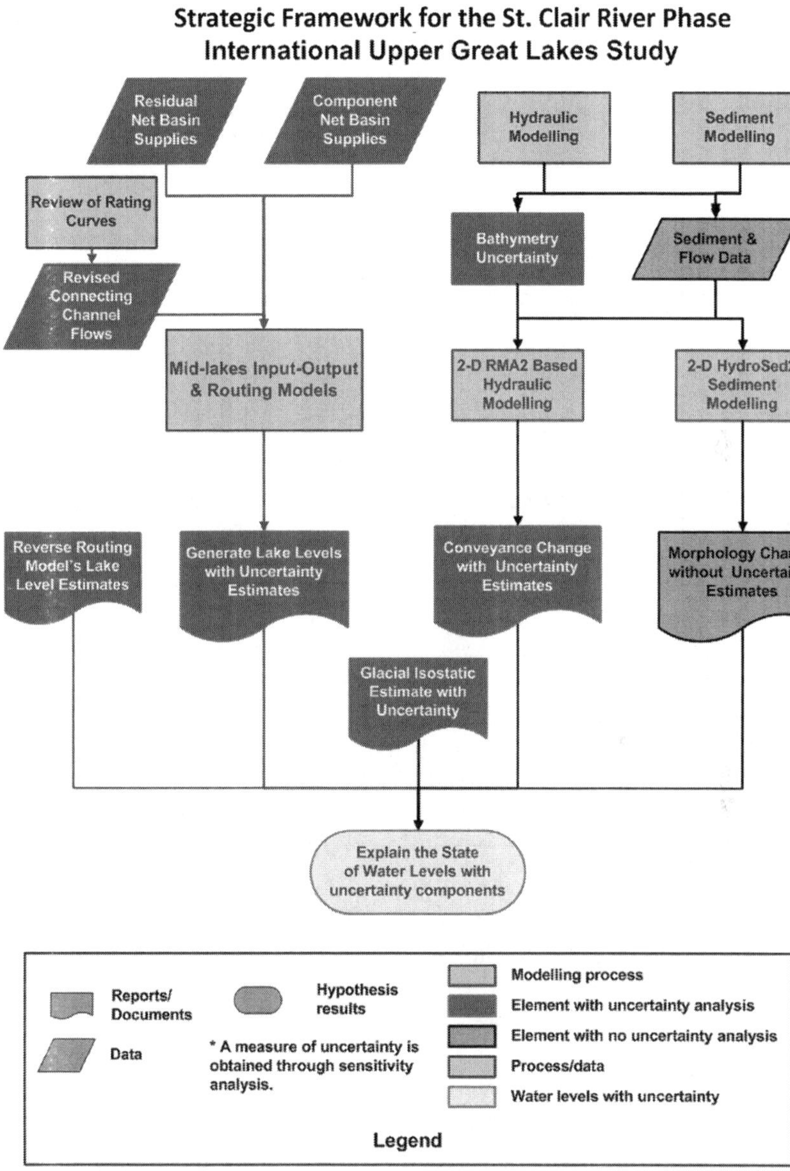

Figure 5.5 Study Strategy of Multi-pronged Approach to Analysis.

all the issues related to the conveyance of the St. Clair River and factors affecting Lake Michigan-Huron levels. The second component considers revisions to the regulation of the Lake Superior through the control works at Sault Ste. Marie.

Table 5.2 Conveyance change calculations from several techniques using Hydraulic Performance Graphs.

Type of Analysis	Water Level Change
Modelling	
1-D Basic HEC-RAS Modelling	10 cm
1-D Conveyance analysis*	2.5 to 3.2%
RMA2 2-D Modelling	12 cm
TELEMAC 2-D Modelling	13 cm
HydroSed 2-D Sediment Modelling	9 cm
Data and Flow Analysis	
Lake-to-lake water level analysis	8 to 10 cm
Gauge-to-gauge water level analysis	Up to 14 cm
HPG analysis	12 cm
Stage-Fall-Discharge equation analysis	8 cm
Hydroclimate Modelling	
Mid-lakes Routing	13 cm
Coordinated Routing - Component	8 cm
Coordinated Routing - Residual	7 cm
Deterministic Mid-lakes Routing	7 cm
** Change expressed as percent change in conveyance since 1971*	

(Source: IUGLS 2009).

The St. Clair River is about 63 km in length and extends from Lake Michigan-Huron to Lake St. Clair. Over this distance the water level falls about 1.5 m. The average annual discharge of the river is about 5,200 m^3/s. Figure 5.1 shows the Study Area with the connecting channels. Lake Michigan-Huron that empties into St. Clair River has a drainage area of 366700 km^2 of which lake surface area accounts of 117,400 km^2 or approximately one-third of the basin. The control works at the outlet of Lake Superior in Sault Ste. Marie and Niagara Falls downstream of Lake Erie form the bounds of this component of the Study. The middle Lakes Michigan-Huron, St. Clair (drainage area 16,895 km^2, water area 1,165 km^2) and Erie (86,830 km^2, 25,800 km^2) form a dynamically linked water body system connected by the St. Clair and Detroit Rivers. Also, features of the lakes like the average depth and surface area come into play for reacting to changes in water supplies and water levels.

With the immense storage capacities of the lakes, in combination with the relatively small capacity of the outflow channels, the upper Great Lakes system is largely naturally regulated. Large variations in water supplies to the basin can be absorbed relatively easily by the lakes, with the outflows remaining remarkably steady compared to the range of flows in other large river systems in the world. The large size of the lakes also means that extremely high or low levels and flows can persist for a considerable time even after the factors that caused them have changed.

The head difference between Lakes Michigan-Huron and Erie based on the long-term average water levels (1918–2006) is 2.30 m. This difference between Harbor Beach on Lake Huron and Cleveland on Lake Erie has steadily declined from a high of 2.9 m in 1860 to 2.1 m in early 2000. This change in head is largely a result of various dredging operations in early 1900s, sand and gravel mining in 1930s and culminated in the last dredging between 1959 and 1962 to maintain 8.02 m navigable depths. In earlier IJC Reference (Lake Reference Study Board 1993) had estimated the drop in the head difference prior to the 1962 dredging. The Study was therefore seeking factors that have left a drop of 23 cm since the last dredging as shown in Fig. 5.3.

St. Clair River Study Objective

As per the directive for conducting the International Upper Great Lakes Study (IUGLS), the IJC was asked by the Governments of Canada and the USA to examine physical processes and possible ongoing changes in the St. Clair River and their impacts on levels of Lake Michigan-Huron and, if applicable, evaluate and recommend potential remedial options. This directive was converted into four primary objective-defining questions. These were supported by several other secondary questions to strengthen the objective. These were:

1. Has the morphology of the St. Clair River been altered since the 1962 dredging?
2. Has the conveyance of the St. Clair River changed since the 1962 dredging and is this causing the declining head difference between Lake Michigan-Huron and Lake Erie?
3. How has climate affected the change in lake level relationship between Lake Michigan-Huron and Lake Erie?
4. What role, if any, does the glacial isostatic adjustment have with respect to changing relative water levels in the upper Great Lakes?

Study Strategy

What is implicit in the state of Lake Michigan-Huron water levels is not only the steep decline in levels since 1997, but also the narrowing of the water level difference between Lake Michigan-Huron and Lake Erie (also known as the head difference). To make a meaningful interpretation of the change in the head difference since the dredging in 1962, the Study used the entire water level data series from 1860 to 2006. Finally, the Study sought to explain the decrease in head difference between Lake Michigan-Huron and Lake Erie of 23 cm observed since 1962 and estimated from linear trend (IUGLS 2009).

Addressing these two closely linked issues of changes in the St. Clair River and the effects of such changes on levels of Lake Michigan-Huron, required a comprehensive understanding of hydraulic, hydrological and geomorphological processes. The water levels in Lake Michigan-Huron depend not only on the connecting channel flows and the basin supplies, but also to varying degrees on the respective conveyance changes in the St. Clair, Detroit and Niagara rivers and on the water level in Lake Erie.

The Independent Peer Review Process

A first for an IJC Reference Study of this magnitude was the continual and ongoing Independent Peer Review (IPR) process designed to ensure the quality of science and engineering delivered was consistent with the needs and requirements of the set objectives. Traditionally, IPR, as practiced by such organizations as the US National Research Council or the Council of Canadian Academies, operates in such a way as to provide an independent, one-time final assessment of a particular project or study, with limited client feedback and interaction with study management. The IUGLS, however, being an operational study, required real-time feedback from experts to permit mid-course adjustments on a series of highly technical issues and choices. The IUGLS required both an independent peer review and a peer-review advisory function. This review function was supplied by the Environmental and Water Resources Institute (EWRI) of the American Society of Civil Engineers (ASCE) and the Canadian Water Resources Association (CWRA).

For the Study, there were four distinct milestones where peer review was sought. The first step was after the development of study methodology looking at problem definition, strategy to answer the science and operational questions posed at the objective setting stage. Once the peer review approval was received, the projects were completed. Key science deliveries at the sub-product level were next reviewed. At the third step, synthesis of a number of project findings were evaluated; and, at the final step, the full report was reviewed. The following Fig. 5.4 captures the stages of the peer review.

Defining the Decrease in Lake to Lake Water Level Differences

The head between lakes is defined as the water level difference between the two master gauges at Harbor Beach on Lake Huron and Cleveland on Lake Erie. The change in the head difference can be considered as function of: the change in the conveyance of the St. Clair River; any net shortfall in the basin supplies on Lake Michigan-Huron or a higher basin supplies in Lake Erie relative to Lake Michigan-Huron; and smaller effects from conveyance changes in the Detroit and Niagara rivers and the glacial isostatic adjustment. This relationship in simple mathematical terms is expressed as:

$$\Delta \boldsymbol{Head}_{(MH-E)} = \boldsymbol{function}\left\{\begin{matrix} \Delta \boldsymbol{Conveyance}_{St.Clair,} \\ \Delta \boldsymbol{NTS}_{MH,}\Delta \boldsymbol{NTS}_{(E-MH),}\Delta \boldsymbol{GIA}_{(MH-E)} \\ \Delta \boldsymbol{Conveyance}_{Detroit/Niagara} \end{matrix}\right\}$$

Where:

$\Delta Head_{(MH\text{-}E)}$ = Change in lake-to-lake fall between Harbor Beach and Cleveland

$\Delta Conveyance_{St.\ Clair}$ = Change in fall from hydraulic property change

ΔNTS_{MH} = Change in fall from lake-wide surplus or deficit from Net Total Supplies

$\Delta NTS_{(E\text{-}MH)}$ = Change in fall due to differential in Net Total Supplies between Lake Erie and Lake Michigan-Huron

$\Delta GIA_{(MH\text{-}E)}$ = Change in fall from Glacial Isostatic Adjustments from Lake Michigan-Huron to Lake Erie

$\Delta Conveyance_{Detroit/Niagara}$ = Change in fall between Erie and Michigan-Huron from Niagara/Detroit River

ε = Rounding errors and unknowns

Note: all units are in centimetres (cm).

For the Study purposes and as noted above, it was decided to rationalize what were the various components noted above that make up for a drop of 23 cm in the head difference between 1962 and 2006. It should be recognized that due to the natural seasonal and supply variability, there are periods within this time domain when the drop in head is substantially greater and highly dependent on the choice of the start and finish years.

Science and Engineering Studies

A complex problem of this nature is best addressed when each of the sub-problems is solved individually and the solutions integrated collectively (Fig. 5.5). The Study applied this approach. All the projects that the Study commissioned were designed to address one or more science questions concerning the components of the change in head equation. One of the unique features of the projects was the concept of designed redundancy. For example, the Study evaluated the particular aspect conveyance capacity from several directions using 1- and 2-D models, inverse modelling, etc.

The Study took a structured approach in addressing the science question. It was important to assess the quality of data being employed by various other studies. For this reason 20 percent of the projects were developed to verify the data and rationalize the drop in the head respecting that both Harbor Beach and Cleveland were subject to Glacial Isostatic Adjustment, a phenomenon that adds a level of complexity in the temporal data series of lake water levels. In this case Harbor Beach is rising with respect to Lake Huron outlet, while, Cleveland is sinking with respect to the outlet of Lake Erie at Buffalo.

In excess of 40 studies were designed to address the science questions listed under Study Objective; these were carried out from 2007 to 2009 over an 18 mon period and co-managed across two countries and multiple jurisdictions. Of these there are 11 projects addressing the science questions in the Hydraulic studies field, 12 projects were designed to conduct analysis and modelling of sediment studies, mobile bed modelling, etc. Ten projects were developed to carry out Hydroclimate modelling and statistical analysis. In total more than Canadian $ 4.2 million were spent for these projects.

Results and Integration

Using the lake-to-lake head difference data and by translating the Glacial Isostatic Adjustment at Harbor Beach a rate of 3.5 cm uplift per century relative to the Lake Huron outlet and at Cleveland a rate of 7.0 cm subsidence per century relative to Lake Erie outlet. The Glacial Isostatic Adjustment therefore accounts for 4.0 cm of the 23 cm drop in the lake-to-lake head.

To evaluate the change in the conveyance capacity of the St. Clair River, it was important to access one set of bathymetry close to the 1962 period and another in 2007. The Study used three data sets for this purpose. The closest in time to the 1962 period was the single-beam bathymetry of 1971, a second set, also single-beam from 2000 and the Study collected a multi-beam bathymetry in 2007. For all practical purposes, 1971 bathymetry was used as a surrogate for 1962. As noted earlier, several rigid and mobile bed boundary models were employed using the same basic bathymetry. As part of evaluating model results, all projects were asked to develop Hydraulic Performance Graphs (HPG) for integrating results from different models. A HPG is a set of curves that relate water levels at the upstream and downstream ends of a channel reach to channel discharge, thus providing a tool for describing and visualizing the backwater profiles of a given river reach under a full range of hydraulic scenarios. Figure 5.6 shows a typical HPG from one of the model runs. Similar HPG from several models on a common platform provides normalization and integration.

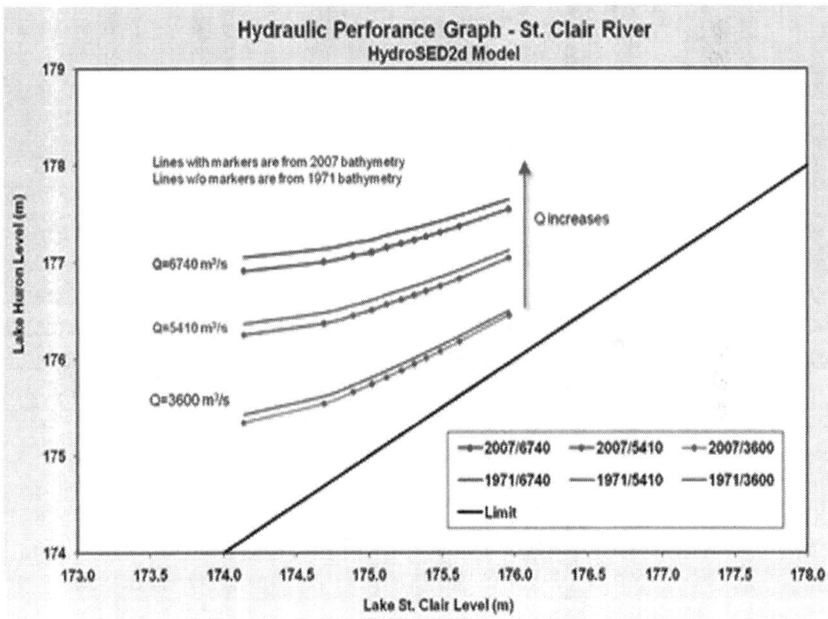

Figure 5.6 Hydraulic Performance Graphs to Detect Conveyance Change.

The Study found that the channel in 2000 and 2007 was deeper in the lower sections of the river when compared to the 1970 bathymetry. This resulted in an increase in the conveyance capacity that was confirmed by the various modelling and data analysis approaches. This is noted in Fig. 5.7. A range of 8 to 13 cm was found for the various rigid and mobile beds, from a variety of hydraulic and sediment models. Of the averaged 23 cm fall in head, thus, physical conveyance capacity change accounts between 8 and 13 cm. The change in the conveyance capacity was also approached

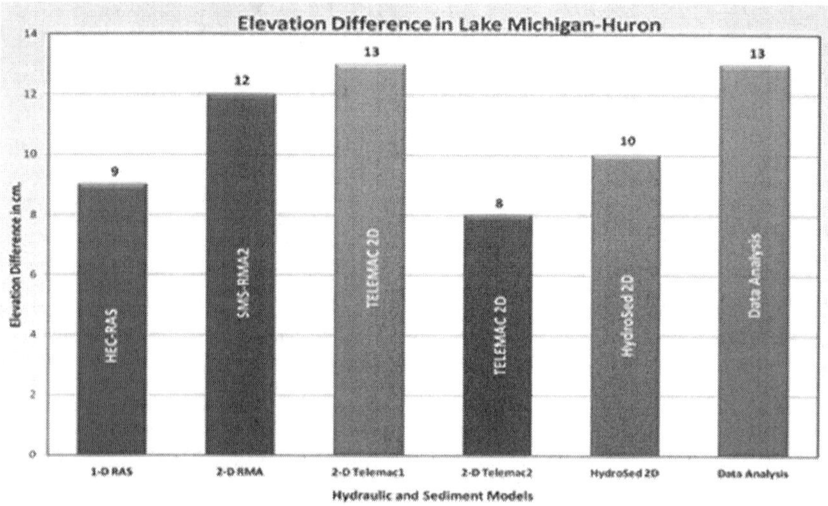

Figure 5.7 Hydraulic and Sediment Modelling Results (Source: IUGLS 2009).

from several other techniques including data analysis, water level/flow measurements and Hydroclimate modelling. These estimates are captured in Table 5.2.

Further hydrologic analysis was carried out by employing the mid-lakes (Michigan-Huron-St. Clair-Erie) routing model. This was to replicate the water levels using hydrology, lake-to-lake fall-discharge relationships, etc. This step involved computing the basin supplies by the two established methods; the component method employs a physical based hydrologic model developed by the Great Lakes Environmental Research Laboratories, an arm of the US National Oceanic and Atmospheric Administration (NOAA); the second method is a mass balance technique, termed residual method uses computed connecting channel, flows, diversions in or out and a change in the lake storage to estimate the basin supplies. The two techniques provided similar results. The general conclusions were that the increase in the channel conveyance capacity and basin hydrology were equally responsible for water level fluctuations from 1962 to late 1980s. The recent drop in the Lake Michigan-Huron water levels is largely a result of the change in water supplies that have averaged about 350 m^3/s-months less between 1987 and 2006 than in the 1963 to 1986 period. From this exercise, an estimate of 9 to 27 cm was obtained in explaining the impact of hydrology on the average in the lake-to-lake head drop.

Explaining the Drop in Head

The IUGLS relied on multiple levels of evidence to explain the factors impacting the state of water levels of Lake Michigan-Huron and further whether there were physical evidences of the change in the conveyance capacity of the St. Clair River. Three major factors impacting the water levels are the Glacial Isostatic Adjustment, an increase in the St. Clair River conveyance capacity and the variability of the Hydroclimate of the region in the time domain of 1962 to 2006.

The Glacial Isostatic Adjustment accounted for 4.0 cm of the overall 23 cm averaged drop in the lake-to-lake head. It was found from hydraulic modelling that the conveyance capacity increased between 1971 and 2000 by impacting water levels by about 16 cm. The change however reversed slightly between 2000 and 2007 when it appears that the conveyance capacity decreased in an amount of 3 cm. The net change when averaged from the various modelling and analytical approaches is 10 to 12 cm. The Hydroclimate modelling concluded that the climate variability amounts to a head drop of 9 to 27 cm. The smaller components of other impacts, notably changes in the Niagara River rating uncertainty/errors and conveyance changes in the Detroit River were absorbed in the results from Hydroclimate modelling. The values will add up to be greater than 23 cm to reflect climate variability and as high as 51 cm from 1986 to 2006 for example. These values are tabulated in Table 5.3.

Table 5.3 Synthesis of Results.

Change in lake-to-lake fall, between Harbor Beach & Cleveland	Components of the Fall	
23 cm	Change in fall from hydraulic property change	7 – 14 cm
	Change in fall from Glacial Isostatic Adjustment	4 - 5 cm
	Change in lake-wide surplus or deficit from Net Total Supplies (NTS)	9 – 17 cm
	Change in fall due to difference in NTS between Erie & Michigan-Huron	
	Change in fall between Erie & Michigan Huron from Niagara/Detroit	
	Rounding errors & unknowns	Negligible

(Source: IUGLS 2009).

Part 2—Lake Superior Regulation Phase

In the entire upper lakes basin, water levels are currently influenced through a regulation plan at the St. Marys River control structures at Sault Ste. Marie (Fig. 5.8), where Lake Superior outflows have been regulated since 1914.

Over long periods of time, a regulation plan can generally affect the balance between Lake Superior and Lake Michigan-Huron (considered as one lake, since they are at the same water level). However, the ability to influence high and low water levels through regulation is severely limited by the natural variation in climate conditions, the risks that climate change could introduce more extreme conditions in the future, and the physical geography of the lakes and connecting channels. Moreover, the natural shifting of the earth's crust has serious implications for both water regulation and coastal interests.

There is a high degree of uncertainty about how climate change will affect future water levels over the next several decades. In response, the Study has undertaken the most comprehensive and balanced analysis ever made of climate change in the Great Lakes basin. The Study concludes that future water levels are likely to remain within a relatively small range around their long-term averages. While lower water levels in the future are likely, water users around the lakes have to be prepared for episodes

Figure 5.8 Lake Superior Control Works at Sault Ste. Marie, Ontario/Michigan. (Source: IUGLS 2012).

of higher levels, too. Any new regulation plan, therefore, must be robust—effective and flexible enough to perform well in an uncertain future.

A new regulation plan also must recognize and balance the needs of the key economic and environmental interests in the upper Great Lakes. Some of these interests, such as recreational boating and ecosystems, were not specifically listed in the original 1909 treaty between Canada and the United States that established a co-management approach to boundary waters. The Study's shared vision planning exercise has allowed representatives of these key interests to provide information regarding their needs and preferences related to Great Lakes water management.

Given the limited ability to regulate the lakes, the Study also has looked at:

- the feasibility of building new control structures to either restore water levels on Lake Michigan-Huron to conditions that existed prior to channel modifications or regulate the entire Great Lakes—St. Lawrence River system; and,
- adaptive management measures, such as strengthened monitoring and information sharing, that can help water managers and property owners know what to expect in terms of changing water levels so that they can take action to reduce risks.

The Study's final report was submitted to the International Joint Commission in March 2012.

Why Review the Regulation of Lake Superior?

A five-year international study is looking at future water levels in the upper Great Lakes and the options for regulating those levels to support the region's economic and environmental interests.

Water levels in the upper Great Lakes basin (Fig. 5.1) of Canada and the United States have a profound effect on the lives of the more than 25 million people who live in the region. Many depend on the lakes or connecting channels for drinking water and electricity, enjoy boating and fishing, or work for industries that rely on the Great Lakes fleet to transport raw materials and finished products. Water levels also are important for maintaining healthy wetlands, fisheries and ecosystems. The Study was carried by connecting various components of the projects into a structured tree, as shown in Fig. 5.9.

In the vast territory of the upper Great Lakes basin, water levels can be affected by regulation at only one location upstream from Niagara Falls: the control structures in the St. Marys River at the twin cities of Sault Ste. Marie in Ontario and Michigan. The release of water from Lake Superior has been regulated by the International Joint Commission, the bi-national

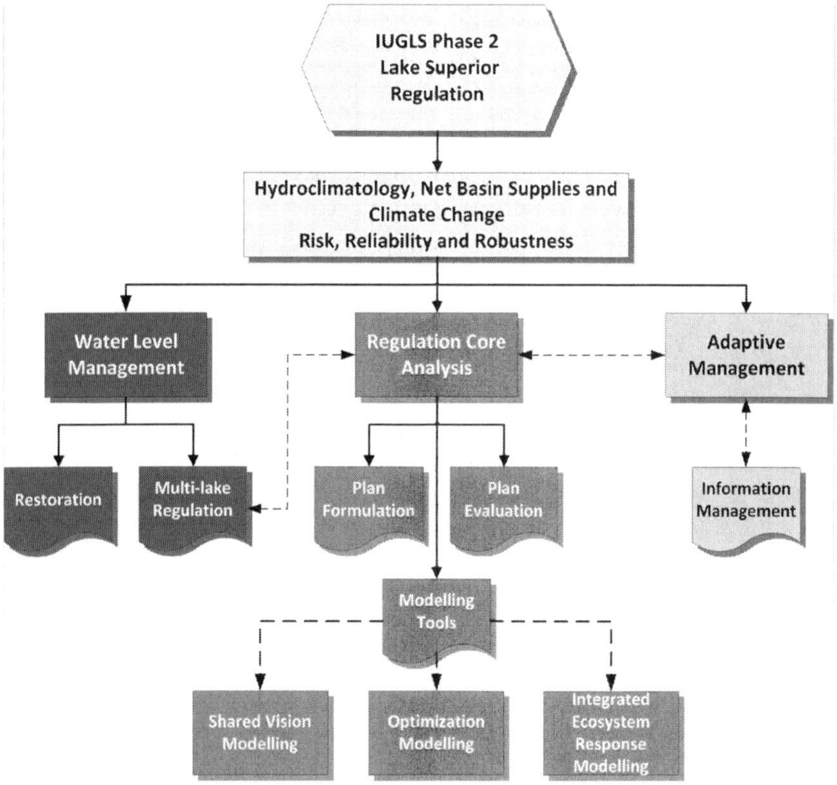

Figure 5.9 Major Elements of Study Strategy.
(Source: IUGLS 2009).

agency established by Canada and the United States under the Boundary Waters Treaty of 1909.

There are limits to the ability to influence high or low water levels in the Great Lakes through a regulation plan. The major factors affecting the water supply to the lakes—precipitation, evaporation and runoff—vary naturally over time and cannot be controlled. Now climate change raises the risk of even more extreme conditions in the decades ahead. In addition, the St. Marys River is small compared to the huge surface area and great depth of Lake Superior and has a limited capacity to move the water downstream. At the same time, apparent water levels are affected by glacial isostatic adjustment, the gradual and uneven tilting of the land as the earth's crust adjusts from the last period of continental glaciation more than 10,000 years ago.

Key Elements in Co-Management

A primary objective of the Study was to develop and evaluate possible new Lake Superior regulation plans to determine if a new plan could improve on the performance of 1977A—particularly in the context of the considerable uncertainty about future climate conditions and corresponding water levels on the upper Great Lakes.

In formulating, evaluating and ranking regulation plans, the Study applied shared vision planning, an iterative and collaborative process through which participants can better understand the implications of any regulatory decision. The Study Board used a shared vision model to undertake "practice" decisions; allowing experts, stakeholders and decision makers a series of opportunities to weigh the results as information developed.

Shared Vision Planning

The Study Board established clear objectives for a new Lake Superior regulation plan—and for the upper Great Lakes basin as a whole—based on the IJC's Directive and feedback received at public meetings:

- To maintain or improve the health of coastal ecosystems;
- To reduce flooding, erosion and shore protection damages;
- To reduce the impact of low water levels on the value of coastal property;
- To reduce shipping costs;
- To maintain or increase hydropower value;
- To maintain or increase the value of recreational boating and tourism opportunities; and,
- To maintain or enhance municipal-industrial water supply withdrawal and wastewater discharge capacity.

Of the hundreds of NBS sequences generated by the Study's hydroclimatic analysis, 13 were chosen as representative of the range of plausible future conditions that could be used to test the limits of any new proposed regulation plan. This suite of NBS sequences allowed the Study Board to test plans for robustness—the capacity to meet particular regulation objectives under a broad range of possible future NBS conditions. The plan formulation and evaluation decision logic is captured in Fig. 5.10.

Study plan formulators generated more than 100 alternative regulation plans, using a variety of scientific approaches, so as to ensure a comprehensive search for new regulation plans. Through the series of

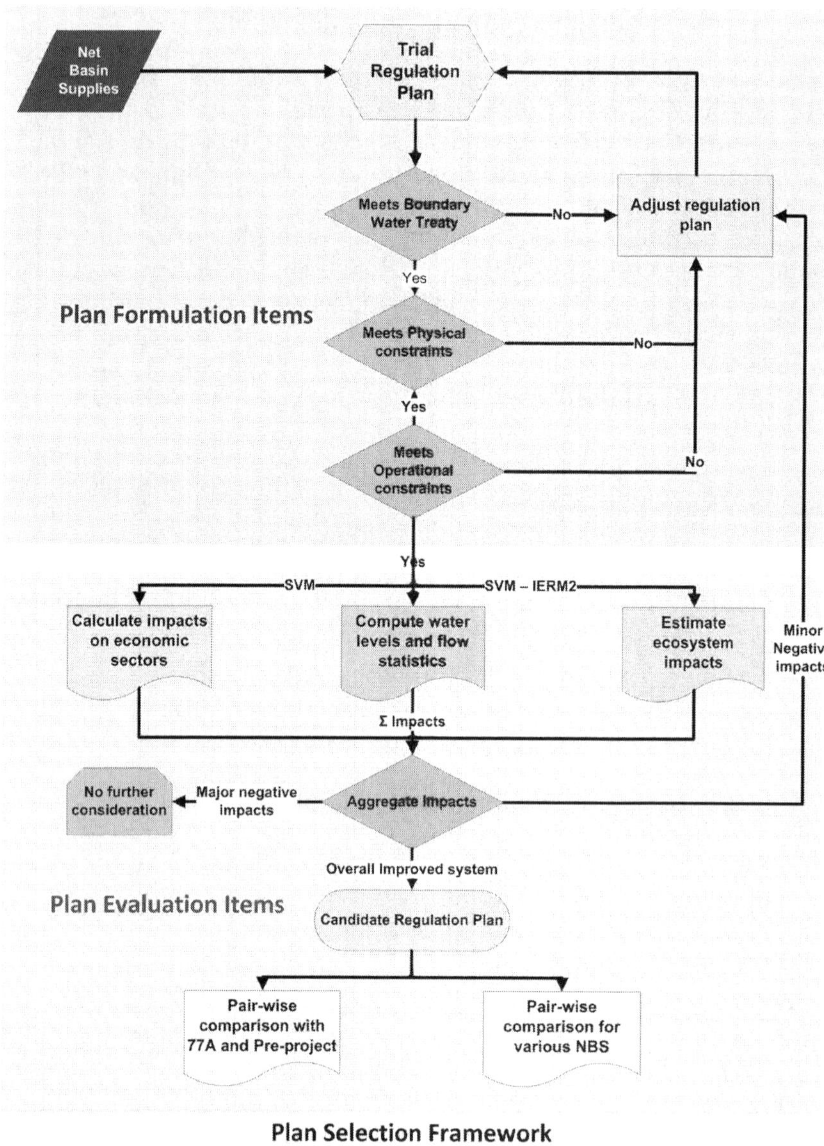

Figure 5.10 Plan evaluation framework.

"practice decisions", the Study Board reduced the list of plans to four. One of the final four plans performed better than or as well as any other regulation plan considered, regardless of the NBS sequence or the decision criterion applied. As a final step in the selection process, plan formulators developed three variations of the preferred plan as part of an optimization analysis. One of the variations was selected as the recommended plan.

The recommended plan will perform in a similar manner as the existing plan if future NBS are similar to those that have been experienced since 1900. However, there is a noticeable difference in plan performance under extreme NBS circumstances. For example, the Study's analysis simulated the high water levels of 1986, when high water levels created serious concerns among coastal zone interests and other interests downstream from Lake Superior. Under these conditions, the recommended plan resulted in levels on Lake Superior that were slightly higher than those under plan 1977A and slightly lower on Lake Michigan-Huron. In simulated wetter NBS conditions, the outcomes were reversed.

Public Engagement and Peer Review

Public involvement was a core element of the Study from the outset. Recognizing the many interests concerned with the future of water levels in the upper Great Lakes, the IJC appointed a bi-national Public Interest Advisory Group (PIAG) to provide advice to the Study Board on issues related to the Study and advice and support in the development and implementation of the Study Board's public information and engagement activities.

Finally, the IJC and Study Board recognized the need to ensure that the Study was scientifically credible and transparent, given the diverse public and private interests concerned about Great Lakes water levels and the uncertainty and debate around some of the scientific issues. As a result, much of the Study's work was subject to a high level of independent scientific scrutiny by external peer reviewers as well as extensive review by internal experts. The peer reviewers operated independently of the Study Board and provided their views directly to the IJC. They reviewed drafts and background studies of several of the Study's scientific and technical chapters. The Study's final report also was reviewed by the co-leads of the independent expert reviewers group. Study Task Team members considered and responded to each comment from the expert reviewers.

A New Regulation Plan to Balance Many Interests in the Upper Great Lakes

As noted before, the Study applied a "shared vision" approach to balancing the many interests of water users in the basin.

The Boundary Waters Treaty establishes an order of precedence for water uses. Under the Treaty, the interests of domestic and sanitary water uses, navigation and hydroelectric generation and irrigation are given preference in the development of regulation plans. No mention was made in 1909 of interests that are now recognized as playing an important role in supporting a healthy and vibrant Great Lakes, such as recreational uses and ecosystems. However, the Treaty does require that the Commission consider impacts on all interests (Article VIII).

With this in mind, the Study is looking at the evolving needs of six key interest groups affected by any new regulation plan:

1. *Domestic, Municipal and Industrial Water Uses*: public and private sector organizations using water for domestic, municipal and industrial purposes, including owners/operators of water and water treatment facilities, power plants, farms relying on irrigation, and large industrial plants, such as mines, paper manufacturers and chemical plants.
2. Commercial Navigation: owners/operators of the U.S. and Canadian fleets of bulk carriers, tankers, barges and other commercial ships transporting goods in the Great Lakes—St. Lawrence Seaway system, as well as ocean-going cargo vessels.
3. *Hydroelectric Generation*: the owners/operators of the three hydroelectric generating stations on the St. Marys River, the stations in the Niagara River and those that use the Welland Canal.
4. *Ecosystems*: the biological components, and the ecological services they provide, of the natural environment of the Great Lakes basin.
5. *Coastal Zone*: individuals and organizations with a direct interest in property along the shorelines and connecting channels of the upper Great Lakes.
6. *Recreational Boating and Tourism*: individuals, companies and associations with a direct involvement in coastal tourism, recreational boating and fishing, marinas and boat retailers, and the commercial cruise ship industry.

To recognize and balance the needs of these different interests, the Study has applied shared vision planning, a proven water management planning technique in which representatives of each interest are directly involved in the development of candidate plans. Through user-friendly computer modelling, participants have been able to learn about potential outcomes under various plans and help the Study identify needs and preferences related to Great Lakes water management. These interests in the upper Great Lakes Region are shown in Fig. 5.11.

Figure 5.11 An integrated view of interests in Lake Superior regulation.
(Source: IUGLS 2012).

A Regulation Plan Dealing with Climate Change and Global Isostatic Adjustment (GIA)

In working toward a new regulation plan for Lake Superior outflows, the Study is undertaking a rigorous analysis of two powerful forces affecting water levels in the Great Lakes basin: climate change and glacial isostatic adjustment (GIA).

In particular, climate change introduces a high level of uncertainty to predicting likely future water levels across the basin, making it difficult to design one regulation plan that will be optimal for all plausible future conditions. Therefore, candidate regulation plans will be evaluated for robustness—their ability to handle a range of conditions in an uncertain future. However, the regulation plan for controlling outflows from Lake Superior cannot be the sole mechanism for dealing with climate change. In this regard, the Study approached the hydroclimatic work in a multipronged approach. Three themes were central to the Study's approach to the hydroclimatic analysis:

1. understanding the water balance (precipitation, evaporation and runoff) of the Great Lakes;
2. assessing the reliability of historical recorded and estimated data, and increasing understanding of potential NBS conditions through the use of paleo-information and stochastic analysis; and,
3. addressing the plausibility and scope of climate change impacts on water supplies through new modelling work.

Climate Change

Given the high degree of uncertainty about climate change and climate variability at the regional level of the Great Lakes basin, the Study has approached the issue carefully to ensure a rigorous and balanced analysis. The Study has used a range of models and data sources, including:

- global- and regional-scale climate models;
- statistical models;
- historical water levels; and
- geological records going back nearly 5,000 years.

The impact of climate on the water levels for past reconstructed from beach ridge information, recent past from recorded water levels in the past century and a potential future are presented in Fig. 5.12. Preliminary findings from the Study's climate research indicate that, compared to current climate conditions, the climate in the upper Great Lakes basin during the next 30 yr is likely to be characterized by:

- an increase in precipitation and possibly more frequent intense storms;
- an increase in evaporation coupled with increased wind speeds that may offset the precipitation increase;
- increased lake temperatures; and,
- slight increases in water supply to the basin during winter/spring accompanied by larger decreases in supply during late summer/early fall, resulting in slight overall annual declines.

The Study's work indicates that, on an annual basis, water levels are likely to remain within a range fairly close to their long-term averages. The 2,000-year paleo record of lake levels shows somewhat higher lake levels

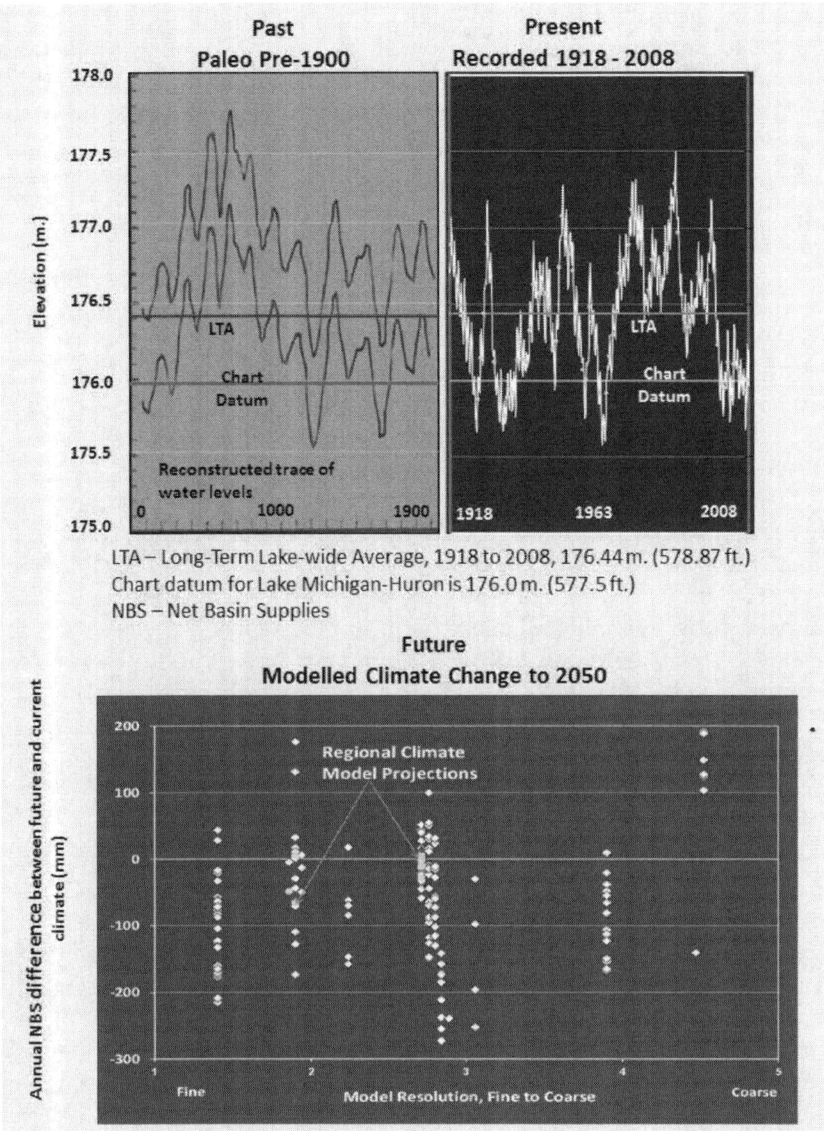

Figure 5.12 Range of water levels for Lake Michigan-Huron from past, present and future climate.

than the recorded 100-year water levels, but with no difference in low water levels. The data are referenced to the long-term average (LTA) and chart datum. Chart datum is the reference level for navigation charts. While lower water levels in the future are likely, we have to be prepared for episodes of higher levels, too. This finding is in contrast to past studies of climate change impacts in which upper Great Lakes water levels, at the extreme, could decline by a metre or more by the end of this century. Modelling results from the Study's two recently developed regional climate models predict that net basin supply will remain at near historic levels, whereas global climate models show much greater variability.

Glacial Isostatic Adjustment

During the last period of continental glaciation, which ended about 10,000 years ago, the tremendous weight of ice that covered most of the Great Lakes region depressed the earth's crust. At the same time, land beyond the edge of the glacier bulged upward. When the glacier retreated, the land that had been covered with ice began to rise while the land that had bulged up began to sink. This rising and sinking process continues today and has the effect of gradually "tilting" the Great Lakes basin over time.

The impact of this "tilting" is particularly noticeable along the shorelines, where features on the rising or sinking land can be compared directly to water levels and near-shore depths. For example, the shoreline of Parry Sound, Ontario, in Georgian Bay is rising at a rate of about 24 cm per century relative to the lake outlet. At the same time, the shoreline around Milwaukee, Wisconsin, is sinking at a rate of about 14 cm per century relative to the lake outlet. Figure 5.13 shows the impacts of GIA in the various geographical areas.

The Limits of Lake Superior Regulation

The Study Board recognized that the dual challenge of limited influence of any alternative regulation plan, particularly on Lake Michigan-Huron water levels, and the potential for exogenous drivers (particularly climate change and GIA) to affect water levels to a greater degree than the effects of regulation necessitated a re-thinking of the strategy. The Study Board concluded that to more fully address changing water levels in the upper Great Lakes basin, there was a need to look beyond the existing system of Great Lakes regulation, and consider alternative approaches for managing and responding to uncertain future water level conditions. These alternative approaches were: restoration of Lake Michigan-Huron water levels; multi-lake regulation of the Great Lakes—St. Lawrence River system as a whole; and, adaptive management.

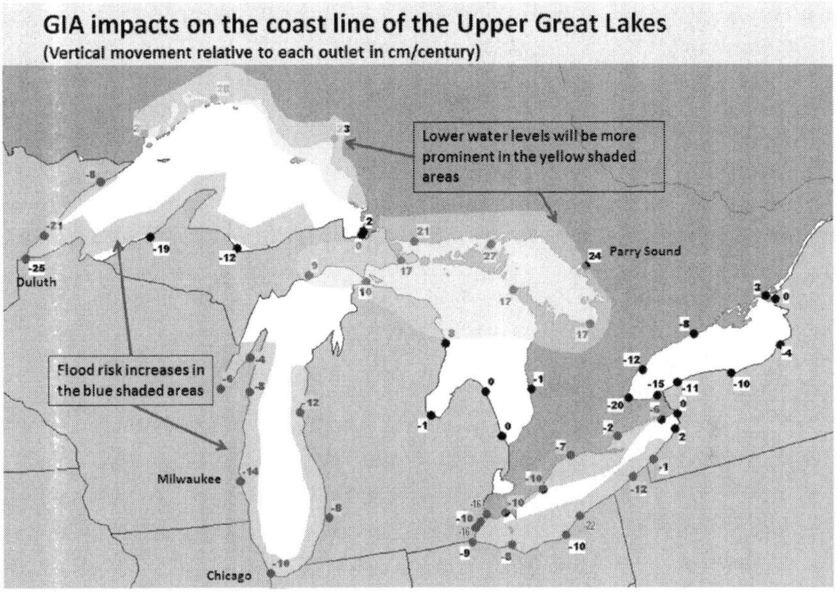

Figure 5.13 GIA impacts on the coast line of the upper Great Lakes (Source: IUGLS 2012).

Restoration and Multi-Lake Regulation

The Study is examining the feasibility of:

- building new control structures in the St. Clair River to restore historical water levels in Lake Michigan-Huron; and,
- using existing and new control structures to help regulate the Great Lakes—St. Lawrence River system on a system-wide basis.

A technical assessment commissioned by the Study concluded that restoration structures, such as submerged sills or weirs, are technically feasible. Submerged sills act as "speed bumps" at the bottom of the river, restricting water flows and raising upstream levels.

However, new control structures would generate a mix of benefits and adverse impacts for various sectors and locations. For example, higher water levels from these structures would likely benefit commercial navigation in the lakes, as well as shoreline property and wetlands in Georgian Bay. But structures to raise Lake Michigan-Huron likely would adversely affect hydroelectric generation and shoreline property and wetlands along Lake St.

Clair and Lake Erie—as well as damage sensitive habitat for five threatened or endangered species in the St. Clair River system, including the Lake Sturgeon. In addition, such structures can be expensive—a series of up to 13 underwater sills in the upper St. Clair River, for example, could cost from $ 70 to more than $ 200 million—and require decades to seek agreements, approvals and funding for construction.

Multi-lake regulation would involve using existing control structures on the St. Marys and St. Lawrence Rivers and building new structures on the St. Clair and Niagara Rivers. The general objective would be to keep the entire Great Lakes—St. Lawrence River system within observed historical extremes on all lakes, even under the more extreme projected climate conditions in the future. Preliminary estimates put the cost of multi-lake regulation at more than $ 8 billion.

Adaptive Management

There are risks to property owners, companies, local governments, ecosystems and other interests whether water levels rise or decline. High water levels can cause significant damage through flooding, erosion, and loss of beaches, recreational lands, and wetlands. Low levels can threaten water supplies, restrict power generation, expose mudflats, limit tourism, isolate wetlands, and severely restrict navigation.

Improved information and modelling of potential changes can help water managers, state, provincial and municipal planners and property owners adapt and improve their ability to cope with high and low levels.

Adaptive management is a process of continuous learning—improving planning decisions as new information becomes available or as conditions in the basin change. Figure 5. 14 illustrates the components as employed in the Study. Building the capacity in this area will require:

- enhanced monitoring and modelling of precipitation and evaporation over the lakes and runoff to the lakes;
- improved tracking of physical changes underway in the lakes and connecting channels, including those caused by human activity; and,
- distributing timely information to individuals, governments and companies in the Great Lakes so that they can plan to reduce possible risks from changing water levels.

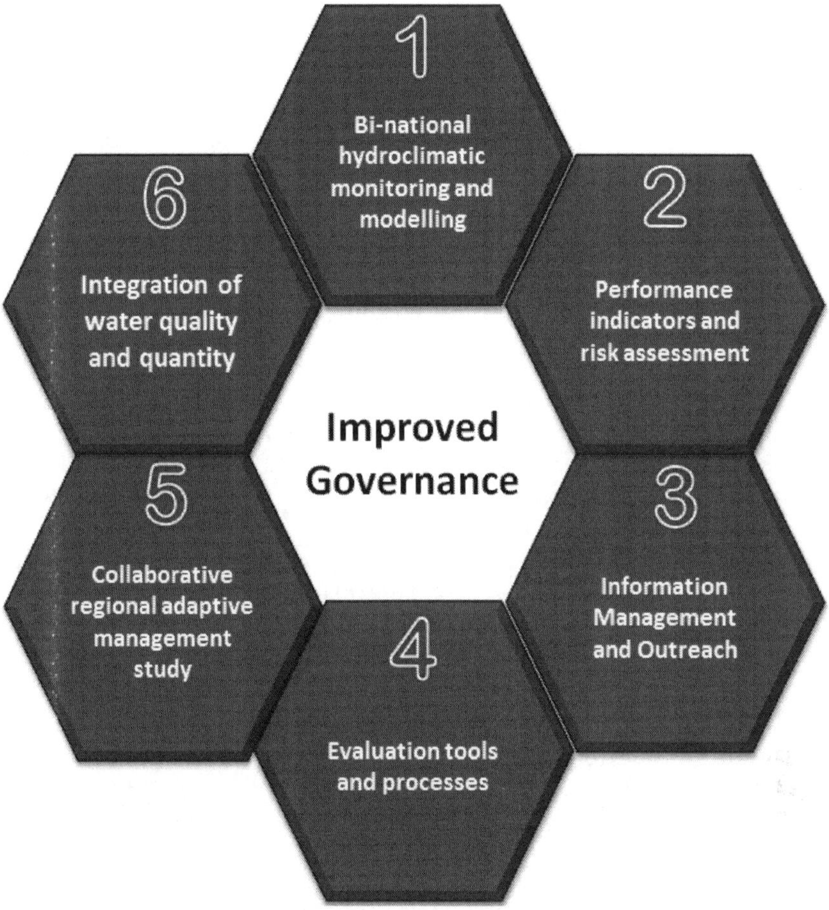

Figure 5.14 Elements of adaptive management in the Study.
(Source: IUGLS 2012).

Key Findings from Lake Superior Regulation Phase

This section summarizes the major findings of the Study.

The Key Interests Served by the Upper Great Lakes System

The Study undertook a comprehensive analysis of the current and emerging conditions and perspectives of six key interests likely to be affected by possible future changes in water levels in the upper Great Lakes basin. Based on this analysis, the Study concluded that:

- Under the Boundary Waters Treaty of 1909, the interests of domestic and sanitary water uses, navigation and hydroelectric generation and irrigation are given this order of precedence in water uses in the development of regulation plans. Today, it is recognized that other interests, such as ecosystems, coastal zone uses and recreational and tourism uses have rights under the Treaty, consistent with the IJC's balancing principle, which provides for benefits or relief to interests affected by water levels and flows without causing undue detriment to other interests.
- All six interests are experiencing major change as a result of broad, underlying economic, social and environmental forces. The decline in heavy industry and manufacturing in the region has put into motion changes such as declines in income, population, and municipal taxes, which in turn affect demand for shipping, energy and recreation. At the same time, the region's economic transition could see the rise of new, more water-intensive industries, such as irrigated agriculture, biofuels and oilsands refining.
- All the interests have a long-established presence in the upper Great Lakes basin, and all represent significant economic value to the region. There are clear expectations across all the interests that water levels will be maintained in the future to support their needs.
- All six interests can be adversely affected by both high and low water conditions. Most of the interests have demonstrated their capacity to adapt to changes in water level conditions that have been within historical upper or lower ranges. However, future water levels that are outside these historical ranges would require some interests to carry out more comprehensive and costly adaptive responses than any undertaken to date.
- For thousands of years, and continuing into the present, many Native American communities and First Nations have relied on the natural resources of the Great Lakes to meet their economic, cultural and spiritual needs. A fundamental ongoing concern of indigenous peoples is the extent to which they are involved in the decisions of federal and state/provincial governments in the United States and Canada with regard to the Great Lakes.

Uncertainty in Future Upper Great Lakes Water Levels

The Study undertook extensive analysis to improve understanding of the hydroclimatic forces at work in the upper Great Lakes basin and their likely impacts on future water levels. Based on this analysis, the Study concluded that:

- Perhaps most striking from the perspective of effective lake regulation is how little the lake dynamics on inter-annual and decadal timescales are understood. Despite best efforts, the lake levels remain almost entirely unpredictable more than a month ahead. In terms of understanding the lakes system relative to lake levels, the unavoidable conclusion is that Great Lakes are a complex system whose dynamics are only partially understood.
- Determination of climate change impacts on NBS using RCM tools provided insights into the dynamics of the hydroclimatic systems that are unavailable with statistical down-scaling. Features such as local feedback and recycled runoff are not captured in any of the GCMs. These aspects advanced scientific knowledge in this area. Due to the limited number of RCM runs, however, the full range of impacts were not computed.
- The NBS series generated by stochastic approaches provides a useful representation of future climate uncertainty in the near-term (i.e., the next 30 yr). Based on the Study's findings, there is no evidence that the statistics of the historical record are not representative of what can be expected within the next 30 yr, the Study's planning horizon.
- The current record of Great Lakes NBS appears continuingly stationary, marked by strong inter-annual and decadal variability, and showing no response that may be attributable to climate change. Increased evaporation and related local precipitation induced by climate change (with loss of ice cover), tend to be compensating each other, resulting in little change in NBS. During the Study's 30-yr time horizon in terms of implementing a new Lake Superior regulation plan, "natural variability" is likely to mask any climate forcing due to greenhouse gas emissions.
- As a result, changes in lake levels in the near-term future may not be as extreme as previous studies have predicted. Lake levels are likely to continue to fluctuate, but still remain within a relatively narrow historical range. While lower levels are likely, the possibility of higher levels cannot be dismissed but rather must be considered in the development of a new regulation plan.
- Beyond the next 30 yr, the predictions of global climate models (GCMs) for more extreme water level conditions in the upper Great Lakes may hold more merit. However, due to the limitations in the GCM projections for the Great Lakes region, it is clear that at present there is no satisfying representation of future climate for the region on that time span.
- Therefore, in terms of water management and lake regulation, the best approach is to make decisions in such a way as there is not great reliance on assumptions of particular future climatic and lake level

conditions. Robustness—the capacity to meet regulation objectives under a broad range of possible future water level conditions—must be a highly prized characteristic of any new regulation plan.

Lake Superior Regulation Plan

The Study developed and evaluated Lake Superior regulation plans to determine if a new plan could improve on the performance of 1977A through a shared vision planning process. Based on this work, the recommended plan will bring several important benefits compared to the existing plan:

- The new plan will do a better job preserving water levels on Lake Superior, with no adverse impacts on the downstream lakes, if future NBS become significantly drier under climate change.
- If future NBS are much drier than historical conditions, then Lake Superior 2012 will still be able to avoid infrequent but serious impacts to the spawning habitat of an endangered species, lake sturgeon, in the St. Marys River. Under 1977A, fish habitat impacts would be more frequent under drier NBS conditions.
- Lake Superior 2012 will provide modest benefits compared to the existing plan for commercial navigation, hydroelectric generation and coastal zone interests, under both wetter and drier NBS conditions. Importantly, under very dry future NBS conditions, commercial navigation through the Sault Ste. Marie locks, as well as hydroelectric generation at the St. Marys River power plants would be threatened with closure under 1977A, but not under Lake Superior 2012.
- Month-to-month changes in flow on the St. Marys River with Lake Superior 2012 would generally be smaller than with 1977A, which will give the St. Marys River a more natural flow relationship to Lake Superior levels. Natural river flow frequencies have been identified as an important factor in sustaining riverine ecosystem health. The smaller changes will also help hydroelectric energy producers.
- The rules for operating Lake Superior 2012 would be much less complex than rules for 1977A, making the new plan easier to manage, maintain and adapt to changing circumstances.

Adaptive Management

Consideration of the role of adaptive management in helping interests in the upper Great Lakes basin better anticipate and respond to future extreme water levels, the Study concluded that:

- Adaptive management has an important role to play in addressing the risks of future changes in water levels in the upper Great Lakes. Lake Superior regulation on its own can do little to address risks of extreme lake levels downstream of Lake Superior. Nor can multi-lake regulation fully eliminate the risk of extreme lake levels outside the historical range. New structures in various parts of the Great Lakes Basin could take decades to implement and cost billions of dollars. Therefore, regardless of the Lake Superior regulation plan adopted by the IJC, ongoing monitoring and modelling efforts will be required to continue to assess risks and address uncertainties and changing conditions.
- Information and education are powerful components of adaptive management. They contribute to both anticipating and preventing lake level-induced damage, particularly when focused on understanding risk, the limits of regulation, inherent uncertainties and system vulnerability.
- Existing legal, regulatory and programmatic efforts related to adaptive management vary considerably from one jurisdiction to the next. Federal, state and provincial governments generally provide the policy and regulatory framework, while site-specific selection and application of adaptive risk management measures is largely a local government responsibility.
- In addition, no bi-national organization exists to oversee an ongoing coordinated adaptive management effort in the Great Lakes basin. Efforts to coordinate approaches and promote consistency across jurisdictions have been limited and generally have focused on accommodating seasonal lake level fluctuations and the occasional extreme high and low water events. Furthermore, little focus has been placed on long-term implications of climate change-induced impacts and the need for new adaptive risk management measures.
- Therefore, application of a comprehensive adaptive management strategy would require a modified governance structure.
- Adaptive management to address future levels in the upper Great Lakes basin has direct relevance to several important initiatives in the Great Lakes—St. Lawrence River system, including:
    - adaptive management efforts in the Lake Ontario—St. Lawrence River part of the system;
    - the Great Lakes Water Quality Agreement; and,
    - the Great Lakes—St. Lawrence River Basin Sustainable Water Resource Agreement.

Acknowledgements

The contents of this chapter are based on the International Upper Great Lakes Study. The Study was managed by a ten-member Study Board appointed by the IJC. The contributing studies and projects were designed, supervised and executed by the Study Co-Managers. The chapter is a summary of work conducted by nearly 200 scientists and engineers under the direction of two task teams. The individual contributions of all Study participants are appreciatively acknowledged; a list of Study participants is noted in the IUGLS report references. The findings and positions noted in this chapter, while drawing from two main Study reports, do not necessarily constitute the positions of IJC or the Study Board.

References

In order to preserve valuable space, only key references are noted. All references can be found in the two Study reports noted below.

Baird, W.F. and Associates. 2005. Regime Change (Man Made Intervention) and Ongoing Erosion in the St. Clair River and Impacts on Lake Michigan-Huron Lake Levels. The report was prepared for the GBA Foundation, June 2005.

Coordinating Committee on Great Lakes Basic Hydraulic and Hydrologic Data. 1970. Lake Superior Outflow 1860–1968 (and post-1968 data published subsequently), June 1970.

Coordinating Committee on Great Lakes Basic Hydraulic and Hydrologic Data. 1976. Lake Erie Outflow 1860–1975 (and post-1975 data published subsequently), June 1976.

Coordinating Committee on Great Lakes Hydraulic and Hydrologic Data. 1977. Coordinated Great Lakes Physical Data.

Coordinating Committee on Great Lakes Hydraulic and Hydrologic Data (1995). Establishment of International Great Lakes Datum 1985.

Coordinating Committee on Great Lakes Basic Hydraulic and Hydrologic Data. 1988. Lakes Michigan-Huron Outflows—St. Clair and Detroit Rivers 1900–1986 (and post-1986 data published subsequently), October 1988.

Great Lakes Commission and U.S. Army Corps of Engineers. 1999. Living with the Lakes: Understanding and Adapting to Great Lakes Water Level Changes.

International Great Lakes Levels Board. 1973. Regulation of Great Lakes Water Levels. Report to the International Joint Commission, December 1973.

International Upper Great Lakes Study Board. 2009. Impacts on Upper Great Lakes Water Levels: St. Clair River.

International Upper Great Lakes Study Board. 2012. Lake Superior Regulation: Addressing Future Water Levels in the Upper Great Lakes.

Levels Reference Study Board. 1993. Levels Reference Study—Great Lakes-St. Lawrence Basin, Report to the International Joint Commission.

Nicholas, J.R. 2003. Status Assessment of Great Lakes—St. Lawrence River Water Resources. Chapter 2, Water Resources Management Decision Support System for the Great Lakes. Great Lakes Commission, 2003.

6

Towards Community-based Management of Water Resources

A Critical Ethnography of Lake and Groundwater Conservation in Pushkar, India

Christina A. Joseph

Introduction

The town of Pushkar (population 16,000), on the edge of the Great Indian Desert in Rajasthan (India), has been a lake-based Hindu pilgrimage site (*tirtha*) for over 2,000 years. Despite this longevity, Pushkar residents have expressed anxiety about imminent lake "extinction" since the 1980s and continually petitioned the state for remediation. Finally, in 2008, a technology intensive solution budgeted at nearly a million US dollars, was sanctioned through the National Lake Conservation Plan (NLCP). This enterprise, the Pushkar Project henceforth, proposed to restore the lake to its "past and pristine glory" (DPR 2008) with massive engineering works.[1] However, subsequent project implementation in the summer of 2009 reduced the lake to "mere spittle… on a powder-dry flatland."[2] While

Department of Anthropology, Baldwin Hall, Room # 250, University of Georgia, Athens, GA 30602, (706)542-3922.
Email: cjoseph@uga.edu

[1]The term Pushkar Project distinguishes it from other NLCP projects nationwide.

the state tried to deflect attention to a failed monsoon season, unappeased locals blamed poor project oversight and corruption, and experts reiterated their objections to desilting the lake.[3] In a complete reversal, the 2010 above-average monsoon rains brought a record breaking 18 feet of water to the lake and project officials seized the moment to declare success.[4]

Grounded in ethnographic and genealogical research, my goal was to examine the twin problems addressed by the Pushkar Project, lake pollution (water quality) and lake sedimentation (water quantity). Two decades of ethnographic engagement with this area provided the close-up lens through which the lake pollution was reframed by me as a cultural problem and not just a technological one. This localized perspective also considered how decentralized initiatives, based on locally relevant religious concepts and legal precedents, could produce the cultural changes necessary for sustainable solutions and community-based management strategies. The genealogical research, based on historical accounts of a discontinuous and contradictory past (Foucault 1984), allowed the focus to shift to a regional level and called into question reductive and linear narratives of desertification as the causal factor of lake sedimentation. This wide-angle perspective also disrupted the silence on the critical but invisible issue of groundwater.

Relevant Literature

This chapter draws on different strands of conservation literature ranging from discursive discussions of religious environmentalism and environmental imaginaries to more empirical research on common pool resources (CPR) and political ecology.

Research on sacred sites in India, like forests and rivers, inevitably focuses on the impact of religious precepts/concepts on ecological conservation (see Alley 1994; Gold and Gujar 1989). While it is not straightforward to use scriptural concepts as privileged sources for environmentalism (Dundas 2002), they do represent a "forceful cultural creativity" that merge local cultural repertoire with global ideas about the environment (Pedersen

[2]Wells, J. 2009, 'Hills are alive with the sound of ... camels', The Star.com, viewed 31 October 2009, from http://www.thestar.com/news/world/india/article/717796-hills-are-alive-with-the-sound-of-camels.

[3]Gaur, K. 2009, 'Sadhus annoyed over "vanished" Pushkar lake', The Times of India, viewed 26 October 2009, from http://timesofindia.indiatimes.com/city/jaipur/Sadhus-annoyed-over-vanished-Pushkar-lake/articleshow/5161871.cms and Mishra, S., 2009, 'Pushkar without a Dip', India Today, viewed 1 November 2009, from http://indiatoday.intoday.in/index.php?option=com_content&task=view&issueid=111&id=68794&Itemid=1&sectionid=114.

[4]2010, 'Pushkar Lake regains lost glory', YouTube, viewed 26 July 2010, from http://www.youtube.com/watch?v=4XJOYFji8CM.

1995). So, eschewing a naturalized link between religion and ecological protection, the strategic deployment of religious concepts to win people over to show concern for the lake's ecological wellbeing were examined by me (Nugteren 2005). Such strategies had been successfully applied in other areas of India. For instance, several environmental projects in India mobilized a religious concept, like *shramdan* or donated labor (Cochran and Ray 2009), to produce the "alchemical shift in interest, beliefs, and actions" necessary for the production of environmental subjects (Agrawal 2005). Sant Seechewal, a respected Sikh leader, used religious appeals to motivate his community to clean and maintain the heavily polluted Kali Bein River in Punjab through donated labor (Nigah 2007).[5] In Udaipur, Rajasthan, volunteers were inspired by the Hindu precept that lakes were to be "respected and worshipped" to voluntarily clean Lake Pichola (Anand 1994).[6] Closer to Pushkar, the neighboring Budha Pushkar Lake was also desilted using shramdan (Sharma and Chouhan 2008). These are powerful examples of the strategic use of "the religious environmental paradigm" to express concern for the environment that may not have worked by the invocation of pure ecology (Pedersen 1995).

Conservation efforts at a sacred center like Pushkar were inevitably mobilized around familiar religious concepts but grassroots groups were also increasingly utilizing CPR-like arrangements nested in many layers. Pushkar Lake was not exactly analogous to the English commons because, not unlike many sacred lakes in India (see Jain et al. 2004), it had no extractive or utilitarian purpose. It was not a source of water (for domestic use or irrigation) or food (fish). Lotus tubers that had once been extracted from the lake had since disappeared with deteriorating water conditions.[7] Leisure activities like swimming and boating in the lake were also forbidden. Yet one could make the argument that the lake was a CPR because it had been common (*shamlat deh*) to the town historically, was still central to its pilgrimage/tourism economy, and most significantly had symbolic capital (Mosse 2003).[8] The literature on managing groundwater resources further suggests that the argument for a CPR could also be extended to groundwater (Feeny et al. 1990). Could CPR arrangements, like a multiplicity of rules, incentives, increased information, monitored use, induced compliance

[5]See also Singh, M. 2008, 'Balbir Singh Seechewal', Time, viewed 24 September 2008, from http://www.time.com/time/specials/packages/article/0,28804,1841778_1841781_1841808,00.html.

[6]*Rajender Kumar Razdan v. State of Rajasthan* (D.B. Writ Petition No. 4271/1999), February 6, 2007.

[7]An 1894 document referred to a bitter dispute between the two neighborhoods of Pushkar over the right to harvest the "abundant" lotus plants. Judgment of District Judge of Ajmer-Merwara, *Brahmin jagirdars of Basti Kalan in Pushkar v. Brahmins of Basti Khurd in Pushkar*, Suit No. 36, 1894.

[8]For *shamlat deh* see *Brahmin jagirdars of Basti Kalan in Pushkar v. Brahmins of Basti Khurd in Pushkar*, Suit No. 36, 1894.

(Dietz et al. 2003 and Ostrom 1999), and co-management by the state and users (Feeny et al. 1990), then protect the lake and conserve groundwater? Given the precedents set by self-mobilized community-based groups (Pretty and Shah 1997) and decentralized local initiatives elsewhere in generating norms of use that constrained self-interested behavior and protected CPRs without external coercion (Bardhan and Ray 2008), a CPR approach was a viable option in Pushkar.

At the regional level, the local environmental imaginary[9] (Peet and Watts 1996) of an advancing desert emerged as a locally salient issue. Desertification as a result of the anthropogenic degradation of land (UNEP Report 1987) had long been exposed as environmental orthodoxy in academic circles (Forsyth 2001). Literature on crisis narratives cautioned that such so-called facts about environmental degradation had less to do with biophysical reality (Guthman 1997) and more to do with competing claims of different political and bureaucratic constituencies (Swift 1996) that were rarely challenged in public policy-making or popular reporting.[9a] Yet development driven projects in Pushkar stabilized the received narrative (Swift 1996) of desertification as a scientific fact. Feedback loops between local discourse and interventions, like the one found by me in Pushkar, raised a red flag about privileging indigenous environmental imaginaries and local knowledge for it too may be "mythology" (Forsyth 1996). One way of countering such mythologies or received narratives was to understand how natural resources, like Pushkar Lake or groundwater, came to be imagined, appropriated and contested through practices of cultural production (Baviskar 2003).

The final strand in the literature deals with civil society strategies in resolving issues of political ecology, like access and control over natural resources. This was significant because emerging grassroots groups in Pushkar emulated successful legal precedents set by civil society action elsewhere. For instance, the most cited precedent in Pushkar was the Public Interest Litigation (PIL) Writ Petition filed by the Udaipur's Lake Conservation Society to hold the state accountable for ecological degradation and negotiate the right to implement locally relevant alternatives (Anand 1994). Local activists followed suit in using PILs and the Right to Information Act to demand public accountability. They also cited the civil society initiatives, like that of the civil activist Anna Hazare in Ralegan Siddhi, Maharashtra, and NGOs operating in Alwar and Udaipur, Rajasthan (namely PRADAN, Tarun Bharat Sangh or TBS and Seva Mandir), that successfully challenged the state's control of water by

[9] The term "environmental imaginaries" refers to the ways in which the natural environment shapes attitudes, discourses, and practices of the people who dwell there.

[9a] For popular reporting see: Team Bhaskar, 2008, 'Pushkar Ksetra Mein Bhi Hain Kai Gap (There are Many Gaps in the Pushkar Region Too)', Dainik Bhaskar, 27 June.

promoting decentralized alternatives and new water management norms leading to demonstrable ecological outcomes over the past 25–35 years (Ahluwalia 1997; Gosling 2001; van Steenbergen and Shah 2003). Such scenarios opened up possibilities for decentralized, self-regulating water (especially groundwater) management systems without quantified rights and without a central regulatory power (Roy and Shah 2003) in Pushkar.

Methods

Fieldwork was conducted during the planning phase of the Pushkar Project in 2008, implementation in 2009, and post-implementation in 2010–11. It included interviews, attending court proceedings, meetings between government/project officials and the community, and observing public behavior at the lakeside. Interviews, informal and formal, were conducted in Hindi, Marwari, and English with Pushkar residents and pilgrims, farmers, local/state officials, research scientists and NGOs. Informal interviews with pilgrims and residents were based on a convenience sample, while formal interviews were conducted with stakeholders. Two key informants and two voluntary groups emerged as community proponents. I drew heavily on these sources as their viewpoints crystallized differing but widely held community perspectives. The use of multiple standpoints prevents the simplification of Pushkar as a "coherent, stable," and undifferentiated community (Li 2002; Cochran and Ray 2009).

I knew the two key informants, well-respected and vocal interlocutors from the dominant, upper-caste Parashar Brahmin community, from previous field trips. As inveterate letter writers and state petitioners, they presented contrasting leadership styles and appealed to different constituencies: the old and the young respectively. The older man, locally referred to as "Ex-Chairman," was a practicing lawyer with a veritable repository of gray documents who worked through a legal appeals process, sit-downs, and negotiations with politicians that he pragmatically termed as the "patronage of politics." The younger man, a self-described "Community Activist," was a hotel owner and popular leader of two local NGOs whose members accompanied him to meetings, interviews, etc. He appealed to religious rituals, like stream/river worship at the onset of the monsoons (*nadi pujan*), and caste authority to rally support for his agenda of change but also demonstrated an unusual mastery of legal and technical details. Interestingly, his brother was a local newspaper reporter and television commentator through whom both the Community Activist and the Ex-Chairman pushed forward their respective media agendas. Both men were consulted extensively during the planning phase of the Pushkar Project, monitored its execution closely, and were central to its future sustainability.

The two groups that shaped the emergent conservation discourse were an Ajmer based NGO named the Common Cause Society and a group of Pushkar based self-styled Good Samaritans. Both had overlapping memberships largely composed of high-ranking retired administrators and locally influential seniors motivated by nostalgic memories of regional forests teaming with wildlife and unpolluted lakes where families picnicked and prayed. Their official connections meant they were consulted for the project and their outsider status shifted the discourse from the lake to the catchment area.

Ethnographic field research over the summers of 2008–09 and the winter of 2010–11 was informed by archival research and virtual follow-ups via online new media resources, like traveler's blogs, Flickr photographs, YouTube videos, as well as vernacular press clippings (courtesy the Ex-Chairman) that tracked the Pushkar Project longitudinally between stints of fieldwork. I relied on this "knowledge of details" and an "accumulation of source material" to chart a more complex course of ecological degradation (Foucault 1984). This multi-textual approach decenters the "field as the one, privileged site of anthropological knowledge" (Gupta and Ferguson 1997) and allows for the supplemental use of "digital" observations. This is not unimportant for a place like Pushkar that has been part of the online news cycle for some time due to networked international tourists who blog and upload photographs/videos in real time. In 2011, I found local youth, like the Ex-Chairman's son, had joined the densely interconnected blogosphere and were using social media to "virtually" shape the conversations on Pushkar.

Setting

Pushkar—The Town

A panoramic view of Pushkar reveals a valley flanked by the rugged Aravalli range and substantial sand dunes with seasonal streams draining into a natural depression to form a lake enclosed by *ghats* (step-like embankments). Less than half a square mile in size, the lake is the dominant topographical feature surrounded by Pushkar town and agricultural fields.

Pushkar is an important Hindu pilgrimage that has also become an international tourist destination in the last two decades due to the aggressive state promotion of mass tourism. It draws approximately 1.5 million pilgrims and 50,000 international tourists annually (CDP 2006). The domestic pilgrims come to visit the sacred place eulogized for its sanctity in sacred Hindu texts, like the Padma Purana and Mahabharata (Bhatt 1988; Van Buitenen 1975). Foreign tourists come for its "hippy vibe" and well-publicized winter camel fair (Joseph 2007). The five-day fair combines commerce and religious activities and averages 400,000 visitors (CDP 2006).

The main inhabitants of the town are Brahmins, many of whom (approximately 2,000) work as pilgrimage priests (*pandas*). Tourism has recently brought in a largely non-Brahmin population employed in hotels, restaurants, and shops along the main street. A striking feature of the social organization of the town is that for centuries it has been divided into two neighborhoods—Big Basti (inhabited by Parashars) and Small Basti—that have feuded bitterly in the streets (Broughton 1892) and in civil/revenue courts for proprietary and hereditary rights to pilgrims, territory, and the lake (Sarda 1914). This infighting negatively impacted the development of Pushkar. In 1933 the British blamed the "deplorable" state of the town on the "warring factions" that "disfigured public life" and "brought about its present ruinous condition."[10] Post-independence the feuding continued in the municipality, holding up projects, like the construction of a bus stop, for decades while the factions wrangled over rights. While loyalties were blurring, especially with a town-wide acceptance of the Community Activist, the social divide is still a potential impediment to community-based actions.

Pushkar—The Lake

Pushkar is acclaimed for its Brahma temple but the ritual nucleus of the tirtha is the lake, locally known as Pushkarraj. Dawn breaks with pilgrims joining residents in song (*aarti*) by the lakeside. Throughout the day pilgrims stream in to perform worship (*puja*), ablutions (*snan*), and circumambulations (*parikrama*), and hear priests praise the lake's power to cure diseases (Zeitlyn 1988) and ensure salvation (Bhatt 1988). In a videotaped interview with project officials, the Ex-Chairman explained the sanctity of the lake to pilgrims: "If you're visiting Pushkar, you are not coming to visit it as any simple or ordinary pond. It has got a special significance. It is a supreme place of pilgrimage in India; tirtha guru we call it."

Prayers, overheard at the lakeside, invoked Pushkarraj who was commonly identified as the divine aspect (*swaroop*) of god Brahma or "god in the form of a lake." On the ghats, a panda ad libbed the Padma Purana to explain Pushkarraj was one form of Brahma (Brahm roopam Pushkar hai). He then tapped the Pushkar origin myth to elucidate the connection between Brahma and the lake: when Brahma's estranged spouse, Savitri, limited his temple rituals to renouncers only, Brahma countered her curse by proclaiming he could be worshipped by everyone at the lakeside as

[10]*Proposal to form a municipality at Pushkar*. File No: D.XI-458 of 1933. Rajasthan State Archives, Ajmer.

Pushkarraj. The pandas at various ghats reiterated this view by stressing that the lake itself was the temple of Pushkarraj. Others like the Ex-Chairman and the Community Activist resisted this identification of Pushkarraj as Brahma; "Pushkarraj is Pushkarraj," they said. All, however, agreed on the centrality of Pushkarraj to local residents; an elderly panda, meditating on a less frequented ghat, explained that Pushkarraj was their livelihood and therefore their tutelary deity (*isht devta*).

The Community Activist referred to the lake as *nabalik* (a minor under 18). Originating in the Rajasthan Tenancy Act 1955 (Section 46(3)), the concept of nabalik referred to the legal position of a temple deity as a perpetual minor who functioned as a juristic person through a guardian, like a temple priest. In other words, the lake was a minor for whom local Brahmins, as the ritual specialists, were self-appointed guardians. To an observer, this spirit of guardianship was evident in the zealous protection of the lake's purity from ritual pollution. Local residents, especially the priests, vociferously reminded visitors to leave footwear 30 feet away from the lake, not to use soap, rinse mouths or spit, or touch the water with unclean hands or, in the case of women, if they were menstruating. While similar proscriptions, based on the *Pravascitta Tatva*, were followed near the sacred river Ganga (Dwivedi 2004), in Pushkar, fishing, swimming, and boating were also banned. Locals claimed these injunctions were made law by the Mughal emperor Aurangzeb but archival evidence suggested that the British introduced them as sanitation rules in 1938.[11] The local administration publicly posted these rules for visitors in 1984; they appear on streets, notice boards, and in hotel lobbies in Hindi, English, Hebrew, and several European languages.

Lake Pollution

Despite this locally protective stance towards the lake, it had a pea soup like appearance in 2008, locals reported skin rashes from ritual ablutions, and fish die-offs occurred on an unprecedented scale. Several town elders, on their daily circumambulation round the lake, sadly contrasted this sorry state of affairs to their own memories of crystal clear water where lotus proliferated and a coin could be seen even in ten feet of water. They fondly reminisced about splashing in the lake as children while their mothers carried home pots of water for household use.[12] They begged forgiveness of Pushkarraj as they confided that now they would only touch the lake

[11]Rules for the improvement of sanitation in certain villages of Ajmer State, 1938. *Misc. correspondence regarding Pushkar municipal committees*. File No: D.XI(g)44, 1952. Rajasthan State Archives, Ajmer.

water with their fingertips. While the Community Activist justified his own regular ablutions in the lake to me by saying, "God has changed his form to test our faith," the Ex-Chairman expressed the more commonly held view: "increasing mud and silt, decreasing water level transformed this whole lake into very muddy and very dirty water which can't even be touched. How to take a holy dip in it?"

Lake water tests conducted by the Pushkar Project confirmed a high bacterial count that made even secondary use, like ritual bathing, unsafe (DPR 2008). Locals attributed this deteriorating water quality to the town sewage overflowing into the lake during heavy rains.[13] But few connected the very rituals that celebrated the sanctity of the lake to its current state. Worship paraphernalia, like vermilion, turmeric, sesame seed, unhusked wheat, coconut, and flowers, were immersed in the lake despite the Public Nuisance law banning it. Pandas urged tourists to "throw flower in holy lake" for blessings and instructed pilgrims to feed the fish to earn religious merit. Even the annual purification rite for the lake (*phul dhol*), that I first observed in 1989 and recently again saw in photographs from the 2010 celebration (on the Facebook community page maintained by the Ex-Chairman's son), involved pouring gallons of milk and bushels of flowers into the lake. Not surprisingly, then, sediment samples taken by the Pushkar Project indicated an accrual of excessive allochthonous materials and organic nutrients that caused eutrophication, cycles of algae blooms and decomposition, depleted dissolved oxygen, and excessive turbidity (DPR 2008). Other prescribed, but problematic, immersions included ashes of the dead consigned to the lake to ensure deliverance of the soul, and large Plaster of Paris statues of gods (*murtis*). The latter released toxic substances that severely impacted water quality, for instance in 2008 the immersion of dozens of murtis during the Ganesh festival led to a massive fish die-off.[14] Lastly, frequent mass religious bathing events resulted in a 10–1,000-fold increase in fecal coliform bacteria (Lal 1996).

This paradox of pollution, despite the centrality of the lake to ritual life and the generally protective attitude towards it, was recognized by the local reporter when he noted, "we live in our own community, we destroy our own community." However, for the average pilgrim the pollution was insignificant because of the inherent purity of the lake. As in the case of the

[12]While the record of customary rights (*wajib-ul-arz*) of Pushkar confirmed this past usage, the lake was no longer a source of domestic or irrigation water. *Manuring in Pushkar Lake*. File No: I-(a)-5, 1939. Rajasthan State Archives, Ajmer.

[13]2011, 'Sewer water enters Pushkar lake, priests up in arms', The Times of India, viewed 11 July 2011, from http://articles.timesofindia.indiatimes.com/2011-07-03/jaipur/29732788_1_pushkar-lake-sewerage-sewer-water.

[14]'Abke Baras Tu Jaldi Aa (This Year Come Early)', Bhaskar News, 26 September, 2008.

river Ganga whose purity was not "defeated" by pollution (Alley 1994), the holy water of Pushkar was considered pure even when polluted. However, even this high level of tolerance for polluted holy water reached a tipping point when the fish started dying in large numbers. In June 2008 two massive fish die-offs lead to angry protests at the Pushkar Municipal Board office that culminated in fish being thrown at the board chairman. Another die-off in September 2008 virtually decimated the fish population in the lake; newspapers reported a "whiff" so bad that tourists fled the town, and even pilgrims avoided bathing in the lake.[15]

These events shattered the complacency about lake pollution and local activists organized a rally on October 14, 2008. School children formed a "human chain" along the street holding placards exhorting everyone to "keep Pushkar Lake clean" and not to "put polluting materials in the holy lake." Marchers chanted refrains like, "the lake is our future, keep it clean," "the lake is our soul, don't hurt it," and "stop the pollution."[16] The Ex-Chairman, a rally organizer, pointed out that the stagnant lake had a limited capacity to absorb polluting materials unlike the flowing river Ganga that carried it away.

To understand this shift in attitude, one needs to understand local beliefs about living things in Pushkar. According to tradition, no living being was to be harmed within the Pushkar *ksetra* (greater Pushkar sacred region); vegetarianism was enforced and reverence for all living beings that was expressed daily through the practice of feeding animals. Specific proscriptions against harming living creatures in the lake went back to the colonial period when instances of Europeans shooting crocodiles to rescue bathing pilgrims met with near revolts (La Touche 1879). The British eventually prohibited shooting and fishing in 1910 in the entire Pushkar ksetra.[17] Post-independence the crocodiles were relocated and the lake proscriptions focused on the fish that, in the words of the Community Activist, could not be "hunted" (*shikar*). Even the mere suspicion of fishing in the lake led to a weeklong protest and hunger strike by a local leader in the 1980s (Joseph 2001). So, to lose almost the entire fish population in one incident was a monumental event for the community.

To mitigate lake pollution, the Pushkar Project proposed culturally relevant changes, like the exclusion of animals from ghats, and limited

[15]Shekhawat, P. S. 2008, 'Preserving Pushkar', Tehelka, viewed 5 December 2008, from http://www.tehelka.com/story_main40.asp?filename=Ws291108preserving_pushkar.asp.

[16]2008, 'Pushkar ecology demonstration', YouTube, viewed 31 October 2008, from http://www.youtube.com/watch?v=9Ey5EfIY7AM.

[17]*Prohibition of fishing and shooting etc. within limits of Pushkar and Anasagar*. File No: XXVIII-81B, 1895. Rajasthan State Archives, Ajmer.

immersions of ashes, murtis, fish food, flowers, grain, and offerings to ancestors (DPR 2008). Expressing opposition to a total ban on immersions and feeding of animals, the Community Activist quoted from the Hindu scriptures, "pilgrims come to serve the cow, the Brahmin, and the tirtha."[18] But he also acknowledged the detrimental effect such ritual acts had on the lake and proposed accommodations, like nets to catch and remove excessive offerings or dedicated immersion tanks. On the other hand, the Ex-Chairman pushed for more purposeful restraint in resource use and actual cultural changes, "If you want to protect that pious lake and keep it alive for long, you should take some preventive steps. Religiously there are certain hard things that have to be swallowed."

On my field trip in 2010, I found that very few of the changes instituted by the Pushkar Project were working. Pilgrims fed pigeons on the ghats, and immersed offerings into the water right under new signs that expressly banned both things; the edge of the lake was littered with offerings like flowers, coconuts, clay lamps, immersed idols, ritually discarded clothes, etc. On the positive side, the lake level at 8–10 feet was far below the level of the newly constructed bathing tanks, thus preventing the allochthonous materials from washing into the lake. The most significant changes had taken place at a ghat controlled by the Community Activist's NGO where no animals were fed and offerings were immersed in a small metal bowl, not in the tanks or lake. Under the leadership of the Community Activist, significant changes had taken place over the two years since the lake crisis.

The Lake as a Commons

Given the sobering fact that the allocation for changing attitudes through public awareness was a mere one percent of the total budget of Pushkar Project, the question was whether it was possible to sustain the momentum created by spontaneous eco-mobilizations, like the rally, and eventually produce alchemical shifts and environmental subjects. Further, could historic usufruct rights of local residents to the lake, that continued to be nested within state ownership post-independence, be promoted to foster polycentric governance systems?[19]

There was no doubt that the informal local norms that previously governed the use of the lake as a commons and prevented a free-for-all

[18]Pilgrims earned religious merit by feeding animals and documentary evidence dated the sale of fish and bird food to the 1800s (Larking 1888; Black 1908). The British defended ash immersion as an established custom in 1936. *Land and ghat adjoining Man Mahal building of Jaipur State in Pushkar*. File No: 167, 1936. National Archives, New Delhi.

[19]Rajasthan Irrigation and Drainage Act 1954 vested all rights to water bodies in the state. Rajasthan Municipality Act 1959 handed the trusteeship to municipalities.

were seriously undermined by the post-Independence transition to *de facto* open access under the *de jure* stewardship of the state (Feeny et al. 1990). Bromley and Cernea (1989) referred to this transition as the "tragedy of open access." This was reflected in the complaints of local priests that if they told pilgrims to follow the new rules, they were accused of harassment. Similarly, administrators complained that political interference stymied their attempts to regulate the sources of pollution. For instance, a local administrator who tried to enforce the ban on the sale of fish food was summarily transferred. While new rules of use were emerging, like a ban on immersions, more public awareness was required to achieve compliance.[20]

The public awareness campaign, developed by the Pushkar Project, drew upon the local devotion to the lake to induce "self-belongingness among people towards the lake and assets created under the project" (DPR 2008). The pilot campaign rolled out during the 2008 camel fair, invoked Pushkar's main deity: "take Brahma's name and start the important work of protecting Pushkar water."[21] This was an example of the kind of "forceful cultural creativity" that merged global ideas about nature with local cultural repertoire (Pedersen 1995).

However, a richer cultural repertoire of religious concepts featured regularly in conversations with locals but was not harnessed by the state public awareness campaign. For instance, the notions of hereditary and historical custodianship (nabalik) and of divine ecology (Gold and Gujar 1989) that predated the colonial era (Brandis 1897), like the protected sacred groves (*dev banis*) in Pushkar valley. Recorded in revenue records as belonging to the Gujjar deity Dev Narayan (*devta ke naam se zameen*), these groves were protected from "cutting" by the common acceptance that transgressors suffered "sorrow and losses" due to the deity's wrath. Such notions served as sustainable "social fences," i.e., used traditional stewardship to protect resources (Borrini-Feyerabend 1997). Another traditional proscription in Rajasthan—reinforced by the Padma Purana injunction, "those who break good wells, lakes, places of water for travelers, or ponds go to hell"—advocated against sullying catchment areas (Agarwal 2000 Deshpande 1990). These religious ideas were deeply ingrained in the community and the pandas, the ever-vigilant custodians of the lake who could be recruited to promote them in addition to ritual rules that they regularly enforced.

The intentional use of many of these concepts by the Community Activists and his NGO members pointed the way forward. However, a

[20]'Jhilon Mein Murti Visarjan, Gandagi Par Rokh (Ban on Idol Immersion and Dirt in Lakes)', Bhaskar News, March 2009 (date not available).

[21]2008, 'Camel Cart - Save Pushkar Lake Awareness', YouTube, viewed 7 January 2009, from http://www.youtube.com/watch?v=DZHkd45uY80.

wider mobilization of such religious concepts required more than just public awareness; Hardin suggested "mutual coercion... agreed upon by the majority of the people affected "(1968). There was some indication that local leaders, like the Ex-Chairman and the Community Activist, exerted such internal coercion—the former by an appeal to the politico-legal nexus and the latter through recourse to religious rituals and language. As powerful members of the community, they mobilized dense networks of interpersonal relations based on kinship, caste, religion, and social standing, and provided leadership in a community with many competing interests and deep divisions. For instance, the Community Activist persuaded squatters to move from a lake feeder by appealing to their devotion to Pushkarraj. Such strategic use of religion was not lost on administrators: at a public meeting, a senior official exhorted project contractors to work with a "missionary zeal... for Pushkarraj." Even local leaders blatantly used religion to appeal to and judge regional and national level political leaders on their "devotion," measured not only in visits to the lake to pray but also in the funds and projects appropriated for the town. Savvy politicians thus actively sought to gain symbolic capital through well-publicized acts of devotion to the lake, especially to offer thanks after being elected to office.

As a case study in the making, Pushkar provides a critical ethnographic road map of what could be possible through the strategic (not naturalized) ecological deployment of religious concepts in a town where religion was the very idiom of the community.

Lake Sedimentation

While lake pollution had become the focus of local ire since 2008, demands for abating lake sedimentation had been expressed since 1907.[22] Desertification was the imaginary of ecological degradation that conceptually underwrote this local discourse on sedimentation and consequently the allocation of remediation resources. My goal here was to investigate how these claims of ecological changes, that so thoroughly imbricate local policy, became naturalized locally.

The sedimentation issue was one of the size and depth of the lake. While toposheets (Survey of India no. 45 J 11, 1928, 1973) and lacustrine deposits beyond the current shoreline (Karanth 1977) indicated that Pushkar Lake had shrunk in diameter to less than half a square mile in size, a reduction in depth was harder to track. In 2007, government surveyors found the lake to be no deeper than six feet at any point (DPR 2008): a contrast to past

[22]*Rough Scheme for the clearance and restoration of the Pushkar Lake* (*correspondence regarding*). File No: 50, 1907. And *Misc. correspondence regarding Pushkar municipal committees*. File No: D.XI(g)44, 1952. Rajasthan State Archives, Ajmer.

hyperbolic claims of a bottomless tank (Beveridge 1868) that went down to the "infernal regions" (Eastwick 1883). Historical accounts more accurately reveal a pattern of inter-annual fluctuations between floods in 1810, 1860 and 1875 and droughts in 1867, 1907 and 1911 (Broughton 1892; Wilson 1877). It was the drought of 1907 that prompted the first recorded demands for desilting by a local group, the Pushkar Jirnoddhar (improvement) Sub-Committee. It successfully raised money from Hindu royal states but never actually desilted the lake due to heavy rains in subsequent years.[23] In recent times, local memories of dry periods, like that of 1974 when the lakebed served as a cricket field, and times of high water levels, like 1975 when the water was so high that kids could dive into it from rooftops, closely correspond to my own recollections of fluctuations since 1985.

Irrespective of this living memory, the lake level was perceived as drastically reduced by aeolian sands from the "advancing desert" (*badhta registhan*) and by fluvial deposits by torrential monsoon streams flowing unchecked through deforested areas. From the 1990s onwards project proposals and reports replicated the global UNEP view that desertification resulted from human impact.[24] They claimed a recent provenance for "active and gigantic sand dunes" that had swallowed "85 percent of the land in the valley" (UNEP Report 1987), ignoring prior evidence to the contrary: namely, archaeological findings that the sand dunes were fossil landforms from the relatively humid Holocene period (Allchin and Goudie 1974); pollen analysis sequences indicating the area had been semi-arid since 400 CE (Singh et al. 1974); and 19th century British accounts of "sand-hills of considerable magnitude" and "immense mounds of shifting sand" (Tod 1983; Rousselet 1875). UNEP activities in Pushkar included film screenings to "depict" triggers for desertification and to train locals to slow/halt this process through eco-regeneration activities, like afforestation, confined grazing, bio-tree guards, and bio-fencing (UNEP Report 1987). This was consistent with the global UNEP view that desertification in arid/semiarid areas resulting from human impact could be arrested by human action, e.g., by planting trees on sand dunes (Pearce 1992).

Some scientific reports went a step further in predicting that desertification would result in the "extinction" of Pushkar Lake unless immediate action was taken (Kar 1986; UNEP Report 1987). They had the unfortunate outcome of reinforcing a cultural suggestibility to desertification narratives based on a local legend about the obliteration of Pushkar Lake by a sandstorm, brought on by the curse of a powerful ascetic denied his

[23]*Rough Scheme for the clearance and restoration of the Pushkar Lake* (*correspondence regarding*). File No: 50, 1907.

[24]See also Integrated Development Plan, 1989–94 and 1992–95; Pushkar Lake Development Project, 1994; and India-Canada Environment Facility, 1997–2001.

customary alms by residents. It was supposedly restored when the King of Mandor accidently discovered a spring with miraculous water and had the area excavated (Zeitlyn 1988). The UNEP "training" in Pushkar, thus, had the unintended consequence of producing subjects vested in a crisis narrative where desertification could only be arrested by human action, like eco-regeneration (UNEP Report 1987).

The local claims of deforestation were equally obscure. While some historical accounts suggested a past of "dense jungles," counter narratives insisted there were "no trees of consequence," and hills were "as bare as one's hands" (Beveridge 1868; Bonavia 1885; Eastwick 1883; Heber 1829; La Touche 1879; Rousselet 1875; Tod 1983). These discrepancies could be attributed to drought pulses and policy changes; for instance, when the British handed forest land to the villages in 1850 a massive felling of trees left "naked hillsides." However, after the 1874 return of management to forest officers, a rapid re-growth was reported (Brandis 1897). In modern times, deforestation was equally hard to substantiate, especially in an area where dune stabilization and afforestation had been a priority for the forest service. Forest cover, on what were bare sand dunes in the 1980s, visibly increased under the soil and dune fixation schemes, including aerial seeding. In August 2008, the Ajmer Conservator of Forests categorically stated that no deforestation had taken place in the catchment area of Pushkar Lake.[25]

Despite the above disagreements, reports since the 1990s agreed on the fact that the flow of surface water to the lake was blocked or diverted. Successive projects repeated a formulaic prescription of desilting, afforestation, and construction of feeder channels and check dams to regulate water flow. Unsurprisingly, the Pushkar Project was more of the same on a grander scale: with an exclusive focus on, in the words of the Ex-Chairman, to feeding "every drop of rainwater which falls on the surface of Pushkar hills" to the lake.[26] A quick breakdown of the budget was telling in its focus on structural features that failed to be sustainable in the past: feeder and water reservoir construction at 75 percent of the budget, lake desilting at 13 percent, and silt excluders and afforestation at 8 percent.

The implementation of the project in 2009 was "haphazard" according to the Ex-Chairman; desilting removed too much soil from the lake, damaging the clay layer that prevented seepage. The remaining water percolated through the sandy bottom, an inconsequential monsoon failed to replenish the lake, leaving a dust bowl in its wake. The anger expressed by residents about a "lake-less Pushkar" forced make-shift measures, like water tankers and tubewells to fill concrete bathing tanks for pilgrims,

[25]Response of the District Collector of Ajmer to PIL, August 2008.

[26]n.d., 'Smile on Every Face', Pushkarguide.com, viewed 13 January 2011, from http://www.pushkarguide.com/pushkar_article2.html.

during the Pushkar fair of 2009 (Ex-Chairman letter to author, December 24, 2009). Overall, the Pushkar Project ignored lessons from the previous failures as well as the real but less visible threat to the fragile ecosystem of the valley, i.e., the depletion of groundwater in an area where the draft exceeded the recharge potential.[27]

Groundwater Depletion

Historically Pushkar valley had a high groundwater table; benchmarked at 14.1 feet below ground level (BGL) in 1966 (Sharma n.d.), it dropped dramatically to 75 feet BGL by 2004 (DPR 2008). Data from the NASA groundwater sensing satellite GRACE indicated that this severe depletion took place despite close to normal annual rainfall in Rajasthan from 2002–08 (Rodell et al. 2009). In Pushkar valley, the depletion was attributed to excessive withdrawals for the railways (Allchin et al. 1972), for water supply to the neighboring town of Ajmer, for the cultivation of high water consumption crops, like roses (Sharma n.d.), and for tourism related expansions, like hotels, restaurants, swimming pools, gardens, etc. The main factor, however, was the enthusiastic embrace of tubewell technology in the late 1960s by the Indian government leading to a 75 percent increase from 1995–2001 in Rajasthan (Birkenholtz 2006, 2008).[28] Rajasthan now has the dubious distinction of being India's driest state (Machiwal et al. 2011).

Noting the rapidly increasing number of tubewells in Pushkar, the Community Activist wrote in a letter to district administrators that they were endangering the lake. This argument visualized a past when the regional groundwater table was higher than the bottom of the lake and "oozed, and sprang up in abundance" through artesian wells according to the Ex-Chairman. However, when the water table fell below the lake bottom, the water seeped from the lake towards surrounding farmland where the water table was below that of the lake (CGWB 2001). An undated Irrigation Department paper, "Conservation of Pushkar Lake," recommended "*raising*

[27]The annual groundwater recharge in Pushkar was 20.2 mcm while the annual draft was 23.4 (CGWB 2001).

[28]Though groundwater-lifting devices, like the Persian Wheel, had existed since 900 A.D., colonial administrators and missionaries (e.g., the Scottish Mission to India who according to anecdotal accounts owned their own rigs to bore wells in the area) borrowed from oil well drilling technology to further develop it. By the late 1800s, diesel driven tubewells became a source of "preventative irrigation" to reduce drought susceptibility. However, tubewells began to proliferate after the Indian government concluded that large-scale surface water irrigation projects were not paying for themselves and shifted the focus to private tubewells funded by international donors, like the Ford Foundation and World Bank (Birkenholtz 2006). Today India is the largest user of groundwater in the world. It uses an estimated 230 cubic kilometers of groundwater per year—over a quarter of the global total (Pahuja 2010).

the water table of the nearby area" (emphasis mine) to conserve the sacred lake.

Official recognition of the seriousness of the problem came in 2001 when the Central Ground Water Authority (CGWA) identified Pushkar valley as a critically depleted groundwater area and in 2003 declared it a Dark Zone where the draft was 85–100 percent of the annual recharge.[29] However, despite this ruling to stop all abstraction, state laws governing groundwater rights remained unchanged (Rodell et al. 2009).[30] A regional groundwater engineer admitted to me that it was a political issue because groundwater in India was a democratic resource that was lawfully attached to land ownership (Roy and Shah 2003). Reluctant to take on the farmer voting blocs that typically impeded any changes to the current dominant heritage principle (Mukherji 2006), Rajasthan chose instead to regulate groundwater use by limiting the supply of electricity to six hours/day (Birkenholtz 2006).

Such state level ambivalence towards regulating groundwater was also reflected locally in the ineffectual implementation of the CGWA notification by the District Magistrate and flagrant violations by farmers. On the outskirts of Pushkar, a farmer admitted to me that with his present well failing, he was having a deeper well bored stealthily at night undeterred by the potential fine of US$ 200. Such actions were tracked by the Community Activist and generated a marked antipathy towards area farmers who were viewed as enemies of the lake for drawing heavily on groundwater, encroaching on feeder channels by cultivating crops there, and diverting water from seasonal streams to their fields.[31]

The governing party's "vote-bank politics" vis-à-vis farmers and the approval of tubewell boring requests by influential tourism related businesses were also the basis of the Common Cause Society PIL in the Rajasthan High Court in 2007 asking that the state be compelled to comply with the CGWA notification. On February 18, 2009, the Rajasthan High Court stayed all construction in the catchment area of several lakes in the district, including Pushkar Lake.[32] While this represented a major victory

[29]CGWA Public Notice No. 24/2001 and CGWA Notification, September 30, 2003. No. 32-1/CGWA/2001(E). Under the first order all existing tubewells had to be registered and new ones required prior approval. The Groundwater Estimation Committee (GEC) 1984 classified groundwater assessment units into four categories according to the level of development (or withdrawal to recharge ratio): white (<65%), gray (65–85%), dark (85–100%) and over-exploited (>100%). GEC 1997 modified the categories to safe (<70%), semi-critical (>70% but ≤90%) and critical (>100%).

[30]Under India's Constitution, state governments were responsible for enacting water regulations to support CGWA notifications.

[31]These farmers are generally not Brahmins. They belong to the Mali and Rawat castes. There is a possibility that the lines of tension are based on caste and consequently livelihood (farmer vs. priest) but needs further research to substantiate the presence of structures of caste authority of the type that Mosse alluded to in South India (2006).

for the activists, the district administration failed to act on the courts orders. The tourist resort specifically cited in the PIL, for encroaching on the Pushkar feeder, was fully operational in 2010 with approval from the Pushkar municipality. There was little evidence of sustained local efforts in terms of groundwater—the connections that the Community Activist and the Ex-Chairman made between the lake level and groundwater remained tenuous at best.

Conclusion

Emergent civil society initiatives in Pushkar have drawn on a toolkit that consists of the law, legal precedents, religious precepts, indigenous knowledge, and customary institutions, like collective stewardship and cooperation, organized around CPR principles rather than relying on state driven techno-fixes alone. These incipient efforts were far from systematic or even coherent and often undercut each other but therein reside the seeds for the sustainable management of ecological resources, such as Pushkar Lake and groundwater in the valley.

Scientists, technocrats, and administrators in Pushkar were lukewarm about this groundswell of civil society initiatives and activism and dismissed them as "too democratic." NGO-led groundwater conservation efforts were similarly scoffed at by a state geohydrologist and countered by a drilling project to find a mythical underground river in Pushkar.[33] However, bureaucratic skepticism about local initiatives was disingenuous given previous civil society victories in Rajasthan that had secured rights to information, food, and work (Right to Information Act 2005; National Rural Employment Guarantee Act 2005) even as the right to decentralized water management was emerging from the praxis of TBS in Alwar (Bhargava 2007). Anthropological literature further indicates that decentralized collective action successfully emerged with greater tolerance by local authorities (Wade 1988) and strong subaltern organization by local groups (Baviskar 2004). The negative state responses also choose to ignore previous successful experiments in Pushkar with community-based "User Committees," like those initiated by the 1997 Indo-Canadian Pushkar Gap Project. The fact that the groups disbanded, when the funding ended, only highlights the problem of sustainability inherent in government projects.

At the regional level, my research indicated that the issues of ecological degradation were defined too narrowly. The blame for the regional

[32]2009, 'Rajasthan HC bans constructions in lake basins in Pushkar', Indlawnews.com, viewed 21 February 2009, from http://www.indlawnews.com/Newsdisplay.aspx?aef53998-7f00-48ca-b404-e077495e1ef7.

[33]2008, 'Pushkar lake dug up', Times of India, viewed 24 December 2008, from http://timesofindia.indiatimes.com/articleshow/msid-3882630,prtpage-1.cms.

degradation was simplistically affixed on desertification and all the attention was focused on the lake while ignoring the severe regional groundwater depletion, which was arguably central to any holistic response. An example of the close nexus between media and the government in perpetuating desertification narratives came in June 2010 when the Union Minister of State for Environment and Forests wrote a letter to Rajasthan's Chief Minister citing a news story titled, "Illegal mining expands desert area."[34] However, when the press reported the letter the title morphed to "Pushkar Turning Into Desert?"[35] It was then picked up and repeated on the Ex-Chairman's sons' website as a declarative statement minus the question mark. Desertification also provided a convenient legal defense, allowing the district administration to contend that the deterioration of lakes in the Pushkar area was due to "nature" and not "any negligent act of the government."[36]

Data from Pushkar suggests that while it was hard to substantiate claims of desertification and deforestation, local representations of nature had "sociocultural fingerprints" in that they derived from the "dynamics of situated social relations" (Williams 2000). By focusing on desertification as a "site of contestation" (Peet and Watts 1996; Baviskar 2003), instead of a naturalized category, competing environmental discourses and situated social relations came into focus. For instance, the 2009 water crisis affected not only the lake but also floriculturists who demanded a water policy that also catered to their needs.[37] Focusing on such issues of political ecology integrates social constructivist approaches with realist questions about the access, control, and management of water resources (Forsyth 2001). The fact that this anxiety over groundwater depletion finds its best articulation within this constituency implies that farmers, who were ignored by the Pushkar Project, also needed to be stakeholders in regional water management. A wider coalition of priests and farmers would be possible by broadening the definition of desertification from advancing sands to land degradation defined by reduced biological potentiality and groundwater depletion. Such shared local responsibility and co-management could also shift the equation from the current over-dependence on the state and technology to indigenous knowledge, like traditional local rainwater harvesting techniques, to recharge wells, groundwater and the lake (Feeny et al. 1990). A percolation tank (*rapat*) near a holy site in Pushkar, sponsored

---

[34]Singh, R. 2010, 'Illegal mining expands desert area', Times of India, viewed 27 June 2010, from http://timesofindia.indiatimes.com/City/Jaipur/Pushkar-turning-into-a-desert/articleshow/6080716.cm.

[35]2010, 'Act on Pushkar: Jairam to CM', Times of India, viewed 27 June 2010, from http://timesofindia.indiatimes.com/city/jaipur/Act-on-Pushkar-Jairam-to-CM/articleshow/6088704.cms.

[36]Response of the District Collector of Ajmer to PIL, August 2008.

[37]2009, 'Water crisis hits rose floriculturists of Pushkar', WebIndia123.com, viewed 14 July 2009, from http://news.webindia123.com/news/Articles/India/20090623/1281093.html.

by the local legislator's fund, indicates that this traditional ecological knowledge still persists in the area (Agarwal and Narain 1997) and if promoted on a larger scale would benefit not only the farmers but also the lake. In the strategic co-management plan that this chapter visualizes, the long-term self-interest of users would provide motivation, while the state could provide coordination over a wide geographic scope at potentially lower rule-enforcement cost (Feeny et al. 1990).

Finally, a word from local lexicon that defined the sacred region of Pushkar, ksetra, was repeatedly brought into play in 2009. The Good Samaritans persuaded district administrators to back an initiative to preserve heritage sites along a traditional 48-mile circumambulatory pilgrimage route (Joseph 1994) around the ksetra. And the Community Activist laid out another proposal to incorporate the entire ksetra under the jurisdiction of the Pushkar municipality to ensure greater control of the lake catchment area. Continued cultural appropriation of the religious concept of ksetra by civil society organizers had the potential to shift the exclusive focus from the lake to include the greater Pushkar region.

In conclusion, my fieldwork suggests that self-initiated mobilization and decentralized strategies require tremendous community-based political will, persuasive leaders, state backing, and, on a sober note given the groundwater crisis in Pushkar, time to succeed. After the plentiful monsoons of 2010 replenished the lake locals noted that with the administration and contractors so absorbed in self-congratulations, the urgency had dissipated and problems, like water loss through lake-bottom seepage, were being ignored.[38] Such identification of early "successes" then had the potential to subvert real attempts at building the slower, institutional processes that could bring about sustainable long-term change (Baviskar 2004). In Pushkar, the alchemical paradigm shift required to sustain the sacred lake and groundwater may well be contingent on another regional surface water crisis.

References

Agarwal, A. Can Hindu beliefs and values help India meet its ecological crisis? pp. 163–179. *In:* C. Chapple and M. Tucker [eds.]. 2000. Hinduism and Ecology: The intersection of earth, sky, and water. Harvard University Press, Cambridge.

Agarwal, A. and S. Narain. 1997. Dying wisdom: Rise, fall and potential of India's traditional water harvesting systems. State of India's Environment Report 4. Centre for Science and Environment, New Delhi.

Agrawal, A. 2005. Community, Intimate Government, and the Making of Environmental Subjects in Kumaon, India. Current Anthropology 46.2: 161–181.

[38]'Kaise Ayega Sarovar Mein Pani, Mitti Ban Gaye Pareshani (How will the Water Come to the Lake when the Sediment Has Become a Problem)', Bhaskar News, 27August, 2010.

Ahluwalia, M. 1997. Representing communities: The case of a community-based watershed management project in Rajasthan, India. IDS Bulletin 28.4: 23–34.
Allchin, B. and A. Goudie. 1974. Pushkar: Prehistory and climatic change in Western India. World Archaeology 5.3: 358–368.
Allchin, B., K.T.M. Hegde and A. Goudie. 1972. Prehistory and environmental change in Eastern India: A Note on the Budha Pushkar Basin, Rajasthan. Man, New Series 7.4: 541–564.
Alley, K. 1994. Ganga and *gandagi*: Interpretations of pollution and waste in Benaras. Ethnology 33.2: 127–145.
Anand, R. 1994. Churning the stilled waters. Down to Earth 3.10: 22–23.
Bardhan, P. and I. Ray. 2008. The contested commons: Conversations between economists and anthropologists. Blackwell Publishing, Oxford.
Baviskar, A. 2003. For a cultural politics of natural resources. Economic and Political Weekly 38.48: 5051–5055.
Baviskar, A. 2004. Between micro-politics and administrative imperatives: Decentralisation and the watershed mission in Madhya Pradesh, India. The European Journal of Development Research 16.1: 26–40.
Beveridge, H. 1868. The memoirs of Jahangir, vols. I & II. Munshiram, Delhi. pp. 254–264.
Bhargava, P. Civil society in Rajasthan: Initiatives and inhibitions. pp. 257–281. *In*: V. Vyasa, S. Acharya, S. Singh and V. Sagar [eds.]. 2007. Rajasthan: The quest for sustainable development. Academic Foundation, Delhi.
Bhatt, G.P. 1988. The Padma Purana, part I. Sterling, New Delhi. p. 402.
Birkenholtz, T.L. 2006. The politics of groundwater scarcity: Technology, institutions, and governance in Rajasthani Irrigation. Ph.D. dissertation, Ohio State University. Ohio.
Birkenholtz, T.L. 2008. Contesting expertise: The politics of environmental knowledge in Northern Indian groundwater practices. Geoforum 39: 466–482.
Black, W. 1908. Winter days in India and elsewhere. James Macklehose and Sons, Glasgow. p. 14.
Bonavia, E. 1885. A trip to Jeypore. p. 140. *In:* The Gardeners' Chronicle: A Weekly Illustrated Journal of Horticulture and Allied Subject, vol. XXIII, New Series. Bradbury, Agnew & Co., London.
Borrini-Feyerabend, G. 1997. Beyond fences: Seeking social sustainability in conservation. IUCN, Gland.
Brandis, D. 1897. Indian forestry. Oriental University Institute, Woking. UK. pp. 2–16.
Bromley, D. and M.M. Cernea. 1989. The management of common property natural resources: Some conceptual and operational fallacies. World Bank Washington, D.C.
Broughton, T. 1892. Letters written in a Mahratta camp during the year 1809: Descriptive of the character, manners, domestic habits, and religious ceremonies of the Mahrattas. Constable and Company, Westminster. pp. 257–260.
CDP. 2006. City development plan for Ajmer & Pushkar. Government of Rajasthan, Ajmer. pp. 217–218.
CGWB. 2001. Groundwater conditions and management strategies in Pushkar Valley. District Ajmer, Rajasthan. Central Ground Water Board, Jaipur.
Cochran, J. and I. Ray. 2009. Equity reexamined: A study of community-based rainwater harvesting in Rajasthan, India. World Development 37.2: 435–444.
Deshpande, N.A. 1990. The Padma Purana, part IV. Motilal Banarsidass Publishers, Delhi. p. 1255.
Dietz, T., E. Ostrom and P.C. Stern. 2003. The struggle to govern the commons. Science 302: 1907–1912.
DPR. 2008. Detailed Project Report (DPR) for conservation and sustainable management of Pushkar Sarovar. Tetra Tech, Delhi. pp. 9–148.
Dundas, P. The limits of a Jain environmental ethic. pp. 95-117. *In:* C. Chapple [ed.]. 2002. Jainism and ecology: Nonviolence in the web of life. Harvard University Press, Cambridge.

Dwivedi, O.P. Satyagraha for conservation: Awakening the spirit of Hinduism. pp. 145–157. *In:* R.S. Gottlieb [ed.]. 2004. This Sacred Earth: Religion, Nature, Environment. Routledge, New York.

Eastwick, E. 1883. Handbook of the Punjab, Western Rajputana, Kashmir, and Upper Sindh. John Murray, London. pp. 134–135.

Feeny, D., F. Berkes, B.J. McCay and J.M. Acheson. 1990. The tragedy of the commons: Twenty-two years later. Human Ecology 18.1: 1–19.

Forsyth, T. 1996. Science, myth and knowledge: Testing Himalayan environmental degradation in Thailand. Geoforum 27.3: 375–392.

Forsyth, T. Critical realism and political ecology. pp. 146–154. *In:* J. López and G. Potter [eds.]. 2001.After postmodernism: An introduction to critical realism. Athlone, New York.

Foucault, M. Nietzsche, genealogy, history. pp. 76–100. *In:* P. Rabinow [ed.]. 1984. The Foucault reader. Parthenon Books, New York. p. 76.

Gold, A.G. and B.R. Gujar. 1989. Of gods, trees and boundaries: Divine conservation in Rajasthan. Asian Folklore Studies 48: 211–229.

Gosling, D. 2001. Religion and ecology in India and Southeast Asia. Routledge, New York. pp. 63–64.

Gupta, A. and J. Ferguson. 1997. Anthropological locations: Boundaries and grounds of a field science. University of California Press, Berkeley. p. 37.

Guthman, J. 1997. Representing crisis: The theory of Himalayan environmental degradation and the project of development in post-Rana Nepal. Development and Change 28: 45–69.

Hardin, G. 1968. The tragedy of the commons. Science 162.3859: 1243–1248.

Heber, R. 1829. Narratives of a journey through the upper provinces of India: From Calcutta to Bombay, 1824–1825, vol. II. Carey, Lea & Carey, Philadelphia. p. 28.

Jain, A., H. Birkumar Singh, S.C. Rai and E. Sharma. 2004. Folklores of sacred Khecheopalri Lake in the Sikkim Himalaya of India: A plea for conservation. Asian Folklore Studies 63.2: 291–302.

Joseph, C.A. 1994. Temples, tourists and the politics of exclusion: The articulation of sacred space at the Hindu pilgrimage center of Pushkar, India. Ph.D. Dissertation, University of Rochester. Rochester, NY.

Joseph, C.A. 2001. Mediated resistance: Tourism and the host community. Annals of Tourism Research 28(4): 998–1009.

Joseph, C.A. Hindu nationalism, community rhetoric and the impact of tourism: The "divine dilemma" of Pushkar, Rajasthan. pp. 203–219. *In:* C.E. Henderson and M. Weisgrau [eds.]. 2007. Raj rhapsodies: Tourism, heritage and the seduction of history. Ashgate Publishing Limited, Hampshire, UK.

Kar, A. 1986. Physical environment, human influences and desertification in Pushkar—Budha Pushkar Lake region of Rajasthan, India. The Environmentalist 6.3: 227–232.

Karanth, K.R. 1977. Geology and groundwater resources of upper Luni Basin, Ajmer District, Rajasthan. Controller of Publications, Delhi. p. 32.

Lal, A.K. 1996. Effects of mass bathing on water quality of Pushkar Sarovar. Indian Journal of Environmental Protection 16.11: 831–836.

Larking, C. 1888. Bandobast and *khabar*: Reminiscences of India. Hurst and Blackett, London. p. 125.

La Touche, J.D. 1879. The Rajputana gazetteer, vol. II. Office of the Superintendent of Government Printing, Calcutta. pp. 48–70.

Li, T.M. 2002. Engaging simplifications: Community-based resource management, market processes and state agendas in upland Southeast Asia. World Development 30.2: 265–283.

Machiwal, D., M.K. Jha and B.C. Mal. 2011. Assessment of groundwater potential in a semi-arid region of India using remote sensing, GIS and MCDM techniques. Water Resource Management 25: 1359–1386.

Mosse, D. 2003. The rule of water: Statecraft, ecology, and collective action in South India. Oxford University Press, Delhi.

Mosse, D. 2006. Collective action, common property, and social capital in South India: An anthropological commentary. Economic Development and Cultural Change 54.3: 695–724.

Mukherji, A. 2006. Political ecology of groundwater: The contrasting case of water-abundant West Bengal and water-scarce Gujarat, India. Hydrogeology Journal 14: 392–406.

Nigah, M. 2007. An assessment of Seechewal initiative in the state of Punjab, India: An example of community-based conservation? Masters Thesis, University of Manitoba. Manitoba.

Nugteren, A. 2005. Belief, bounty, and beauty: Rituals around sacred trees in India. Brill Academic Publishers, Boston. pp. 365–366.

Ostrom, E. 1999. Coping with tragedies of the commons. Annual Review of Political Science 2: 493–535.

Pahuja, S. 2010. Deep wells and prudence: Towards pragmatic action for addressing groundwater overexploitation in India. World Bank Report 51676. World Bank, Washington, DC. p. IX.

Pearce, F. 1992. Mirage of the shifting sands. New Scientist, 12 December: 38–42.

Pedersen, P. Nature, religion and cultural identity: The religious environmentalist paradigm. pp. 258–276. *In:* O. Bruun and A. Kalland [eds.]. 1995. Asian perceptions of nature: A critical approach. Curzon Richmond, Surrey.

Peet, R. and M. Watts. 1996. Liberation ecologies: Environment, development, social movements. Routledge, New York. p. 263.

Pretty, J.N. and P. Shah. 1997. Making soil and water conservation sustainable: From coercion and control to partnerships and participation. Land Degradation & Development 8: 39–58.

Rodell, M., I. Velicogna and J. Famiglietti. 2009. Satellite-based estimates of groundwater depletion in India. Nature, 20 August 460: 999–1002.

Rousselet, L. 1875. India and its native princes: Travels in Central India and in the presidencies of Bombay and Bengal. Chapman and Hall, London. p. 213.

Roy, D. and T. Shah. Socio-ecology of groundwater irrigation in India. pp. 307–335. *In:* M.R. Llamas and E. Custodio [eds.]. 2003. Intensive use of groundwater: Challenges and opportunities. Balkema Publishers, The Netherlands.

Sarda, H.B. 1914. The Pushkar case: Judgment. Small Cause Court: Ajmer.

Sharma, D.C. (n.d.). Report by Chief Hydrogeologist. Ground Water Department, Jodhpur. pp. 3–4.

Sharma, K.C. and C.S. Chouhan. Ecology and restoration of Lake Budha Pushkar – A threatened water body of Ajmer, Rajasthan. pp. 1757–1764. *In:* M. Sengupta and R. Dalwani [eds.]. 2008. Proceedings of Taal 2007. The 12th World Lake Conference. Rajasthan, India.

Singh, G., R.D. Joshi, S.K. Chopra and A.B. Singh. 1974. Late Quaternary history of vegetation and climate of the Rajasthan Desert, India, Journal Philosophical Transactions of the Royal Society of London 267.889: 467–501.

Swift, J. Desertification: Narratives, winners and losers. pp. 73–90. *In:* M. Leach and R. Mearns [eds.]. 1996. The lie of the land: Challenging received wisdom on the African environment. James Currey Ltd., Oxford.

Tod, J. 1983. Annals and antiquities of Rajasthan, vol. I, M.N. Publishers, New Delhi. pp. 606–608.

UNEP Report. 1987. A plan of action to combat desertification: A case study in environment management: Regeneration of Pushkar (Ajmer, India) Lake Valley ecosystem. UNEP, Bangkok.

Van Buitenen, J.A.B. 1975. The Mahabharata: The book of the assembly hall and the book of the forest. University Press, Chicago. p. 374.

van Steenbergen, F. and T. Shah. Rules rather than rights: Self-regulation in intensively used groundwater systems. pp. 241–256. *In:* M.R. Llamas and E. Custodio [eds.]. 2003. Intensive use of groundwater: Challenges and opportunities. Balkema Publishers, The Netherlands.

Wade, R. 1988. Village republics: The economic conditions of collective action in India. University Press, Cambridge.
Williams, D.M. 2000. Representation of nature on the Mongolian steppe: an investigation of scientific knowledge construction. American Anthropologist 102.3: 503–519.
Wilson, J. 1877. Indian caste, vol. I. Times of India Office, Bombay. p. 169.
Zeitlyn, S. 1988. The construction and reconstruction of Pushkar. South Asian Studies 4: 41–50.

7

Governance Reform to Secure Resilience in the Nearshore Waters of the Great Lakes Basin

Gail Krantzberg

Introduction

Prime Minister Pierre Trudeau and President Richard Nixon signed the Great Lakes Water Quality Agreement (GLWQA) in 1972 (United States and Canada 1972). This Agreement expresses the commitment of Canada and the United States to restore and maintain the chemical, physical and biological integrity of the waters of the Great Lakes Basin Ecosystem (United States and Canada 1972). The GLWQA has had substantial influence on the cleanup and restoration of the region. The progress made since 1972 is evidenced by the documentation by scientists early in the 21st century (for the first time since 1916), of the presence of spawning lake whitefish and eggs in the Detroit River, the resurgence of cormorant populations, the rediscovery of sturgeon populations, and the return of nesting and fledging bald eagles (Krantzberg 2008).

As Manno and Krantzberg (2008) explain:

> "The Great Lakes Water Quality Agreement was negotiated pursuant to the 1909 Boundary Waters Treaty between the United States and British Canada that had created the International Joint

Professor and Director, Centre for Engineering and Public Policy, McMaster University, 1280 Main St. W., Hamilton, Ontario, L8S 1K4.
Email: krantz@mcmaster.ca

> Commission (IJC) to help resolve problems including pollution that was causing injury to health or property crossing the binational boarder. The IJC and the institutions added to it … were based on the principle of bi-nationalism (two countries collaborating on achieving a set of shared goals) rather than bi-lateralism (two countries negotiating with each other in an attempt to balance interests and protect each others rights)."

For four decades the Great Lakes regime has invoked the GLWQA as the mechanism for binational cooperation on programs and policies to enhance and protect the integrity of the Great Lakes. Many advances in water quality have lead to unquestionable improvements in ecosystem quality, habitat and biodiversity, and water infrastructure. The 2009 State of the Lakes Ecosystem Conference reported that "[r]eleases of targeted bioaccumulative toxic chemicals have declined significantly from their peak period in past decades and, for the most part, no longer limit the reproduction of fish, birds and mammals. Concentrations of contaminants in the open waters are low, and many contaminants are further declining" (SOLEC 2009).

While acknowledging progress towards meeting the purpose of the GLWQA, Great Lakes scientists have issued compelling evidence that the ecological health of the basin ecosystem is at significant risk and could be approaching a tipping point. Bails et al. (2005) contend "[t]here is widespread agreement that the Great Lakes presently are exhibiting symptoms of extreme stress from a combination of sources that include toxic contaminants, invasive species, nutrient loading, shoreline and upland land use changes, and hydrologic modifications… Factors such as the size of the lakes, the time delay between the introduction of stress and subsequent impacts, the temporary recovery of some portions of the ecosystem, and failure to understand the ecosystem-level disruptions caused by the combination of multiple stresses have led to the false assumption that the Great Lakes ecosystem is healthy and resilient."

The Great Lakes continue to be challenged and the capacity of desired ecosystem states to cope with events and disturbances is not readily predicted (Jackson et al. 2001; Paine et al. 1998). Folke et al. (2004) emphasize that humanity strongly influences biogeochemical, hydrological, and ecological processes, at many geographical scales. Environment is understood to be more variable than previous management schemes imagined, with greater uncertainty about how ecosystems will respond to human use and pressures (Steffen et al. 2004). This degree of complexity requires a management change from the traditional paradigm of command-and-control that aims to stabilize preferred conditions, to a governance paradigm based on managing resilience in uncertain environments to secure essential ecosystem services (Holling and Meffe 1996; Ludwig et al. 2001). Here, resilience is defined as a measure of the ability of a complex adaptive system to self-organize and the degree to which the system can build and increase the capacity for learning and adaptation (Carpenter et al. 2001b; Levin 1999). Folke et al. (2004) clearly illustrate that regime shifts in ecosystems are, to a large extent, driven by human actions. "A combination of top-down impacts, such as fishing down food webs and losing response diversity and functional groups of species, and bottom-up impacts, such as accumulation of nutrients, soil erosion, or redirection of water flows, as well as altered disturbance regimes, such as suppression of fire and increased frequency and intensity of storms, have shifted several ecosystems into less desired states with diminished capacities to generate ecosystem services." The management of ecosystem resilience, biodiversity, and shifts in ecosystem states require adaptability among the actors involved in ecosystem management (Berkes et al. 2003), where adaptability is defined as the capacity of actors in a system to manage resilience in the face of uncertainty and surprise (Gunderson and Holling 2002).

The tradition of top down command and control of the Great Lakes basin ecosystem is no longer suited to the diversity of human-induced drivers of change, particularly evidenced by Bails et al. (2005), in the nearshore waters. Governance reform that engages a multitude of actors to effect change at a watershed scale is proving to be a desirable strategy for the newly renegotiated GLWQA. In fact, Hall (2009) points out that the GLWQA relies "heavily on citizens to ensure compliance and implicitly recognize that the two federal governments may have more in common with each other than with citizens and other stakeholders on both sides of the border when it comes to environmental protection and harm."

The nearshore waters connect the land to the receiving waters and are critical zones for many species supplying food and critical habitats. The significant past and present anthropogenic impacts threaten the health of the Great Lakes ecosystem, the provision of ecological services, economic sustainability, and quality of life. Reoccurrence of severe algae outbreaks in the lakes in recent years point to the need for new forms of interventions. For example, summer blooms of the planktonic toxic cyanobacterium *Microcystis aeruginosa* have become more frequent in western Lake Erie since the mid-1990s (Conroy et al. 2008). Rinta-Kanto et al. (2005) report that in high-bloom years, surface scums of *Microcystis* may stretch for hundreds of square kilometers and produce concentrations of microcystin toxin that exceed World Health Organization guidelines for human consumption. To illustrate the watershed loading, Bridgeman et al. (2011) demonstrate the proximity of *Microcystis* and *Lyngbya* blooms to the inflow of the Maumee River and Maumee Bay, suggesting that nutrients loaded from the river influence the development of algal blooms in western Lake Erie.

Measurements during the bloom of 2008 indicated that *Microcystis* growth rates were greater in the vicinity of Maumee Bay than in offshore waters and that *Microcystis* was most frequently phosphorus limited.

Over the past decade, scientists have been discussing excessive growth of Cladophora in the Great Lakes. Joose and Baker (2011) observe that management of non-point or diffuse sources of phosphorus will be more important in the future in order to address symptoms of eutrophication in the nearshore zone. They call for a renewed focus on managing non-point source tributary loads. Changes that have occurred in the lakes and tributaries in the past 15 years indicate a greater need to focus on non-point sources, whether urban or rural. (Joose and Baker 2011), mandating a focus on the land-lake interface, and actions at a local or regional scale.

Governance

Water governance refers to the processes and institutions through which decisions are made about water. This includes the range of political, organizational, and administrative processes used to make and implement decisions, as well as how decision makers are held accountable. This is different from water management, which refers to the operational, on-the-ground activity to regulate the water resource and the conditions of its use (NRTEE 2011).

Effective collaborative water governance requires the involvement of a broad range of stakeholders. To stay engaged and committed, stakeholders need incentives and solid, attainable outcomes. As explained by NRTEE (2011) "To encourage participation in collaborative water governance, governments need to demonstrate strong leadership and act on the recommendations provided by the collaborative process." Collaborative water governance requires time and dedicated resources, as well as clear rules and guidance from governments. Figure 7.1 depicts many of the elements of successful collaborative governance. This is a well-defined problem that can be easily dealt with by a government department.

Governments at all levels (federal, state/provincial, regional and local) struggle to create lasting, effective, responsive and representative regulatory and institutional arrangements for water resources management, which is complicated by multiple functions, mandates and goals of various orders of government. Mandarano and Paulson (2011) review recent efforts that have focused on developing regional institutions that foster collaborative planning and management approaches. They conclude that collaborative partnerships can be effective institutional arrangements for managing complex regional environmental problems, such as those in the nearshore zone.

An ecosystem approach to restore and protect the Great Lakes is fundamental to the GLWQA, and has led to the emergence of integrated

Figure 7.1 Word Cloud depicting elements of successful collaborative governance models.

water resource management (IWRM). The Global Water Partnership (GWP 2000) defined IWRM as "a process which promotes the coordinated development and management of water, land and related resources, in order to maximize the resultant economic and social welfare in an equitable manner without compromising the sustainability of vital ecosystems." Mitchell (2005) reflects on the need to promote the coordination of management initiatives for water, land, and related resources. Analysis and decisions are required to determine how the connections among these resources are to be made. It is important to explicitly integrate water and land-based systems for management purposes. Watersheds, defined as areas of land draining into a common body of water (USEPA 2008), are an accepted unit for water governance and management (Baril et al. 2006; Koehler and Koontz 2008). Cohen and Davidson (2011) summarize the many proponents who have touted the advantages of using watershed boundaries over their jurisdictional predecessors (e.g., Montgomery et al. 1995; McGinnis 1999), and list a number of more recent scholars who have questioned the benefits of this approach to water governance and have identified significant challenges with its implementation (e.g., Draper 2007; Warner 2008; Norman and Bakker 2009). They characterize the emerging debate surrounding specific elements of the watershed approach, particularly participation and accountability.

There are consequently implications for the design of institutional arrangements related to public agencies responsible for water and land management, as well as other resources. For the governance framework to work (see Fig. 7.2), the predisposition of resource-based agencies to act independently needs to be overcome, and the model established that organizations with shared interests and shared responsibilities.

The benefits of inclusiveness and empowerment of place-based stakeholders in watershed planning, decision-making, and implementation often surround the benefits of decentralized decision-making. Cohen and

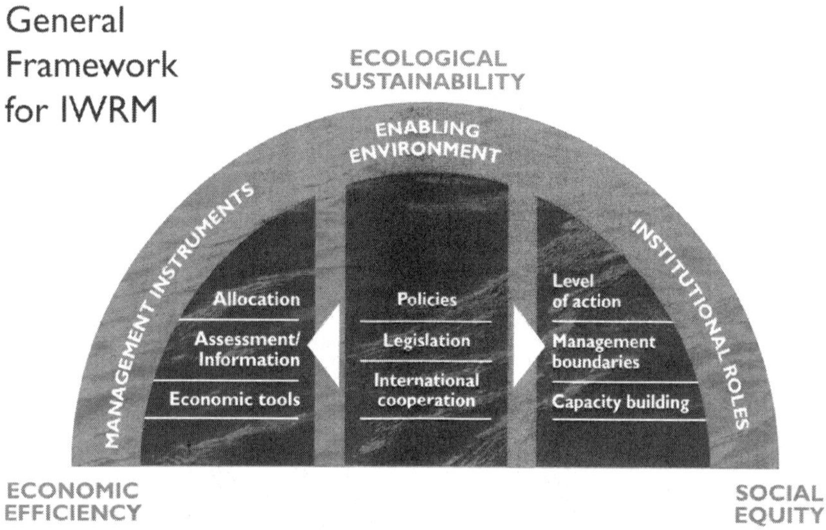

Figure 7.2 A generalized framework for Integrated Water Resource Management: from http://www.gwp.org/The-Challenge/What-is-IWRM/IWRM-pillars/.

Davidson (2011) summarize the frequently-cited benefits of decentralization including: increased proximity between decision-makers and those affected by governance decisions, an increase in sub-national level democratic participation, greater access to local knowledge and expertise, heightened responsiveness to citizen needs and concerns, and empowerment of local communities (Paehlke 2001; Rabe 2006; Hill et al. 2008). Norman and Bakker (2009), however, point out that despite significant strides in Canada-US transboundary water governance, higher orders of government have not loosened their grip on their decision-making power and local groups have not been empowered notwithstanding apparent inclusiveness. For the nearshore zone of the Great Lakes to respond with resilience to watershed-based anthropogenic pressures, the ability of different orders of government to recognize the potential capacity for beneficial changes and empower local stakeholders to build that capacity is imperative.

On the matter of accountability, besides that of governments, the issue is one of delegating decision-making authority to non-elected collaborators, typically extra-governmental participants. In the case of the watershed approach, the accountability challenge can be seen as a function of the process through, and the degree to which, participants and stakeholders have been involved in the decision-making process and are legitimate representatives of the community and society pressures on the resource.

Notwithstanding the challenges, Holley and Cunningham (2011) conclude that a quiet revolution is taking place in the ways in which citizens and governments are seeking to engage with complex social and environmental issues. They cite numerous "experiments" which seek to address problems through mechanisms that supplement and in some cases supplant conventional regimes of regulation, administration and enforcement. For example, teamwork is considered to be a mechanism to more effectively mobilize stakeholders to improve performance. By reorganizing the group co-management processes to accommodate task interdependencies, and then leaving particants with some autonomy in determining how to handle interdependencies, teamwork results in a satisfying experience for stakeholders and more effective in achieving the project goals. Another such "experiment" on collaborative watershed management example is that of the Grand Traverse Bay Watershed Initiative (GTBWI) on Lake Michigan, established in 1990 after the realization by numerous agencies and organizations that collaboration would enable the managed programs that had the purpose of enhancing and sustaining the bay's watershed. Organizers of the GTBWI developed a partnership agreement that included private and public institutions, all orders of government, businesses, agriculture, civic, and environmental organizations. Accordintg to Konisky and Beierle (2001), "[t]he partnership agreement provides a decentralized, noncontractual but coordinated management framework to assist local agencies and organizations to collaborate on watershed protection projects". The aim is to manage public problems through localized collaborations and nonbinding agreements.

The term "governance" reflects that these forms of social steering are not necessarily dependent on formal legal regulation or other interventions by the nation state to drive them. The new governance literature does not derive from a single legal or socio-legal theory but rather has a diversity of terms that have been used to describe it. As summarized by Holley and Cunningham (2011), prominent amongst these are "experimentalism", "modular regulation", "collaborative governance", "multilevel governance" and "regional collaboration". The significant commonalities include a focus on collaboration, integration, participation, decentralization, deliberative styles of decision-making and flexibility.

Collaborative watershed partnerships can take many forms. Some are comprised of diverse stakeholders including government, industry, environmentalists, farmers, and local citizens. Others are grass-roots nonprofit organizations that operate with little government support. According to Koontz and others (2004), a common theme necessary for a management initiative to be considered collaborative is interaction among or between different stakeholder groups as a means for airing diverse viewpoints and generating information that will address increasingly complex environmental problems. Hardy (2010) notes that the membership composition of collaborative efforts often correlates with their goals, activities, and outcomes: "Groups made up primarily of government personnel have been shown to excel at highly complex issues, while citizen-dominated groups appear most appropriate when issues are more diffuse in scope and there is an overarching need for community support." The lesson is that different types of groups have unique internal structures and functions, and may be more appropriate for different watershed management activities. No prescription for structure is necessarily appropriate, nor are the slate of stakeholders to be engaged in a co-management process.

That said, collaborative governance could be successfully applied if:

- input from multiple stakeholders into decision on vision, goals, and outcomes are required;
- long-term commitment from diverse sector of society is established;
- public policy decision making directives are required;
- watershed plans are being developed and implemented;
- responsibilities, mandates, and terms of reference are clear;
- common objectives and benefits can be determined;
- stable funding is available to support the collaborative process; and
- participants share a commitment to horizontal rather than hierarchical structures to achieve sustainable water resources.

Some Promising Potential

Citizen engagement has been central in the Great Lakes regime, and relying exclusively on national governments for compliance ignores the potentially powerful role that citizens can and do play in environmental law and policy (Hall 2007).

After years of top-down policy implementation by public agencies, collaborative partnerships made up of diverse stakeholders are increasingly taking shape to address environmental problems in the watersheds. In some cases, collaboration takes place primarily among government agencies, with local stakeholder involvement relegated to public hearings and comment periods. In others, collaboration primarily involves actors in the private and nonprofit sectors, with public agencies providing the resources necessary for collaboration (Koontz et al. 2004).

The primary commonality of collaborative watershed groups, in particular is their focus on implementation or active co-management. Collaborative watershed management efforts typically move beyond visioning and planning exercises to the implementation of concrete initiatives, ranging from cleanup and restoration efforts to resource use issues. This feature of these processes represents devolution of decision making from a government agency to a group of multi-interested stakeholders working jointly with agency officials to codetermine the management of resources in a watershed (Konisky and Beierle 2001).

Effective sustainable management of a transboundary watershed system requires coordinated actions among governments, particularly when the transboundary waters reside in the nearshore zone for which many jurisdictions may have compatible or competing objectives. Chen (2008) states that this inter-state approach is important yet inadequate. "Policies and management plans developed by formal inter-state processes eventually rely on the implementation at local sites; hence community-based actions are critical to the effectiveness of policies." Chen advocates integrating community-based actions in watershed management, which will be complicated if citizen engagement is superficial and limited. While a consensus and willingness to cooperate among governments and extra-government participants is central to management of the Great Lakes watersheds, implementation of programs and plans must take place at the local level by enabling community engagement. Chen (2008) contends that it is impractical and inefficient for all interventions to be made centrally to protect ecosystem integrity.

Conclusions: Requisites for Change

A number of recent and significant voices agree that governance reform in the Great Lakes is critical to future ecosystemic recovery and well-being in the Basin and that implementation of the renegotiated GLWQA should produce substantive changes in the governance structure in the Basin (e.g., Krantzberg and Manno 2010; Jackson and Kraft Sloan 2008).

Botts and Muldoon (2005) called for "significant and rapid changes, the Great Lakes Agreement" or it will be "at the brink of irrelevancy." Further, they contend that "the Great Lakes themselves [are] subject to an onslaught of existing and new threats without a binational regime in place to deal with them", consistent with the findings of Krantzberg and Manno (2010).

Although there is still a need for governance at the ecosystem scale, many policy makers recognize that some threats, such as persistent organic pollutants are a global problem that required a global response. The appropriate scale for the hands-on work of restoring the Great Lakes ecosystem, however, is at the local level where thousands of "Friends of" organizations, local conservancies, beach stewards, and so on, represent a substantial and knowledgeable constituency actively engaged in clean-up and maintenance (Manno and Krantzberg 2008).

Markell (2005) points out that while the GLWQA lacks legally enforceable domestic status, it has given citizens an increased role in shaping policy to address transboundary pollution in the Great Lakes (also Hall

2007). A renegotiated GLWQA could increase the opportunity for public participation in decision-making. It is unlikely that a new agreement would be given treaty status, hence, as noted by Hall (2007), increased public participation would help to insure increased accountability on the part of both federal governments to comply with their joint responsibilities under the GLWQA. The GLWQA has helped create an informed and engaged citizenry on both sides of the border, which could result in an increased role for citizen enforcement.

Creating the conditions that facilitate self-organization, and particularly cross-scale institutional linkages, is the major challenge facing attempts to initiate adaptive co-management (Cundill and Fabricius 2010). Factors requiring greater attention in the restoration and protection of the nearshore zone of the Great Lakes include community perceptions of support from outside agencies, access to long-term funding for adaptive decision making, and access to reliable information about changes in natural resources and legal options for the formation of decision-making bodies. Long term and well-funded social facilitation is central to achieving this.

References

Bails, J., A. Beeton, J. Bulkley, M. DePhilip, J. Gannon, M. Murray, H. Regier and D. Scavia, 2005. Prescription for the Great Lakes http://www.miseagrant.umich.edu/downloads/habitat/PrescriptionforGreatLakes.pdf.

Baril, P., Y. Maranda and J.Baudrand. 2006. Integrated watershed management in Quebec (Canada): A participatory approach centred on local solidarity. Water Science and Technology 53(10): 301–307.

Berkes, F., J. Colding and C. Folke [eds.]. 2003. Navigating Social-Ecological Systems: Building Resilience for Complexity and Change. Cambridge University Press, Cambridge

Botts, L. and P. Muldoon . 2005. Evolution of the great lakes water quality agreement. Michigan State University Press, Michigan.

Bridgeman, T .B., J.D. Chaffin, D.D. Kane, J.D. Conroy, S.E. Panek and P.M. Armenio. 2011. From River to Lake: Phosphorus partitioning and algal community compositional changes in Western Lake Erie. Journal of Great Lakes Res. doi:10.1016/j.jglr.2011.09.010.

Carpenter, S.R., B. Walker, J.M. Anderies and N. Abel . 2001. From metaphor to measurement: Re- silience of what to what? Ecosystems 4: 765–81.

Chen, S. 2008. From Community-based Management to Transboundary Watershed Governance. Development 51: 83–88.

Cohen, A. and S. Davidson. 2011. An examination of the watershed approach: Challenges, antecedents, and the transition from technical tool to governance unit. Water Alternatives 4(1): 1–14.

Conroy, J.D., D.D. Kane and D.A. Culver. Declining Lake Erie ecosystem health—evidence from a multi-year, lake-wide plankton study. pp. 369–408. *In:* M. Munawar and R.T. Heath [eds.]. 2008. Checking the Pulse of Lake Erie.

Cundill, G. and C. Fabricius. 2010. Monitoring the governance dimension of natural resource comanagement. Ecology and Society 15(1): 15. [online] URL: http://www.ecologyandsociety.org/vol15/iss1/art15/.

Draper, S.E. 2007. Administration and institutional provisions of water sharing agreements. Journal of Water Resources Planning and Management 5(1): 446–445.

Folke, C., S. Carpenter, B. Walker, M. Scheffer, T. Elmqvist, L. Gunderson and C.S. Holling. 2004. Regime Shifts, Resilience, and Biodiversity in Ecosystem Management: Annual Review of Ecology, Evolution, and Systematics 35: 557–581.

GWP. 2000. Integrated water resources management, TAC Background Paper 4, Global Water Partnership Secretariat, Hantverkargatan 5, SE-112 21 Stockholm.

Gunderson, L.H. and C.S. Holling [eds.]. 2002. Panar chy: Understanding Transformations in Human and Natural Systems. Island Press, Washington, DC.

Hall, N.D. 2007. The Evolving Role of Citizens in United States-Canadian International Environmental Law Compliance Environmental Law Compliance 2007: 131–160.

Hall, N.D. 2009. Introduction: Canada-United States Transboundary Environmental Protection; 26 Windsor Rev. Legal & Soc. 1 (2009): 1–6.

Hardy, S.D. 2010. Governments, Group Membership, and Watershed Partnerships. Society and Natural Resources 23: 587–603.

Hill, C., K. Furlong, K. Bakker and A. Cohen. 2008. Harmonization versus subsidiarity in water governance: A review of water governance and legislation in the Canadian provinces and territories. Canadian Water Resources Journal 33(4): 315–332.

Holley, C. and N. Gunningham. 2011. Natural Resources, New Governance And Legal Regulation: When Does Collaboration Work? New Zealand Universities Law Review 24: 311–336.

Holling, C.S. and G.K. Meffe. 1996. Command and control and the pathology of natural resource management. Conserv Biol 10: 328–37.

Jackson, J.B.C., M.X. Kirb, W.H. Berher, K.A. Bjorndal and L.W. Botsford. 2001. Historical overfishing and the recent collapse of coastal ecosystems. Science 293: 629–38.

Jackson, J. and K.K. Sloan. 2008. A way forward: strengthening decision-making and accountability under the great lakes water quality agreement. A report sponsored by Great Lakes United.

Joosse, P.J. and D.B. Baker. 2011. Context for re-evaluating agricultural source phosphorus loadings to the Great Lakes. Can J Soil Sci 91: 317–327 doi:10.4141/CJSS10005.

Koehler, B. and T.M. Koontz. 2008. Citizen participation in collaborative watershed partnerships. Environmental Management 41(2): 143–154.

Konisky, D. and T. Beierle. 2001. Innovations in Public Participation and Environmental Decision Making: Examples from the Great Lakes Region. Society & Natural Resources: An International Journal 14: 9, 815–826.

Koontz, T.M., T.A. Steelman, J. Carmin, K.S. Korfmacher, C. Moseley and C.W. Thomas. 2004. Collaborative environmental management: What roles for government? Resources for the Future Press, Washington, DC.

Krantzberg, G. 2008. The Great Lakes, A 35th Year Anniversary: Time to Look Forward. Electronic Green Journal Issue 26 Spring 2008 ISSN 1076–7975.

Krantzberg, G. and J.P. Manno. 2010. Renovation and Innovation: It's Time for the Great Lakes Regime to Respond. Water Resources Management 24: 4273–4285.

Levin, S. 1999. Fragile Dominion: Complexity and the Commons. Perseus Books. Reading, MA.

Ludwig, D., M. Mangel and B. Haddad. 2001. Ecology, conservation, and public policy. Annu Rev Ecol Syst 32: 481–517.

Ludwig, D., B. Walker, C.S. Holling, L. Mandarano and K. Paulsen. 2011. Governance capacity in collaborative watershed partnerships: evidence from the Philadelphia region. Journal of Environmental Planning and Management 54: 1293–1313.

Manno, J. and G. Krantzberg. Rediscovering and Revitalizing the Great Lakes Governance. pp. 159–170. *In:* K. Bosselmann, R. Engel and P. Taylor [eds.]. 2008. Governance for Sustainability—Issues, Challenges, Successes. IUCN, Gland. Switzerland.

Markell, D.L. 2005. Governance of International Institutions: A Review of the North American Commission for Environmental Cooperation's Citizen Submissions Process, 30 N.C. Jour. Int'l L. & Com Reg 659–760.

McGinnis, M.V. 1999. Making the watershed connection. Policy Studies Journal 27(3): 497–501.

Mitchell, B. 2005. Integrated water resource management, institutional arrangements, and land-use planning. Environment and Planning A 37: 1335–1352.

Montgomery, D.R., G.E. Grant and K. Sullivan. 1995. Watershed analysis as a framework for implementing ecosystem management. Journal of the American Water Resources Association 31(3): 369–386.

Norman, E. and K. Bakker. 2009. Transgressing scales: Water governance across the Canada—U.S. Borderland. Annals of the Association of American Geographers 99(1): 99–117.

NRTEE. 2011. Charting a Course—Collaborative Water Governance. http://nrtee-trnee.ca/charting-a-course-collaborative-water-governance.

Paehlke, R. Spatial proportionality: Right-sizing environmental decision-making. pp. 73–124. *In:* E.A. Parson [ed.]. 2001. Governing the environment: Persistent challenges, uncertain innovations. University of Toronto Press, Toronto, Ontario.

Paine, R.T., M.J. Tegner and E.A. Johnson. 1998. Com- pounded perturbations yield ecological surprises. Ecosystems 1: 535–45.

Rabe, B. Power to the states: The promise and pitfalls of decentralization. pp. 34–56. *In:* N.J. Vig and M.E. Kraft [eds.]. 2006. Environmental policy: New directions for the twenty-first century. CQ Press, Washington, DC.

Rinta-Kanto, J.M., A.J.A. Ouellette, G.L. Boyer, M.R. Twiss, T.B. Bridgeman and S.W. Wilhelm. 2005. Quantification of toxic *Microcystis* spp. during the 2003 and 2004 blooms in western Lake Erie using quantitative real-time PCR. Environ Sci Techno 39: 4198–4205.

SOLEC 2009. State of the Great Lakes 2009 Highlights. http://binational.net/solec/sogl2009/sogl_2009_h_en.pdf.

Steffen, W., A. Sanderson, J. Jaiger, P.D. Tyson, B. Moore III et al. 2004. Global Change and the Earth System: A Planet Under Pressure. Springer-Verlag, Heidelberg.

United States and Canada. 1972. Great Lakes Water Quality Agreement. www.ijc.org

USEPA. 2008. What is a watershed approach. www.epa.gov/owow/watershed/framework/ch2.html (accessed 2 January 2012).

Warner, J. The beauty of the beast: Multi-stakeholder participation for integrated catchment management. pp. 1–19. *In:* J. Warner [ed.]. 2007. Multi-stakeholder platforms for integrated water management. Ashgate, Cornwall.

8

Co-Generated Knowledge for Co-Management of a Mobile Resource

Anishinabek/Ontario Fisheries Resource Centre

Tom Whillans,[1,*] *Cliff Meness*[2] and *Ed Desson*[3]

Introduction

Co-management of natural resources, as summarized by Armitage et al. (2007), involves the sharing of power and responsibility between governments and communities through the actions of their individual members. Natural resource management in complex ecosystems consists of a set of nested activities, including the initial inventory of resources and their potential uses, assessment of the sustainability of those uses in the context of external stresses, the allocation of rights of access, development and implementation of biophysical and policy options, monitoring of trends in resource condition and adaptive response to these. It aspires to

[1]Environmental and Resource Science, Trent University, Peterborough, Canada K9J 7B8.
Email: twhillans@trentu.ca
[2]Algonquins of Pikwakanagan, 1657A Mishomis Inamo, Pikwakanagan, Golden Lake, Canada K0J 1X0.
Email: councillor.cliff@pikwakanagan.ca
[3]Anishinabek/Ontario Fisheries Resource Centre, 755 Wallace Road, Unit #5, North Bay, Canada P1A 0E7.
Email: edesson@aofrc.org
*Corresponding author

the efficient and effective combination of human effort, equipment, facilities and finances.

In theory, the co-management of a natural resource by two parties where each party has a vested interest and a claim to the rights of partial jurisdiction over the natural resource can be a useful pre-negotiation platform for compromise between the parties about any future differences in opinion about rights and responsibilities regarding the management of the shared resource. Yet those same differences of opinion can interfere with the management of the natural resource. One danger of a well-functioning arrangement is that it may be viewed suspiciously by third parties that have interests in the natural resource.

Various models of co-management have been developed in order to lower the probability of failure due to soured working relationships or defensive intervention by external parties. This chapter documents a unique innovation in co-management of fisheries—the creation and operation, under formal agreement between two co-managing parties, of an independent institution for the generation and assembly of scientific and traditional ecological knowledge.

In 1993, The Union of Ontario Indians, representing the Anishinabek Nation, and the Ontario Government signed the Anishinabek Conservation and Fishing Agreement (ACFA). The parties agreed to establish an independent centre of knowledge for the co-management of fish populations and habitat of shared interest. Since 1996 the Anishinabek/Ontario Fisheries Resource Centre (A/OFRC) has fulfilled this function. The independent mandate in this case is to uncover the truth about fishery problems without the taint of real or perceived bias that might cloak investigations undertaken by either of the signatory organizations. The A/OFRC offers the opportunity to either of the parent organizations of requesting and receiving knowledge that is needed to make decisions. At the same time, its existence minimizes the co-optive peer pressure that both co-managing parties might feel, if they were directly in control of the knowledge generation. The balance is delicate—the role of ownership in knowledge generation in order to achieve trusted deep understanding weighed against the respected impartiality that is accorded to independent inquiry.

Knowledge Generation, Social Learning and Co-Management

The generation of knowledge is cited by Armitage et al. (2007) as core to effective co-management of natural resources. They argue that pluralistic generation and application of knowledge is one of the defining characteristics of adaptive co-management. Anticipating uncertainty, collaborators need to negotiate a flexible approach to learning and research. Berkes (2009) explains how the complexity of socio-ecological systems, that is the root of these inherent uncertainties, limits the ability of any single institution to direct and realize learning.

The same complexity creates the context for an alternative approach—social learning through the first hand involvement of individuals and their distributed understanding within multi-group or community situations. Plummer and FitzGibbon (2007), Pahl-Wostl et al. (2007), Pahl-Wostl and Hare (2004) argue for multi-level social learning through the planned interaction of organized stakeholders from individuals to collective levels. Eventually this can lead to the emergence of communities of practice or the quasi-professionalization of information sharing (Wenger 1998). It has been recognized for some time that in natural resource management, the key ingredient for adaptive problem solving is iterative feedback to collaborating actors (Lee 1993). This social and ecological feedback enables social learning that builds flexibility in co-management (Berkes 2007). In one sense co-management is a knowledge partnership (Berkes 2009) among academics and non-academics, community and government, scientists and non-scientists, formal organizations and informal networks (Pohl et al. 2010). This is an open partnership in which the actors strive to understand each other in order to properly interpret the knowledge that is generated by each and incorporate it into management.

Davidson-Hunt and O'Flaherty (2007) focused on knowledge generation in "context-specific" or "place-based" learning communities. In his case study of watershed-scale resource management at Shoal Lake, Ontario, he described dialogic networks between researchers and indigenous people. Knowledge generation involved decolonization of community members' roles in research and the perceived value of their contributions, as previously publicized by Smith (1999) and later generalized by Pohl et al. (2010). This involved, for example, preparing mutually supported research protocols by making commensurate the university ethics of harm reduction and community ethics of aiming for social good. However well-meaning the efforts toward mutual learning, they may not be not wholly embraced by participants. Fishers in Chile, for example, who shared knowledge, appreciated the resultant exposure of information bias, the sharing of biological observations and organizational advice, but were critical of the added effort and cost of transaction, the relinquishing of exclusive control, and the sense of duplication of effort (Schumann 2010).

Bridging and Boundary Organizations

One model for generating knowledge for co-management involves the creation of "external agents" of impartial knowledge, agents that sometimes may only be involved temporarily in a resource management situation, thereby avoiding perception of vested interest (Pomeroy 2007;

Pomeroy et al. 2001). With respect to fisheries co-management, Pinkerton (1999) recommended a role for "credible third parties" and later (2007) "information clearinghouses". Schumann (2010) reported on the role of independent biological consultants who are employed in fisheries regulation by fishers as part of their access agreement with the government regulators. They assess stocks and calculate quotas. Their role, however, extends beyond this to the education of fishers about fishery science and the building of respect for conservation-related regulation. Schumann (2010) refers to this as social bridging between the government and fishers.

Pahl-Wostl et al. (2007) argued that bridging organizations are needed between groups or networks of actors in order to interpret information. Labelled as "boundary organizations" by Guston (2001), they are ideally accessible without loss of identity by either users of a natural resource or regulators. Knowledge is co-produced (Pohl et al. 2010) within boundary organizations, thereby involving actors from both sides and perhaps mediated by independents. These organizations do not have direct political power to implement findings, but this frees them somewhat from the rigidity of previous commitments in order to recommend innovative alternatives. They may collectively impart weight to those alternatives through communities of practice (Pohl-Wostl and Hare 2004). Davidson-Hunt and O'Flaherty (2007) recognizes the merits of "dialogic networks" involving researchers and indigenous peoples who are working in "place-based learning communities". Boundary or bridging organizations ensure open access to information and open processes of knowledge creation (Pohl-Wostl et al. 2007). It is important that the openness involves a two-way flow of information between the generators and users of knowledge (Berkes 2009).

Pinkerton (2007) observed that fragmented knowledge is one barrier to co-management. Perhaps the most widely recognized form of fragmentation is between scientific knowledge and traditional ecological knowledge (TEK). This dichotomy strengthens the perceived need for boundary organizations. Spak (2005), for example, lamented with respect to caribou that science tends to be the backbone of management, with traditional ecological knowledge being supplementary. She explains this in part by the fact that TEK information providers have an advisory role to science-driven managers. Berkes (1999), Moller et al. (2004) and others argue for more balanced, complementary use of the two forms of knowledge. Boundary organizations have the potential to solicit and co-generate information from the generally more science-based government and the more TEK-based communities and then filter the combined information in the form of an independent interpretation of its contribution to the understanding of the natural resource. The range of knowledge practices that might be undertaken or proposed by boundary organizations includes knowledge gathering, sharing, integration, interpretation and application

(Armitage et al. 2011). The FAO Department for International Development (2005) identified four informational products for fishery co-management: evaluation indicators of local management effectiveness, documentation of management initiatives, formulation/improvement of management plans, and formulation/evaluation of the success of policy or plans.

Knowledge development for co-management is as much about process as it is about product. The process of knowledge creation is key to the co-managers' trust of the knowledge products. When that creation involves experiential learning by the participants that blends their respective conventional roles, then they are more likely to accept the result and may even realign their broader social relationship (Armitage et al. 2011; Berkes 2009; Schumann 2010). Boundary organizations thus have dual purposes as third party providers of trusted knowledge to co-managers and as builders of capacity or social capital within those constituencies to do the same (Berkes 2009; Pomeroy et al. 2001). That capacity can cover the range of planning research, resource and resource use monitoring, communication of results, brokering fair interpretation by co-managers with differing values and beliefs, and development of consensus about action (Ayles et al. 2007; Charles 2007). An argument could be made that over the long-term, a boundary organization's success would be punctuated by its eventual redundancy and dissolution.

Boundary organizations are not political actors. Their utility thus depends on political decision-makers receiving the results of their efforts and applying them. In this respect the boundary organizations are referent bodies with no direct control over the management of a natural resource, such that their impact on management may well be limited by the management context (Pahl-Wostl and Hare 2004).

A Boundary Organization for a Shared Fishery: Establishment of the A/OFRC

The Supreme Court of Canada's decision on the Sparrow case (Sparrow 1990), reinforced later by the Badger case (Badger 1996) was interpreted by the Government of Ontario to mean that governmental management of fisheries needed to recognize Aboriginal rights of access to fish and be collaborative with First Peoples. It prompted discussions between the Government of Ontario and the Union of Ontario Indians (UOI) about how this could occur within the territories fished by 42 (now 39) Anishinabek First Nations (Fig. 8.1). The result was in June 8, 1993, the signing of the Anishinabek Conservation and Fishing Agreement (ACFA). The Minister of Natural Resources, Hon. B. Wildman, and the Minister Responsible for Native Affairs, Hon. H. Hampton, signed on behalf of the Ontario

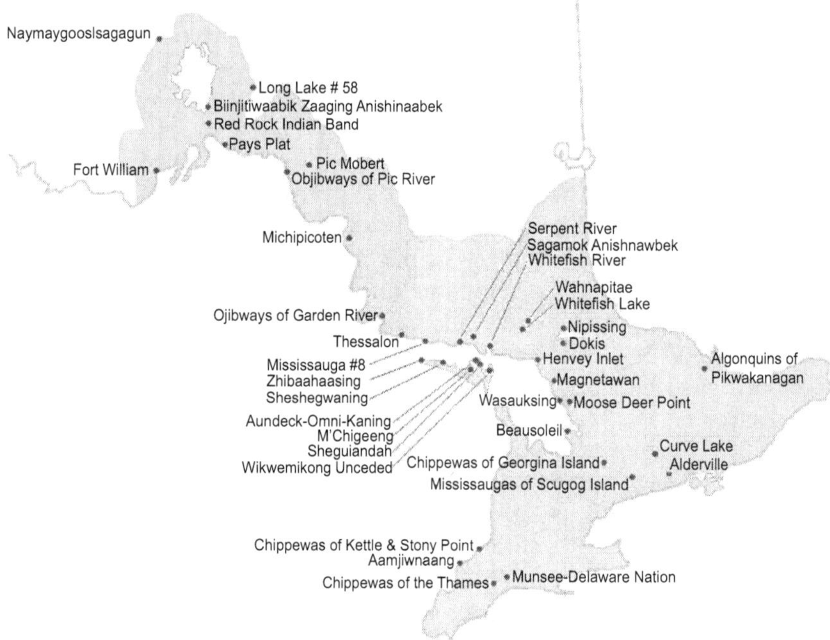

Figure 8.1 First Nations served by the Anishinabek Conservation and Fishing Agreement.

Government and the Chief of the Grand Council, Chief J. Miskokomon, signed on behalf of the UOI. Subsequently the main responsibilities on the part of the Ontario Government have been assumed by the Ministry of Natural Resources (MNR).

The ACFA committed the parties to the establishment of a Fisheries Resource Centre *to act as a central and independent source of information upon technical matters relevant to fisheries conservation and management issues* (ACFA 1993). A follow-up Fisheries Resource Centre Agreement in 1995 (FRCA) between the UOI and the Ontario Government, represented by MNR, specified the functions of the Anishinabek/Ontario Fishery Resource Centre (A/OFRC):

a) *Collect, produce, provide and evaluate information,*
b) *Advise and make recommendations to the Minister and the Grand Council Chief, Anishinabek Nation,*
c) *Promote new technology and other scientific advances,*
d) *Encourage co-operation amongst those interested in the fisheries resource and provide a forum for discussion,*
e) *Identify, assess and advise on issues arising from agreements made under the Agreement,*

f) *Engage in other activities consistent with these objectives, as may be assigned from time to time by the Minister or Grand Council Chief* (FRCA 1995).

An appendix to the FRCA detailed the operating principles of the Centre, including an outline of its roles:

1. *To report on stock status by data collection, inventory, monitoring, and evaluating the impacts of use and other environmental stresses on fish populations and their habitats;*
2. *to make available and promote the use of state of the art science, technology, and techniques; and*
3. *to provide a forum for information sharing and participation with stakeholders* (FRCA 1995).

The appendix specifies that the Centre *will have no jurisdictional authority and its activities will be limited to making non-binding recommendations to ensure the sustainability of aquatic ecosystems, particularly fish stocks* (FRCA 1995). The Centre was incorporated not-for-profit, with a 13 person Board of Directors, six of whom were appointed by each party and the Chair jointly. The Board was later reduced to nine in order to minimize costs.

In order to achieve their respective interests it was necessary for the MNR and the UOI to co-manage fisheries and to develop a mutually respected and sufficient knowledge base for that co-management. Like the water that they inhabit, fish are mobile and do not respect jurisdictional boundaries. Fish managed under the umbrella of the ACFA are potentially accessible to any person with commercial fishing or angling rights in the waters inhabited by the fish. All users of the resource need to believe that their interests are represented. This was addressed in part by ensuring that appointees to the Board of Directors represented a wide range of user interests. Appointees of the UOI covered the geographic range of UOI territory. In the case of those appointees of the MNR, only one was an employee of the Ministry—the others were members of the general public representing a mixture of interests in the fishery. The Board is ultimately accountable for the A/OFRC, but receives or seeks advice from the parent organizations and its clients. It receives advice annually from its parent organizations when presenting the A/OFRC's business plan, occasionally when the organization undergoes external review, periodically when undertaking strategic planning initiatives and *ad hoc* when needed.

The Board oversees the action and reporting on an annual core allocation from the Ministry of Natural Resources that has remained at $ 800,000 (Canadian) since inception. The funding allocation was originally expected to increase as the A/OFRC developed its field units and expanded its focus to address broader habitat concerns; however, the Government of Ontario underwent province-wide cuts in funding, leaving the A/OFRC to determine how to grow, without additional core funding. The organization thus acted on its freedom to solicit funds from other funding organizations in order to better deliver projects that are initiated through its core funding, to undertake additional projects with UOI members for external sponsors, or occasionally, outside of its regional responsibility, to undertake projects that require A/OFRC capability and will generate income that could be applied to other UOI projects. Major supplementary funding has been received from the Department of Fisheries and Oceans, Trillium Foundation, Indian and Northern Affairs Canada, Northern Ontario Heritage Foundation, FedNor, the Aboriginal Inland Habitat Program, and fee-for-service contracts such

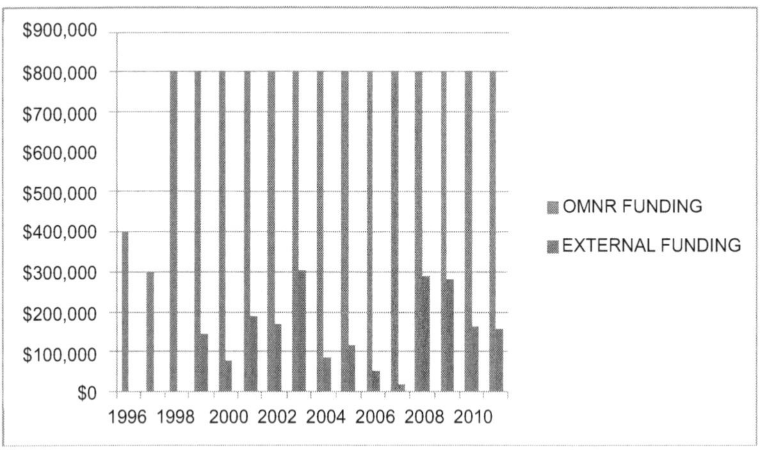

Figure 8.2 A/OFRC funding by source, 1996–2011.

that the total budget over the period 1996–2011 was $ 13.94 M. Figure 8.2 summarizes the core and external supplementary budget through the years.

A/OFRC Operational Logistics

The A/OFRC generates knowledge for its two parent organizations mainly through local-level interaction. Each year in the fall its staff facilitates the preparation of applications for research and capacity-development projects by UOI member First Nations. The applications identify the potential contribution to co-management of the fishery. They are received and assessed first by staff and then by the Projects Committee of the Board and finally approved for funding by the Board as a whole. Often equipment needs for projects are met through the loan of A/OFRC-owned equipment. Each proposal is vetted through the appropriate local office of the Ministry of Natural Resources in terms of its relevance and unique contribution to fishery knowledge. The projects take many forms, including:

- research that is undertaken collaboratively by staff of a First Nation and A/OFRC
- research that is undertaken by a First Nation, with advice from the A/OFRC
- research that is undertaken primarily by the A/OFRC , in consultation with a First Nation
- research that is collaborative amongst a number of First Nations and coordinated by the A/OFRC
- any of the above, with in-kind contributions from the Ministry of Natural Resources of staff and/or equipment
- any of the above with training components
- any of the above with cash and sometimes in-kind contributions from external sponsors
- training events or programs, with research contributing experientially.

Products of the projects are equally diverse. They include full research reports with recommendations, data reports, advisory reports, training summaries and contributions to external integrated resource management documents. Occasionally, in cases where the type of knowledge generated is traditional, as obtained from elders or other community members, then the raw information is archived and retained by the community and the A/OFRC report summarizes the overall findings. Communications for communities are prepared for all projects.

The A/OFRC focuses on the waters around the 39 First Nations of the UOI, a territory spanning over 1200 km east-west and similarly north-south. The original physical plan for the A/OFRC was that it would have a central head office and several regional field offices. This would enable close contact with the individual First Nations, yet take advantage of the efficiencies of central administration, shared equipment and other resources. The central office was initially in Sudbury, but throughout most of its history has been located nearby in North Bay where the UOI offices are also located. Soon after it began, the A/OFRC created a western field office in Rock Bay First Nation on Lake Nipigon, because that First Nation had emerged as a regional leader in fisheries investigation. A second field unit was later initiated at Nipissing First Nation and another was contemplated in the south. However, before further action could be taken, the A/OFRC determined that the cost of running field offices exceeded the fiscal capacity of the organization, especially if it was to expand its focus to encompass habitat and training without additional budget. Thus, the Nipigon field office was closed. Co-ordination of Nipigon and Nipissing field operations was assumed by the central office and by the higher capacity field units of particular First Nations that had developed by this time. Given the great distances to the extremities of the UOI territory and especially the cluster of First Nations and projects in the N.W., the A/OFRC is still considering the possibility of a smaller office presence in those areas, perhaps complemented by a mobile field unit. When the co-generation of locally respected knowledge and related capacitation of First Nation personnel are goals, then field units offer distinct advantages in terms of service and communication with FN's that are the farthest from the A/OFRC's North Bay headquarters.

The FRCA outlined three phases of development of the A/OFRC, with staffing complements increasing from 12, to 21 and then 31 full-time employees. This was never realized, because of funding constraints. Also, in the spirit of capacity-building, the Board decided instead to distribute staffing funds to many short-term project-specific positions in individual First Nations. The A/OFRC operates with 8–10 full-time employees based from its North Bay office and funds the employment of many part-time community-based employees for its projects, perhaps 30–40 in a typical year. Its physical plant is comprised of vehicles (trucks, cars, ATV), boats of various size, including one large lakes fishing tug, nets of various types and sizes, electro-fishing equipment, a BASS unit for bathymetric analysis, fish and water quality sampling gear, and state-of-the-art information technology. Field equipment is routinely loaned to First Nations for A/OFRC-sponsored projects and for other worthy First nation projects. Some technical services such as aging of fish and DNA analysis are out-sourced.

Co-Generating Knowledge: The General Experience of the A/OFRC

The number of projects undertaken per year varies, depending on the complexity of projects, the amount of A/OFRC staff assistance that is requested by First Nations, the size of the staff complement, the need to spread the projects and associated demand for equipment and staff throughout the field season, the size of the supplementary budget, and the interference of other work undertaken by the A/OFRC (e.g., strategic planning, report writing, training courses for First Nations). In the early years the number was also influenced by the tentative quality of some of the proposals; however, the quality of proposals improved markedly through time as the A/OFRC staff worked with First Nations to improve their proposal-writing skills, such that quality of proposals no longer limits the number of projects approved. Figure 8.3 portrays the total number of proposals submitted and the number of First Nations submitting them from 1996 to 2012.

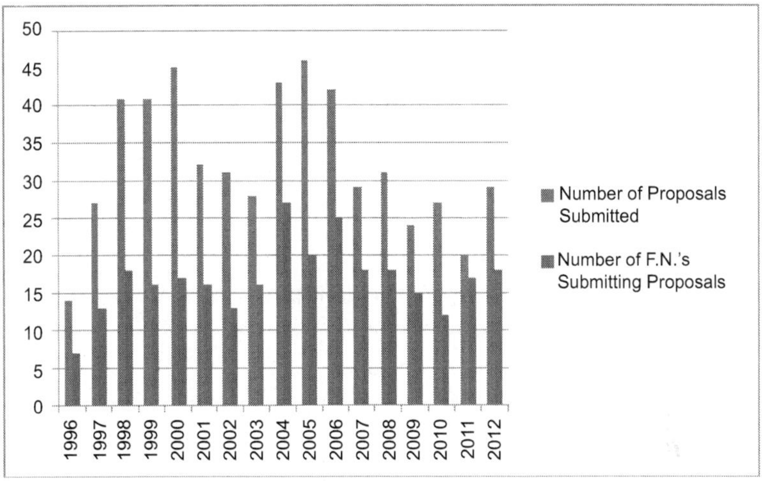

Figure 8.3 The number of project proposals submitted in total and the number of First Nations (F.N.'s) submitting them, 1996–2012.

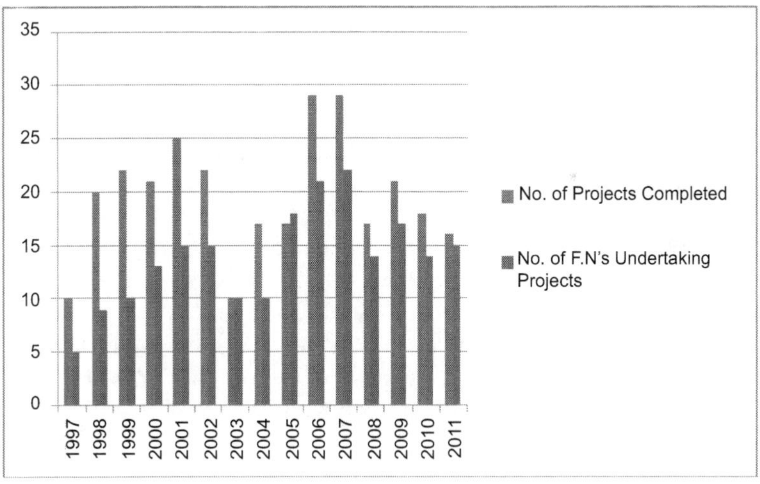

Figure 8.4 The number of projects completed in total and the number of First Nations (F.N.s) undertaking them for the years 1996–7 to 2010–11.

The number of projects actually completed through the A/OFRC and the number of First Nations involved during the same period is summarized in Fig. 8.4. The number of projects averages 19. This represents approximately 59 percent of the projects proposed. Some First Nations will propose and undertake more than one project in a given year, thus the number of participating First Nations averages 13, representing about 76 percent of the First Nations applying on average in a given year. Some 25 percent of the projects were collaborative among First Nations. The total number of projects undertaken from 1996–2011 was 294, involving 37 of the original 42 ACFA First Nations.

The majority of the projects in the initial years focused directly on fish: that is harvests, abundance indices, distribution, fish species diversity, surveys of water bodies, growth rates, general condition of fish, and external stresses on fish. Investigations usually involved scientific methods and about one-third used traditional ecological knowledge (TEK). A few studies were wholly dependent on TEK. The A/OFRC was not funded in the first phase to undertake habitat investigations, but when it became evident that additional funding for this would not be forthcoming, the focus was expanded to include habitat because of its importance to the fishery. The habitat work included general shoreline and bathymetric surveys, studies of reproductive habitat, impact assessment, estimation of rehabilitation potential, water quality measurement, and substrate analysis.

Another complementary focus that emerged was training. The 1993 ACFA had referred to the likelihood that this would be required, but had not proposed the Fishery Resource Centre or any other organization as the delivery agent. Soon enough it became clear that training of community personnel was necessary hand-in-hand with the implementation of many projects, in order to enable First Nations to undertake the projects cooperatively with the A/OFRC and to develop First Nation capacity to generate knowledge. Training was undertaken in several ways. Most frequently, the A/OFRC staff would provide field experiential training while a project was being initiated. In the cases of training that would benefit a number of First Nations and A/OFRC staff, workshops were convened, supplemented at times with external funds. Subjects included assembling TEK, sustainable fisheries, gill net fishery practices, fish habitat, fisheries assessment techniques, working on ice, and introduction to field methods. In 2011 a summer camp program was initiated for First Nations children, emphasizing environmental and natural resource experiences. This and a program of school visits helped to cultivate a culture of fishery awareness within communities.

One of the frustrations when working with a large number of First Nations, only a portion of which undertake knowledge co-generation projects in a given year, is the mobility of community-based expertise. In order to apply or extend their knowledge after a 6-wk or otherwise short-term project, graduate trainees often need to go elsewhere for similar employment or advanced training. The A/OFRC, though effective in its role in cultivating the professional interest, is then faced with the continuous need to train new recruits. One innovation since 2007 has been the operation of an internship program in which the participants work one to two years with the organization on a range of projects in a number of First Nations. Internships are paid apprenticeships that effectively provide advanced training locally and retain expertise in communities for longer periods.

Large Co-Management Territory, Integrative Knowledge

One of the difficulties in undertaking co-management of a fishery for 39 First Nations, many more communities, 55,000 people, and an area that covers over one million km^2 is the identification of shared concerns and co-ordination of related actions. The A/OFRC in its role of knowledge generation has at times led investigations that, by serving multiple clients, enabled more efficient co-management. The largest such undertaking was the 3-yr whitefish stock assessment, initiated in 2002 along the Canadian coasts of Lakes Superior and Huron. Supplemented by funding from FedNor and the Northern Ontario Heritage Fund the A/OFRC co-ordinated the first integrated inventory of lake whitefish (*Coregonus clupeaformis*) stocks within a vast area of coastal waters (Seyler 2002). The shared concern was the state of the most important commercial and subsistence fish species in those waters and the suitability of the quotas, both in terms of conservation and food fish supply. The fish is highly mobile, yet standard information on its abundance across its range did not exist.

The A/OFRC led the collaborative assessment of the stocks by 10 First Nations. Fishery science on large lakes is difficult because of the specialized and expensive vessel and equipment needs, the rigor required for statistically significant samples, and the skills for field and analytical delivery. In addition to its own staff, the A/OFRC drew upon the expert scientific advice of the Ministry of Natural Resources' Fisheries Assessment/ Management Units for Lakes Huron and Superior, the collective experience, traditional knowledge and efforts of fishers from the collaborating First Nations, and significant guidance by First Nation personnel from Lake Nipigon. The results improved measurably the understanding behind rational co-management of whitefish through mutually negotiated quotas in those waters. The A/OFRC was not involved in the negotiations, only the science that informed them.

A different type of integrative effort was led by the A/OFRC in the 2002–2003 period with supplementary funds from the Ontario Trillium Foundation. Entitled "Towards Harmony and Sustainable Use among the Ontario Fisheries Community", the project involved deliberations between First Nation communities and non-First Nation stakeholders on the issues and resource needs for fisheries management (Gillies et al. 2003). It became a pilot project in the N.W. Lake Huron region for information-sharing among stakeholders in the fishery. The outcome was the validation of the need for a standing group to discuss co-management concerns, the initiation of such a group in that region and the recognition of the more general need for regional communication mechanisms.

These types of integrative knowledge generation and dissemination are entirely appropriate for a boundary organization that serves a broadly distributed clientele; yet they represent difficult challenges for the A/OFRC. They are expensive and draw heavily on the human resource base of the organization. The above example exercises required supplementary funds. The demand from individual First Nations for local field-based projects usually exceeds the funds available and is limited by A/OFRC staff availability, such that a large integrative project would severely limit the capacity of the A/OFRC to participate in many local projects that for the individual proponent communities are more important than the integrative ones. The task of balancing the big picture longer-term initiatives with the immediate application parochial ones is delicate, both logistically and in terms of client perspective.

Mixed Knowledge, Dynamic Context

The literature on mixed knowledge generation for co-management exposes the scientific knowledge–TEK dialectic. Often the two types of knowledge are respected unequally by decision-makers, with the greater weight attributed to scientific knowledge. The A/OFRC has endeavoured to broker mutual respect between the generators of each type of knowledge, while struggling itself in specific cases to understand and resolve paradoxes between their findings. One example of the difficulty occurred on Lake Nipigon in the late 1990's. Based on index gill netting, the Ontario Ministry of Natural Resources'(OMNR) Lake Nipigon Fisheries Assessment Unit recommended the quotas for lake whitefish. The local First Nations, led by Rocky Bay F.N., had traditionally recognized an additional stock of whitefish that they felt was unaccounted by the index netting and thus unrepresented in the quota allocation. Working with the First Nations and the Assessment Unit, the A/OFRC modified the index netting protocol to include the waters in which TEK placed the additional stock. The results substantiated the TEK and the modified protocol upon which it was based became the norm that was subsequently applied collaboratively on the lake.

A comparable case occurred on Golden Lake in eastern Ontario. The walleye (*Sander vitreum*) population was perceived to be depressed, having declined over a 10 yr period. In 1997 this came to a head when subsistence fishers from the Algonquins of Pikwakanagan First Nation expressed their alarm, as did local recreational anglers. In this case TEK was the basis of both opinions—that of the longstanding subsistence fishers arising from their declining yields and that of the anglers substantiated by their perceptions and decades of private records of catches kept by a resort owner. Occasional surveys of the fishery through the past decade by the OMNR had indicated steady, but low natural production of walleye. In 1997 at the invitation of the First Nation and with the agreement of the OMNR the A/OFRC undertook near-shore community index netting, spawning stock assessment, creel survey and fall index netting. The research confirmed the low density of adult walleye and the presence of a small, but functional spawning population. The results were presented and deliberated in public meetings. The co-management response was a 5 yr moratorium on walleye fishing in the lake, in order to determine if the spawning potential would be realized in the absence of fishing pressure such that the species would recover.

In both the Lake Nipigon and Golden Lake examples, the knowledge that was generated through the A/OFRC about controversial issues was accepted and used as a basis for action. The A/OFRC was able to deliver the results to the receptive regional ears of its parent co-management organizations. This has not consistently been the case. When after the moratorium on walleye fishing in Golden Lake the walleye apparently had not recovered, a decision was made to stock the lake with walleye without further study by the A/OFRC and perhaps at the peril of the remnant walleye population. The decision was less knowledge-based than political. In Lake Nipigon, apart from the A/OFRC, a disagreement amongst the local First Nations and the UOI resulted in the withdrawal of several from the UOI. Without membership in the parent organization, they ceased to have a relationship with the A/OFRC and its knowledge-generation activities on the lake also ceased. As an apolitical boundary organization, the A/OFRC responds to the requests of its co-managing parent organizations. If those organizations experience political adjustments that mitigate against its involvement in issues, then it must patiently await the resolution of the underlying problems. With 39 First Nations and numerous local offices of the OMNR, it should be no surprise that the relationships with the A/OFRC ebb and flow in response to background social changes. Knowledge is generated within a dynamic, shifting co-management context.

Sensitive Knowledge

In 2010, responding to a regulation problem on the Mississagi River, the OMNR declared a province-wide moratorium on the harvesting of lake sturgeon (*Acipenser fulvescens*). The species has declined markedly throughout most of its range over the past 150 yr because of over-exploitation and habitat degradation and is the subject of recovery plans currently in preparation, as mandated under the provincial and federal endangered species/species-at-risk legislation. The moratorium annoyed First Nations, given the perpetual dependence on the fish species for food. One of the prominent reasons for this is the contradictory TEK indicating that some of the populations of sturgeon appear to be reasonably healthy and that sightings of juvenile sturgeon are frequent indicators of reproductive resilience in a species that does not reach maturity for about 20 yr.

A difficulty in preparing recovery plans for lake sturgeon is the paucity of information on the species, especially in Ontario, and the slow generation of information because fishing for the species is prohibited. Both before and since the moratorium many First Nations proposed projects and co-generated knowledge on sturgeon through the A/OFRC. The Centre in turn obtained special permission from its parent OMNR to mount the research. As a consequence the A/OFRC's body of knowledge and expertise on sturgeon has accumulated and been presented in various international meetings. During the moratorium it has been one of the few sources of primary knowledge on the species in Ontario and certainly the most active source.

The A/OFRC research verifies low sturgeon populations in contrast with those historical, but it also suggests that the species has persisted in low numbers across its range within the UOI area. Telemetry and netting studies in the Pic River, for example, document a healthy spawning population and healthy reproductive habitat (Ecclestone 2011). Substantial numbers of sturgeon were rediscovered in the adjacent White River, where they had been listed as extirpated. The research raises the possibility that one reason why sturgeon numbers are thought to be low is that few studies have actually searched for the species. The assumption of low abundance begs local testing.

One merit of a knowledge-generating boundary organization is that it may move comfortably into an arena that might be difficult for either parent organization. In the case of sturgeon, the regulatory case that provoked the moratorium involved a First Nation fisher within the UOI territory. Thus, approval of First Nation research on the species in the territory might invite criticism. Research undertaken by OMNR might similarly suggest that the knowledge base for the moratorium was inadequate. In either case the knowledge itself might be questioned in terms of its impartiality, given the politicization of the species. The established international reputation of the boundary organization provides the foundation for the generation of respectable knowledge.

Evolving, Adaptive Boundary

Turner et al. (2003) raise the notion that ecologic and social communities that exist on the edge of persistent ecosystems or social systems, respectively, are more capable of adapting to change and thereby impart resilience to those ecosystems or social systems. Intuitively, a knowledge-generating boundary organization should need to be more adaptable than, say, a research organization sitting wholly within the boundaries of one of the parent organizations. It will be less well-understood by the parent bodies, less easily guaranteed funds from either body, more subject to socio-political externalities because of exposure to two socio-political regimes, more pressured by peer review to maintain independence, and propelled more by healthy scepticism about the status quo.

The experience of the A/OFRC is that of an organization on the edge. As illustrated by the Lake Nipigon and Golden Lake experiences, it has the potential to be affected by the socio-political vagaries of the two parent organizations, 39 First Nations and numerous local offices of OMNR. By definition the A/OFRC cannot exist outside of the UOI and OMNR context, thus must react to or anticipate the demands of those organizations. Projects are frequently affected by local personnel, governance, social and other factors. The organization itself has perpetuated adaptation in the form of frequent strategic reviews, one external and four internal during its short period of existence. Its priorities are revised annually in response to proposals by its partners or resourcing opportunities.

Adaptations currently being developed would have the A/OFRC adding as priorities the investigation of food fish quality and climate change effects on fish communities. These foci emerged from the frequent community-based proposals for projects on the former and the interest expressed by the parent organizations in the latter. To the degree that any new priority is covered by the original purposes of the A/OFRC, it would be possible to allocate resources within the existing funding umbrella. Climate change, for example, represents a type of stress on fish populations and habitat that is embraced by the original focus of the organization, though one that could well attract supplementary funding support from untapped sources. Food fish quality was not identified originally as relevant to the A/OFRC, thus would necessitate supplementary support.

The Missing Link

The Lake Nipigon and Golden Lake cases illustrate how co-generated knowledge can find co-management application. Although a co-management application existed in those cases, many A/OFRC projects, although articulated partly in terms of management needs, have not been applied. They were undertaken at the request of First Nations, with advice from the local office of OMNR, but have not yet contributed to active co-management. This is in part because management is a non-linear practice. A number of factors need to fall into place and knowledge is only one of them. When both co-management parties agree that a project could provide useful information for future co-management, is in itself an achievement and worthy of follow-up. Of course sometimes the outcome of a project is the recognition by one of the parties that no further action is necessary or desirable. Nevertheless, although the A/OFRC was created in order to co-develop the information base for co-management decisions, the fact that it is not directly involved in management *per se* means that the results of its projects sometimes follow the path of curiosity-based research rather than applied research. They contribute to a body of knowledge, parts of which may find application in the future by the co-managers.

Missing is a UOI-Ontario Government co-management organization that would process knowledge communicated from the A/OFRC and other sources and apply it. Large integrative projects such as the whitefish stock assessment on Lakes Huron and Superior, the cumulative knowledge from many smaller projects such as those on lake sturgeon, and the implications of future projects on regional issues such as climate change need this sort of forum for decisive follow-up. Also missing is a co-management body that could request of the A/OFRC larger knowledge co-generation projects to address broader issues such as these.

In the late 1990's the UOI and OMNR developed a Resource Management Council (RMC), with three Working Groups, one of which was on Fish and Wildlife Issues. The Fish and Wildlife Working Group consisted of six members, three representing the Anishinabek Nation and three representing OMNR. Viewed from the A/OFRC's perspective it had the potential to fill the niche of a co-management organization that could recommend, receive and act on the knowledge generated. Since it was created, the RMC has deliberated on various issues relating mainly to policy, treaty rights and communications between the co-management parties. A reporting relationship from the A/OFRC to the RMC was never established, although for a period one of the Anishinabek Directors from the A/OFRC Board was coincidentally a member of the Fish and Wildlife Working Group.

Retrospection: Boundary Organizations for Knowledge Co-Generation

Within the overall field of co-management the creation of a boundary organization for the co-generation of knowledge is a relatively rare phenomenon. The experience of the A/OFRC includes both striking successes and lessons from growing pains and it raises intriguing issues for future consideration. The organization has developed into a widely respected authority on freshwater fisheries, co-managed or otherwise, not withstanding the challenging path toward sustainability. The model that it represents may indeed be applicable and replicable for other co-management situations.

The conditions under which a knowledge-generating boundary organization might serve co-management are worth considering. Fundamental is agreement by the co-managing parties that the organization will have an apolitical role as an independent source of information. A corollary to this is that the boundary organization needs to understand this principle of its creation and have the patience to await the resolution of any political problems within its parent bodies. This does not mean that it must refrain from political analysis, but it must do so internally and quietly. In fact the Board of the A/OFRC has needed many times to undertake such analyses in order to understand the need for patience in particular situations. The particular situation of the A/OFRC in serving a wide range of First Nations and local OMNR jurisdictions has translated into the acceptance that one characteristic of its operational model is the ebbs and flows of local involvement because of matters well-beyond the control of the A/OFRC.

Also fundamental is the sustained provision of core resources from the parent organizations that can be used as the basis for and even a lever for soliciting complementary resources from other providers. A characteristic of a mobile resource such as fish is that it often requires management considerations that are ecosystemic in breadth. The overlap with the interests of other government agencies and non-governmental stakeholders exposes the opportunities and constraints of collaborative information gathering. In the A/OFRC's case the enlightened willingness of the parent organizations to sustain the core and magnanimously entertain other partnerships has been a foundation for adaptive investigation. Territoriality would be dangerous.

The final fundamental is communication, respectful communication. In the A/OFRC's case the need is relentless. The capacity to communicate will never be sufficient to meet all demands. No simple communication mechanism can facilitate the preparation of project proposals, enable smooth collaboration in widely distributed field projects, explain the results and implications to communities and governmental managers, maintain clear understanding among the tripartite core organizations, integrate supplementary supporters, educate other stakeholders and ensure intra-organizational harmony. A boundary organization must constantly reinvigorate its communications strategy and will likely always feel that it could do more.

References

Anishinabek Conservation and Fishing Agreement. 1993. Anishinabek Conservation and Fishing Agreement between: Anishinabek Nation as represented by the Union of Ontario Indians and Her Majesty the Queen in Right of Ontario as represented by the Minister of Natural Resources and the Minister Responsible for Native Affairs.

Armitage, D., F. Berkes and N. Doubleday. Introduction: moving beyond co-management. pp. 1–18. *In:* D. Armitage, F. Berkes and N. Doubleday [eds.]. 2007. Adaptive co-management: collaboration, learning and multi-level governance. UBC Press, Vancouver.

Armitage, D., F. Berkes, A. Dale, E. Kocho-Schellenberg and E. Patton. 2011. Co-management and the co-production of knowledge: learning to adapt in Canada's Arctic. Global environmental change 21: 995–1004.

Ayles, B.G., R. Bell and A. Hoyt. Adaptive fisheries co-management in the western Canadian Arctic. pp. 125–150. *In:* D. Armitage, F. Berkes and N. Doubleday [eds.]. 2007. Adaptive co-management: collaboration, learning and multi-level governance. UBC Press, Vancouver.

Berkes, F. Adaptive co-management and complexity: exploring the many faces of co-management. pp. 19–37. *In:* D. Armitage, F. Berkes and N. Doubleday [eds.]. 2007. Adaptive co-management: collaboration, learning and multi-level governance. UBC Press, Vancouver.

Berkes, F. 2009. Evolution of co-management: role of knowledge generation, bridging organizations and social learning. J. Environmental management 90: 1692–1702.

Charles, A. Adaptive co-management for resilient resource systems: some ingredients and the implications of their absence. pp. 83–104. *In:* D. Armitage, F. Berkes and N. Doubleday [eds.]. 2007. Adaptive co-management: collaboration, learning and multi-level governance. UBC Press, Vancouver.

Davidson-Hunt, I.J. and R.M. O'Flaherty. 2007. Researchers, Indigenous peoples, and place-based learning communities. Society and Natural Resources 20: 291–305.

Ecclestone, A. 2011. Movement patterns, habitat utilization, and spawning habitat of Lake Sturgeon (Acipenser fulvescens) in the Pic River, a northeastern Lake Superior tributary in Ontario. M.Sc. thesis, Trent University. Peterborough.

Fisheries Resource Centre Agreement. 1995. Fisheries Resource Centre, Ontario-Anishinabek Conservation and Fishing Agreement, including attached Statement of Operating Principles.

Food and Agriculture Organization Department for International Development. 2005. Information for fisheries co-management. FMSP Project R8462.

Gillies, M., J. Seyler and P. McLeod-Shabogesic. 2003. Towards harmony and sustainable use among the Ontario fisheries community, Taking action: 2003. Final Report. Anishinabek/Ontario Fisheries Resource Centre, North Bay.

Gutson, D.H. 2001. Boundary organizations in environmental policy and science: an introduction. Science, technology and human values 26(4): 399–408.

Kaplan, I.M. and B. McCay. 2004. Cooperative research, co-management and the social dimension of fisheries science and management. Marine policy 28: 257–258.

Lee, K.N. 1993. Compass and gyroscope: integrating science and politics for the environment. Island Press, Washington.

Moller, H., F. Berkes, P. O'Brian Lyver and M. Kislalioglu. 2004. Combining science and traditional ecological knowledge: monitoring populations for co-management. Ecology and society 9(3): 2 (online).

Pahl-Wostl, C. and M. Hare. 2004. Process of social learning in integrated resources management. J. Community and applied social psychology 14: 193–206.

Pahl-Wostl, C., M. Craps, A. Dewulf, E. Mostert, D. Tabara and T. Taillieu. 2007. Social learning and water resources management. Ecology and society 12(2): 5 (online).

Pinkerton, E. 1999. Factors in overcoming barriers in implementing co-management in British Columbia salmon fisheries. Ecology and society 3(2): 2 (online).

Pinkerton, E. Integrated holism and segmentalism: overcoming barriers to adaptive co-management between management agencies and multi-sector bodies. pp. 150–171. *In:* D. Armitage, F. Berkes and N. Doubleday [eds.]. 2007. Adaptive co-management: collaboration, learning and multi-level governance. UBC Press, Vancouver.

Plummer, R. and J. FitzGibbon. Connecting adaptive co-management, social learning, and social capital through theory and practice. pp. 38–61. *In:* D. Armitage, F. Berkes and N. Doubleday [eds.]. 2007. Adaptive co-management: collaboration, learning and multi-level governance. UBC Press, Vancouver.

Pohl, C., S. Rist, A. Zimmerman, P. Fry, G.S. Gurung, F. Schneider, C.I. Speranza, B. Kiteme, S. Boillat, E. Serrano, G.H. Hadorn and U. Wiesmann. 2010. Researchers' roles in knowledge co-production: experience from sustainability research in Kenya, Switzerland, Bolivia and Nepal. Science and Public Policy 37(4): 267–281.

Pomeroy, R.S., B.M. Katon and I. Harkes. 2001. Conditions affecting the success of fisheries co-management: lessons from Asia. Marine Policy 25: 197–208.

Pomeroy, R. Conditions for successful fisheries and coastal resources co-management: lessons learned in Asia, Africa, and the wider Caribbean. pp. 172–190. *In:* D. Armitage, F. Berkes and N. Doubleday [eds.]. 2007. Adaptive co-management: collaboration, learning and multi-level governance. UBC Press, Vancouver.

Schumann, S. 2010. A tenuous triumvirate: the role of independent biologists in Chile's co-management regime for shellfish. Marine Policy 34: 133–138.

Seyler, J. 2002. Assessment of Lake Whitefish stocks in Lake Huron and Lake Superior: Year 1 Progress report. Anishinabek/Ontario Resource Centre, North Bay.

Smith, L.T. 1999. Decolonizing methodologies: research and indigenous peoples. Zed Books, London. Spak, S. 2005. The position of Indigenous knowledge in Canadian co-management organizations. Anthropologica 47: 233–246.

Turner, N.J., I.J. Davidson-Hunt and M. O'Flaherty. 2003. Living on the edge: ecological and cultural edges as sources of diversity for socio-ecological resilience. Human Ecology 31(3): 439–461.

Wenger, E. 1998. Communities of practice: learning, meaning, and identity. Cambridge University Press, Cambridge.

9

Taking the "Co" out of "Co-Management"

The Delegitimization of Fishing Communities on Lake Victoria, E. Africa

Ted Lawrence

Introduction

Socio-political institutions, such as rules of fishing created through the relationship between central governments and local level resource users, called co-management, have been created to manage fresh water fisheries in developing countries. Despite the creation of co-management institutions, many factors inhibit these institutions from being effective (Pomeroy and Berkes 1997; Kaimowitz and Ribot 2002; Jul-Larsen 2003). Of the factors that are researched, *legitimacy* and *accountability* are cited as being critical aspects of a successful fishery management institution. Legitimacy and accountability include state accountability to the local level (Ribot 2002a); legitimacy of local-level crafted rules by higher-level government (Ostrom 1999); legitimacy of the whole institution as viewed by all stakeholders; and legitimacy of the rules by the users. Understanding legitimacy, however, is

Great Lakes Fishery Commission, 2100 Commonwealth Blvd. Ste. 100, Ann Arbor, Michigan 48105; and School of Natural Resources and Environment, University of Michigan, 440 Church Street, Ann Arbor, Mi, 48109-1041., 734.669.3008.
Email: ted@glfc.org

weakest when determining the legitimacy of local level organizations (the lowest political enforcement entity) by the resource users. In the course of study on institutions, legitimacy of local level organizations within the larger fishery co-management program does not receive the attention necessary to develop a suitable explanation of why these institutions fail.

While legitimacy is a necessity at all levels of the fishery co-management institution, the local-level/resource-user interface is an essential level for legitimacy to exist because it is where management authority has been transferred from higher level government and where the majority of enforcement and management activities occur. Legitimacy, however, is often negatively affected at the local level by higher levels of government because central governments transfer insufficient judicial authority and financial resources to lower levels of government (Ribot 2002a,b; Ribot et al. 2006).

On Lake Victoria, East Africa a fishery co-management institution has been created, defined by the shared responsibility, authority, and resources between the central level governments and established local level organizations called Beach Management Units (BMU). Despite the creation of Lake Victoria's fishery co-management program (LVFCP),* illegal fishing continues on the lake (Mkumbo and Marshall et al. 2009; Ogwang' et al. 2009). The 2010 biennial, lake-wide survey indicates that, while fisher (resource extractors) populations and number of boats have decreased since 2008, use of illegal gear and methods continues on the Nile perch fishery (LVFO 2010a); I documented illegal fishing in over 50 percent of my study sites. Data shows that not all individual fishers around Lake Victoria comply, or are able to comply, with the management rules designed on Lake Victoria; data from this research indicates that illegal fishing by individuals is due to weakness at some BMUs due to mid-level government entities either interfering with, or not providing enough support to, the BMUs. This causes delegitimization of the BMUs, decreasing the ability of the BMUs to clearly and consistently enforce rules and conduct management activities.

To understand the notions of legitimacy within the LVFCP, the relationship between higher levels of government and community-run fishery organizations were examined by me. Higher levels of government include the ministries of fisheries, departments of fisheries (DOF), and other non-local enforcement agencies; local level organizations are the BMUs, and are community-run fisheries management organizations consisting of people engaged in fisheries-related activities. Using data collected on Lake Victoria during 2009–2010, this chapter describes two major factors that undermine the legitimacy of the BMUs: a lack of support from higher

*For ease of reference, the informal term LVFCP has been created and used throughout this paper. The LVFCP refers to the efforts of fishery co-management on Lake Victoria and includes central level government entities, fishers and local community organizations, all other fishery stakeholders, and the rules and roles created to change harvest behavior. The LVFCP is not a single institution, but a complex set of institutions and collaborations described on P. 179.

political authorities—in the form of unclear lines of authority—and, a lack of financial returns from the fishery and higher political authorities (a tax "black hole"). The lack of clear judicial authority leads to the delegitimization of BMUs when mid-level enforcement entities (police, army, DOF officers) interfere with BMU committee enforcement, creating inconsistent procedures, consequences, or unrecognized or confusing authority. The inability of fishermen to determine who is the authority leaves offenders in the system to continue illegally fishing and the act of allowing fishermen to continue fishing illegally (defined as using illegal-sized gear or practices or harvesting illegal-sized fish) undermines and delegitimizes the BMU's authority, thus rendering the BMU impotent. The lack of financial resources is a weakness in that: 1) BMUs lack the ability to conduct patrols, administer their functions, or conduct other business because of a lack of money, which 2) leaves offenders in the system to continue illegal fishing, thus 3) undermining BMU legitimacy and rendering the BMU impotent. Judicial weakness and financial insufficiency by higher level government, undermines the BMUs' authority over the fishers, thus rendering the BMU ineffective at implementing fishery management duties.

It was therefore suggested by me that insufficient and unclear power transfer by the central level government leads to the delegitimization of the BMUs. I argued that while the members of the BMUs have the *will* to administer fisheries management, they lack the *ability*, precisely because of insufficient and unclear judicial authority and insufficient financial support. I demonstrated that effective fishery co-management institutions require a strong positive relationship between the national level government and the local level. Within the LVFCP, relationships between those who are governed and those who govern are strengthened by legitimacy, or fishers' acceptance of the regulations, rules, and authority that governs the fishery; and legitimacy is created by the strong positive relationship between the national level government and the local level.

In the below sections legitimacy is defined and its conceptual basis given, its place within co-management, and its importance at the local level are detailed and characterized. Then the concepts of community participation and co-management and how each inform the LVFCP and the historical process and factors that led to its creation, including current theoretical notions of cooperation between central level government and local level resource users are described. Data is then presented and synthesized to determine how legitimacy of the BMUs is affected through the current relationship with higher levels of government on Lake Victoria and how these relationships affect the way fisheries management is conducted.

Methods

Studying legitimacy at the local level on Lake Victoria requires that a) numerous BMUs are visited to provide the ability to compare those BMUs that are "successful" to those that are "unsuccessful" and determine the factors which may cause each, and b) BMUs are visited in each of the three countries to determine differences between the relationship of each of the three countries central level governments and their respective BMUs. Because of the prominence and importance of the newly emerged Nile perch—the species that has most influenced the creation of the LVFCP[1]—study sites were selected based on fishing beaches on which the majority of fishing activity targets Nile perch. There are 1007 total BMUs (Muhoozi 2008) around Lake Victoria, with 318 BMUs in Uganda, 252 in Kenya, and 437 in Tanzania. To represent the population of BMUs around the lake for this study, a 10 percent sample by the total number of BMUs in each of the bordering countries was selected: 32 in Uganda, 35 in Kenya, and 44 in Tanzania, for a total of 111 BMUs.

Data were collected using semi-formal qualitative interviews and a combination of ethnographic research methods including quantitative participant surveys and site observations. Research began informally during multiple short visits to the three partner countries during 2005–2009, but the majority of data collection occurred during a structured six month period during 2009–2010 when the 111 BMUs were visited and interviews conducted.

Interview subjects at each BMU were selected under two categories: boat owners (those who make the decision on fishing practices within fishing boats), and BMU committee members—those who administer fishing enforcement and other BMU activities. An attempt to establish a random sample of fishing boat owners to be interviewed about legal and illegal fishing practices was made. The process of choosing boat owners, however, was dependent on the BMU committee chair's willingness of involvement with the study. It was often up to the chairperson to choose the participants. The number of boat owners, in every case except one,[2] was two members (Ug, 32 x 2 (+3) = 67; Ky 35 x 2 = 70; Tz 44 x 2 = 88) for a total of 225 boat

[1]The LVFCP also creates rules for other fisheries on Lake Victoria.

[2]At the first site visited in Uganda, as the methodology was being created, five boat owners were interviewed. All of their data is included in the analysis.

owner respondents. Additionally, two members of each BMU committee were interviewed[3] (Ug, 32 x 2 (–2) = 62; Ky 35 x 2 = 70; Tz 44 x 2 (–2) = 86) for a total of 218 Committee Member respondents, the chairperson and the secretary or treasurer. The total number of respondents representing the 111 BMUs is 443. The two sets of respondents' data—boat owners and BMU committee members—were collected both for comparative analysis as well as corroborative data depending on the questions being asked.

A structured interview-survey instrument was created based on the analysis of institutions, covering multiple variables. The survey instrument is primarily quantitative, with a short section of open-ended interview questions to gain richer qualitative data. The qualitative data were particularly useful in revealing important information about authority, funding, and legitimacy within the BMUs. Participants were asked: "What changes need to take place at your BMU for a healthy fishery?", "Mention the most important problems at your BMU", and "How would you fix these problems?" All qualitative data were entered into Nvivo computer software and coded for numerous variables that emerged from primary data analysis. Data from the qualitative data were used to inform this study.

Several analytic tools were used, including constant comparison analysis, text search and comparative matrix queries (Leech 2011). The qualitative data indicates that the delegitimization of the BMUs is occurring because of interference by higher level authorities, and word search and comparative matrix queries were conducted to determine the number of respondents stating these problems and the reasons why these problems occurred. Results from coding of the accumulated qualitative data establish clear patterns of financial and authoritative weaknesses emanating from higher level government.

Legitimacy in Institutions

Fishery management relies on the effective enforcement of regulations (Nielsen 2003). Legitimacy, in this chapter, is defined as the acceptance of the rules and regulations by fishers and the amount of loyalty acquired by management leaders which then gives them the authority required to govern effectively (Tyler 1990 as cited in Nielsen 2003). Legitimacy, as the power or authority at the interface of the enforcer (managers) and the enforced (resource harvester), is one of the most critical aspects of resource management institutions. Without legitimacy, users will have no reason

[3]Due to unforeseen circumstances, at two sites in Uganda and two sites in Tanzania committee members of the BMU refused to show up or were unable to meet. Each BMU is therefore represented, but for two of the BMUs in Tanzania and two of the BMUs in Uganda, only one member is represented.

to assume ramifications for circumventing the rules. Legitimacy has two important components: personal experiences by the resource users and procedural fairness (Nielsen 2003).

Personal experience, in the form of participation in the regulation-creating process, leads to resource users who are more likely to adjust and follow those regulations (Nielsen 2003). Acceptance of the rules is created through users' perceptions of legitimate rules and rules-making. Co-management programs often include numerous mechanisms for all stakeholder inclusion, as is the case on Lake Victoria. In particular, democratic decentralization allows resource users to participate in elections of local level organizations, and therefore engage in the management process of their natural resources. Personal experience and involvement by the resource users is essential to improving legitimacy through consent of particular power relationships and of fisheries regulations (Jentoft 1989; Ribot 2002a).

Procedural fairness includes transparent and consistent procedures (Nielsen 2003), most commonly concerning fair and consistent enforcement activities and punishment as perceived by the resource user. Procedural fairness is necessary so that breaking the rules results in administrative or judicial action. The system defines what is right and wrong, and consequences or rewards are observed (Nielsen 2003). If executed correctly, the management program is perceived as legitimate because authority through action has been executed by the relevant (legitimate) authority. The relationship between procedural fairness and personal experiences by the resource users is such that procedural fairness often determines the personal experiences of the resource users, and therefore procedural fairness can be adjusted to change the personal experiences, and thus the resource extraction behavior, of fishers. Procedural fairness, then, is used in determining legitimacy.

When attempting to change behavior in a natural resource management program, incentives to change behavior are essential. Most incentives in natural resource management involve punishment and enforcement. Punishment is the direct consequence of harvesting natural resources outside of established regulations, and enforcement is the actions conducted to catch rule-breakers. Procedural fairness is defined by the ability to enforce rules of illegal resource harvesting. Transparency and consistency is determined by the perceived judicial authority that the enforcement entity has and the financial resources available for that entity to execute the enforcement procedures authorized for them to administer. Procedural fairness, therefore, is the determining factor in the success of many co-management programs, and is influenced, in large part, by the strength of the central level government's willingness to provide resources and clear lines of authority to local level entities, thus allowing the local level organizations themselves to conduct transparent and consistent procedures.

Legitimacy and Authority

Without authority, local level governments will not be downwardly accountable—the obligation or responsibility to report or justify actions to higher level authorities—and will have no legitimacy to effectively represent users (Ribot 2002a). Without legitimacy, resource users will often not follow the rules and, in fact, will try and find ways around the rules. In this case, coercion becomes the only basis for local level involvement in natural resource management (Therkildsen 1992 and Ribot 2002a). Within a strong, legitimate system, actors are less likely to try and find ways around the rules or to cheat the system.

In this research, the notion of authority is described in two ways. First, as the legal (judicial) authority bestowed upon an entity or individual to administer punishment or reward upon others, and second, as the *perceived* authority that the entity or individual has over others. The first is provided through legal arrangements and can be implemented instantly and has been done so on Lake Victoria. The second requires the recognition of that authority by the people upon whom it is imposed.

Judicial authority is the power pertaining to the administration of deserved punishment or reward. Here, judicial authority is framed as the authority to administer deserved punishment where it has been legally granted to the lowest-level of fishery management organization that will implement patrols and enforcement—the action arm of fishery management. In the case of Lake Victoria, the BMUs are the lowest level political authority. Legitimacy of the BMUs requires that there are clear lines of authority between the resources users (those who can be punished) and the BMU committee (those who enforce the rules and punish offenders) allowing recognition of consistent and transparent authority.

The perceived authority of the BMUs by the fishers is of greater concern, because having the legal authority to punish resource users does not necessarily mean having the *ability* to use that authority. As will be shown below, although the BMUs have the legal authority to punish offenders vested in them by the central government, many of the fishers do not always recognize this authority because of a lack of legitimacy resulting from higher levels of government.

Legitimacy and Financial Resources

Sustaining the fishery resources on Lake Victoria largely depends on how the BMUs are funded. Without proper, sustainable financial resources, it is difficult or impossible to administer management activities, especially those activities which create legitimacy and allow rules to be enforced. Therefore, financial resources are a key component to legitimacy, especially at the local level where the administration of enforcement activities occurs (using personnel, boats, engines, fuel, and safety equipment).

Therkildsen (1993) takes the ability of local authorities to collect tax as an indication of the local authorities' legitimacy: "... if local authorities cannot mobilize local revenues it indicates a lack of legitimacy which, in turn, constrains their ability to be effectively involved in NRM [natural resource management] on a consensual basis." Moore (1997) makes a complementary argument that taxation engages the user with the acting authority, creating a basis on which to legitimately expect the authority to provide services (in Ribot 2002a). Similarly, allowing the local level organizations to collect taxes on their own, demonstrates that authority has been given by higher level government and trust to execute those activities exists.

On Lake Victoria, the primary source of funding for BMUs is collection of licensing, taxes, and fines at the local level through legal authority vested in the BMU committee. As will be shown below, BMU funding is not only dependent on the will of a higher-level government authority, but is also insufficient. This lack of funding prevents BMUs from conducting routine procedures, and the government's unwillingness to allow BMUs to collect it, compromises the legitimacy of the BMU.

Co-management: Local Level Fishery Management and Community Participation

During the early 1900s, Lake Victoria's fisheries were managed by central level government who exercised their authority and limited conveyance of discretion to individual fishers (Gaden et al. 2012). Weaknesses of state-centered policies resulted from limited intercourse with citizens and from coercing citizens' resource use through unpopular discipline measures, faulty designed management programs, inefficient implementation, and corruption (Agrawal and Gibson 2001; Bwathondi et al. 2001). On Lake Victoria in particular, this distrust of the central government by the resource users created a culture of fishing activity that contributed to the fisheries' exploitation at more than two times its sustainable level (Hecky 2003), devastating the resource and diminishing the livelihoods of those who depend on it most.

In part due to failures of centralized government approaches and the subsequent realization that community participation is essential for effective management (see Kaimowitz and Ribot 2002), decentralization—where central level authorities formally cede power to actors or organizations at lower political or administrative levels (Ribot et al. 2006)—has prevailed as a model for natural resource management in developing countries. Advocates for decentralization believe that in addition to improved natural resource management through greater efficiency and equity, community participation results in socially sustainable development through "ownership" of local decisions and projects (Ribot et al. 2006). The inclusion of local level resource users provides better knowledge of local needs; further, when granted appropriate powers, users, in the form of local level organizations, are "more likely to respond to local aspirations" (Ribot et al. 2006). Ribot et al. (2006) continue:

> The belief in greater responsiveness is based on the assumption that local authorities have better access to information about their constituents, and are more easily held accountable by local populations. Transfer of significant powers and 'downward accountability' of local authorities are thus central to this formula.

Democratic decentralization is based on local organizations being representative of and accountable to local user-populations and on these populations having a secure and autonomous domain of powers to make and implement meaningful decisions (Ribot 2002a). Successful decentralization with community participation, therefore, must include the transfer of sufficient and appropriate political and judicial powers, financial resources, accountable representation, (Ribot 2002b) and an understanding as to what extent fiscal, administrative, and political control is transferred to those actively managing the resource (Pomeroy and Berkes 1997; Schneider 2003; Larson and Soto 2008).

Co-management programs incorporate the concept that active citizen participation at all levels of fisheries governance can protect the diverse interests of those affected by environmental problems (Lemos and Agrawal 2006; Nunan 2006); that natural resources are a source of livelihood and income, and therefore are already being managed by the local people (self-organized) (Kaimowitz and Ribot 2002; Ostrom 2009); and, that rather than being an expense for central governments, natural resources become a major source of revenue when the appropriate property rights and management schemes are instituted (Kaimowitz and Ribot 2002) and transaction costs lowered at the planning and implementation phase because local level personnel provide resources for fisheries management (Sen and Nielsen 1996).

Citizen participation, however, is only one component of a co-management relationship. Co-management requires a strong and clear commitment by the government to share power, authority, and other resources with local governments and delegated local level fishery organizations (Pomeroy and Berkes 1997). Indeed, co-management is defined, in part, as the "sharing of responsibility and authority between the government and the community of local fishers" in the management of the fishery (Pomeroy and Williams 1994 as cited in Pomeroy and Berkes 1997). The relationship in a co-management program ranges from government control of management, to local level control (Jentoft 2003; Ogwang' et al. 2009); true co-management, therefore, rests between these two points (Jentoft 2003; Ogwang' et al. 2009) with a sharing of power, financial and judicial resources, and responsibility between the central government and communities. Co-management can also be a mechanism for economic development that directs benefits to the community from successful management of the resource (Pinkerton 1989). A co-management institution, therefore, can create a balance between government resources and support to ensure local level administration of fishery management activities.

The objectives of the LVFCP are to enable "all stakeholders to work together in a collaborative and cooperative partnership for sustainable fisheries management and improved livelihoods of fishing communities" (Ogwang' et al. 2009). The LVFCP seeks and incorporates the views of the "fishing communities on the design and implementation of management interventions [to increase] the legitimacy of any actions that are subsequently taken" (Ogwang' et al. 2009).

A key goal of the LVFCP is to improve resource users' compliance of fishing rules that limit resource harvest to a sustainable level. Ogwang' et al. (2009) state the principles:

a) Democracy, transparency, accountability and sustainability in systems, processes and objectives;
b) Power sharing between government, communities and other stakeholders;
c) Partnership between government, fisheries communities and other stakeholder groups; and,
d) Subsidiarity, with management authority being delegated to the lowest possible organization.

Each of the principles addresses the relationship between the lowest-level of authority—in the case of Lake Victoria, the BMUs—and higher levels of government. Each of the principles of co-management addresses the nature of power sharing and thus, the components critical to creating legitimacy through consistent and transparent procedures.

Challenges of Decentralization and Co-Management Programs

Despite the positive notions of inclusiveness and cooperation inherent in co-management, weaknesses within the approach have been observed and often lead to failure of natural resources management. It is widely documented how central governments insufficiently or inappropriately transfer discretionary power, downward accountable representative authority, and resources to the local level (Pomeroy and Berkes 1997; Ribot et al. 2006; Larson and Soto 2008; Lawrence and Watkins 2012). The reasons for the inappropriate or insufficient transfer of power, authority, and resources include flawed program design, resistance from a variety of stakeholders, poor implementation of well-designed programs, lack of trust of local level entities by the government, and inadequate attention to the establishment of administrative and policy structures, all of which erode the effectiveness of such programs (Pomeroy and Berkes 1997; Ribot et al. 2006; Larson and Soto 2008; Lawrence and Watkins 2012).

Ribot et al. (2006) state that central government priorities often differ from the goals of decentralization and rarely provide downward accountability to individuals who have effective authority at the local level. "Instead, central governments often devolve obligations rather than meaningful powers, without adequate resources, to traditional authorities that are not downwardly accountable, to private entities, or to upwardly accountable institutions" (Larson and Soto 2008).

The legitimacy of actors is an integral component of successful fishery co-management. Without the proper resources to implement decisions and create authority, "then discretionary powers have not been effectively transferred" (Ribot et al. 2006). Successful co-management depends on the political commitment of the central government of each partner country, including the appropriate legislation and adequate technical and financial resources (Bwathondi et al. 2001; Lawrence and Watkins 2012). "Seldom, however, is adequate attention given to the establishment of administrative and policy structures that define the legal status, rights and authorities essential for the effective performance of local organizations" (Pomeroy and Berkes 1997). Central-level government shortcomings in the LVFCP reduce the legitimacy of the BMUs, undermining the intent of their existence.

Lake Victoria's Fishery and Co-Management Institution

Lake Victoria is the second largest lake in the world, by surface area, and the largest in Africa; it supports millions of people through vibrant and diverse fisheries. Lake Victoria is also a common-property resource—making it difficult to exclude users. The lake is shared by three political states—Uganda, Kenya, and Tanzania where ownership is 45, 6, 49 percent, respectively (Njiru et al. 2008). The lake has three important commercial fish species, the non-native Nile perch (*Lates niloticus*) and Nile tilapia (*Oreochromis niloticus*), and the indigenous sardine-like fish dagaa (Ug), omena (Kn), or mukene (Tz) (*Rastrineobola argentea*). Fishing pressure on

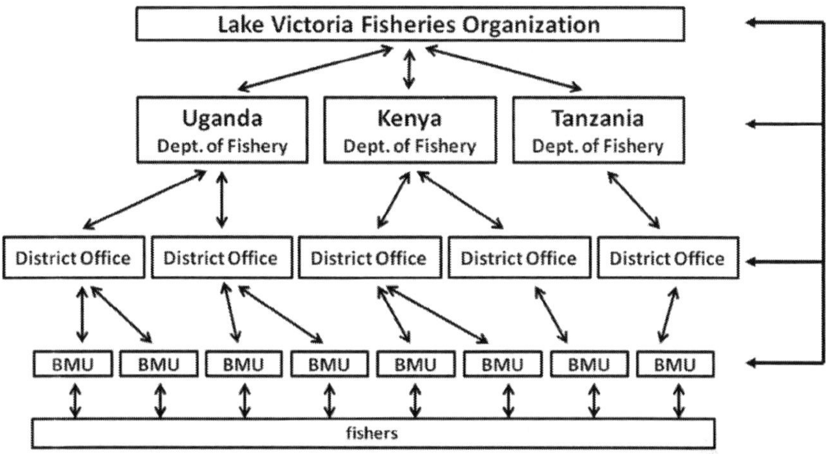

Figure 9.1 A Simplified map of institutional linkages of the LVFCP. The Lake Victoria Fisheries Organization coordinates fishery policies between the three partner states and ensures harmonization of these policies between higher level government and the community level. Note that police and village-level authorities are not included. Adapted from LVFO (2005).

Lake Victoria is high, with more than 1,400 landing sites, 194,000 fishers, and 64,000 fishing craft targeting Nile perch, Nile tilapia, and *Rastrineobola* (LVFO 2010b data (unpublished)). Fishery management of the Nile perch, Nile tilapia, and *Rastrineobola*, is conducted through the formal fishery co-management institution, defined by the relationship between the national level governments and the BMUs, and is meant to include the input of local level resource users in the management of the fishery.

The LVFCP was created to address the challenges of over-harvest of resources faced by fishery management on common property resources, using the current theoretical notions of decentralization, co-management, community participation, and socio-political institutions. These design elements are intended to motivate and reinforce legal fishing behavior and therefore a sustainable fishery and sustainable development (Bwathondi et al. 2001).

The LVFCP consists of organizations, agencies, sets of rules, and regulations to guide fisher's harvest behavior on the lake. Inclusive are the national fishery research institutes of each partner state, regional department of fisheries officers, police, army, and numerous local level actors including fish traders, net makers and menders, fishers, and fish sellers. All of these actors have a role, whether formal or informal, in the management of the fishery and fall under the guidance of the Lake Victoria Fisheries Organization (LVFO), a regional organization responsible for managing fishery resources under the East African Community (EAC). The

LVFO is charged with developing conservation and management measures for sustainable fisheries and coordinating fishery management through fostering cooperation among the partner states and between the partner states and BMUs (LVFO 2011a).

The Lake Victoria Fisheries Organization: An Interjurisdictional Organization

The Lake Victoria Fisheries Organization (LVFO) was formed through a convention signed in 1994 between the three EAC partner states which share Lake Victoria: Uganda, Kenya, and Tanzania. The LVFO is the fisheries management arm of the EAC and coordinates co-management of the fishery by facilitating the partnership between each of the central government's department of fisheries (DOFs) and their respective local level fishing communities (BMUs). The LVFO secretariat is located in Jinja, Uganda, is the executive organization of the Lake Victoria Fisheries Organization, headed by an Executive Secretary and staff who ensure that the policies and decisions created by the Council of Ministers is adopted (LVFO 2011b). Under fisheries policies, the DOFs are mandated to promote, guide and support the fisheries sector in their respective countries. The LVFO facilitates the partnerships between the partner countries through a structure designed to ensure equitable policies, harmonized rules, and a sustainable fisheries co-management program with the local level. The LVFO is comprised of the following organs and organizations: the Council of Ministers—made up of ministers of the partner states' ministries responsible for fisheries—is the supreme body of the LVFO and adopts measures for management and conservation of fisheries resources (LVFO 2011b) and is informed by the following committees: the Policy Steering Committee—consisting of permanent secretaries responsible for fisheries—submit management recommendations to the Council of Ministers; the Executive Committee is comprised of directors of fisheries departments and fisheries research institutes in each country to review management and scientific activities and report to the Policy Steering Committee; the Fisheries Management and Scientific Committees are made up, respectively, of the heads of fisheries departments and heads of fisheries research institutions, both of which report to the Executive Committee to provide advice on management and conservation measures and to convey the appropriate scientific fisheries information needed to inform policy; and other sub-committees and working groups (LVFO 2001; LVFO 2011b).

The objective of the LVFO as an organization, is to harmonize fishery rules and regulations between each of the partner states around the lake, and thus address the challenges that common-property, multi-jurisdictional fisheries present. The fishing regulations created within the LVFCP outline the use of fishing gear and practices, including restrictions on fish size, fishing gear, and fishing methods. These rules are directed at the local user—fishermen—who make the ultimate decision of how they will harvest the resource. The LVFO Secretariat and its partners are charged with teaching the fishing communities (BMU committees and members of BMUs around the lake) about rules and how those rules create sustainable fisheries and community development, the functions and responsibilities of the BMUs and of the fishers, and other functions and responsibilities of the fishing community.

Central Level Government

Within the LVFCP, central level government is defined as the partner states' central authority: the Ministry of Agriculture, Animal Industry and Fisheries (Uganda), Ministry of Fisheries Development (Kenya), and Ministry of Livestock and Fisheries Development (Tanzania) which oversee their respective Department of Fisheries Resources (Ug) or Fisheries Department (Kn, Tz) referred to as department of fisheries (DOF). Each central government around Lake Victoria—under trust law—is obligated to hold natural resources in trust for the people of their countries (United Republic of Tanzania 1977; Ugandan Ministry of Agriculture Animal Industry and Fisheries 2004; Republic of Kenya 2008). Legal ownership of fisheries resources is vested in the states' central government authority as trustee and, therefore, the state is obliged to manage the resource in the interest of the beneficiaries (the people who depend on these resources) (Naluwairo 2005).

The guidelines for fishery management on Lake Victoria state that each partner state's department of fisheries promote, support, and guide the BMUs ability to function, especially under "circumstances where local capacity alone will not be sufficient to safeguard the livelihoods of people depending on fisheries resources" (Ugandan Ministry of Agriculture Animal Industry and Fisheries 2004).

The central level government, primarily through the respective Minister's participation on the LVFO's Council of Ministers, and informed by policy and scientific committees, creates fishery policy. The department of fisheries of each partner country informs lower-level (regional/district) fisheries offices of those policies, where policies will be closer to implementation.

Mid-level Entities

Mid-level political entities are numerous and their charges and authority are either poorly defined or not included under the guidelines of the LVFCP. Mid-level entities include the police, marine police, army, regional and district fisheries offices and officers, and village committees. The literature, including BMU guidelines and other guiding materials for the LVFCP, are not clear to the level or function of enforcement authorities such as the police, marine police, army, or of village committees. Confusion, therefore often exists as to who has the authority in fishery matters. Reports of police and army intervention, both helpful and hampering fishery management, have been reported at BMUs around the lake. It is unclear whether these enforcement entities are required to report to, or collaborate with, local BMUs.

The authority of the police, in any country on Lake Victoria, states that they can arrest any offenders of any law, but many deal solely with duties assigned to them (not necessarily fisheries related). With regard to fisheries in East African Countries, it is the marine police (water-based police force with boats) that work closely with the departments of fisheries on fishery enforcement. Notwithstanding the LVFCP the DOF officers work with both district police commanders and officers and with BMU personnel. Additionally, there are sometimes police posts in or near the fishing villages to facilitate enforcement activities at the village and BMU level, though these police posts are not fishery related unless requested to serve in such capacities. The LVFO advises that BMU committee members, when conducting patrols, to include police personnel because fishers conducting illegal activities may be dangerous (Kirema-Mukasa 2011; unpublished data, Lawrence); BMU committee members also indicate that the use of police during BMU-conducted patrols is necessary because illegal fishers resist arrest with violence or weapons. A Kenyan BMU committee member summed up the dangers by simply stating "If we patrol without the police, the fishermen can kill [us]" (K23-CM-03, QU3). A Tanzanian fisherman explained that those who conduct patrols are not safe, that "some illegal fishers are armed and can kill, so people fear going out at night" (T07-BO-02, QU3).

Contrary to the need for police assistance is the problem with the lack of consistent line of authority and the amount of corruption that takes place at the mid-level, where police and DOF officers are the source of controversy and corruption. Enforcement of fishing rules at the community level is more successful when partnerships between law enforcement officials—police and a DOF officer—occurs. When police and DOF officers, however, usurp BMU authority, either through contrary actions (releasing offenders, taking bribes, or not coordinating with BMU leadership) then the BMUs' authority is delegitimized. Unclear lines of authority, however, have reduced the BMUs ability to conduct enforcement activities and will be discussed below.

The Local Level Fishery Management: Beach Management Units

Beach Management Units are the foundation of the LVFCP. They are legally empowered, local level organizations that are responsible for the majority of fishing management activities, such as patrolling, enforcement of rules, and—to a lesser degree—punishment, on Lake Victoria. Each BMU locates around pre-existing, local, fish-landings (DFR 2003). The BMUs consist of all users engaged in fisheries-related activities, including "boat owners, fishing crew members, fish mongers, artisanal fish processors, local gear makers and repairers, boat builders, fishing input suppliers, and industrial fish processors' agents" (DFR 2003). Following notions of democratic decentralization, the fishing community must elect a BMU Committee of 9–15 members to be drawn from their population and be inclusive of all stakeholder groups and including at least three women (LVFO 2005; LVFO 2011c). The BMU committee operates within pre-determined geographic boundaries, and share policy development, rules enforcement, and administrative duties with each national government's fisheries department. The BMUs: create and enforce their own local bylaws—governed by LVFO (lake-wide) guidelines—for sustainable fisheries management; serve as resource-data collection points for better fisheries management and monitoring; and, increase local users' capacity to manage their finances (Ebong et al. 2004).

Fishery management, such as enforcement of rules, patrolling for illegal gear and activities, and tax collection, are executed by the BMU committees and in "collaboration with the relevant authorities" (LVFO 2005). It is at the local level, at the interface between the BMU committee (regulators) and the fishers, that fisheries extraction is regulated and tax collection is conducted.

On enforcement of fisheries rules, BMUs major charge is specifically to "ensure compliance with local and national regulations ... formulate and enforce community bylaws at the local level; [and] monitor fishing activities within their localities" (Ugandan Ministry of Agriculture Animal Industry and Fisheries 2004; LVFO 2005). BMUs are therefore considered the action arm of fisheries management on the lake, designating the national departments of fisheries to supporting roles. BMU committee members have legal authority to conduct enforcement and the authority to arrest offenders, but BMU committees cannot prosecute offenders. BMU committee members can transfer arrested offenders and their illegal gear or fish to fisheries officers or staff or the police. In Kenya and Tanzania, the authorized officers (fisheries officer, staff, or police) then need a court injunction to ban or dispose of the illegal gear or fish. In Uganda, BMU committee members are allowed to destroy (usually through burning) illegal gear or fish by an authorized officer of the BMU committee (Kirema-Mukasa 2011).

The central government also bestows the authority of tax collection to the local level. Guidelines for BMUs, however, do not automatically allow BMU committees to collect taxes, instead, BMUs in Tanzania must apply to be the tender for tax collection; otherwise other entities (not the BMU committee) are in charge of tax collection. The purpose of tax and fine collection is so that fishery management operations, such as patrols for illegal activity, consistently function. When the central government bestows the authority to the BMUs to collect taxes there is a level of trust that is demonstrated the local level to conduct fisheries management operations; when this activity is usurped unclear authority of the BMU is often observed.

Summary of Lake Victoria Fishery Co-Management

Each partner state on Lake Victoria has a BMU statute defining the powers of the BMU committee as explained above. At the local level of governance, legitimacy is critical to their success.

The LVFCP was developed to give authority and ownership to the local level users while reducing costs of enforcement to the government. Each BMU on Lake Victoria is bound by the same set of fishing rules and guidelines, informed by all partners. The LVFO guides the BMUs in implementing the fisheries management policies. The LVFO ensures that each BMU has the appropriate committee and structure outlined in the BMU guidelines, and that each of the BMUs has the appropriate capacity (through trainings) to organize and follow established rules and guidelines for sustainable fishery management. The mechanisms for allowing BMU committees to collect taxes, however, are insufficient because mid-level governmental entities interfere with what would normally be a clear and consistent tax collection operation by the BMU committee, thus disallowing the BMU committees the ability to administer clear and consistent operations, rendering them illegitimate by the fishers. Of the 111 BMUs, respondents from 106 BMUs (95 percent) indicated that a lack of funding of the BMU and lack of equipment were the biggest problems to successful operation of the BMU and successful management of the fishery. To be effective, money and equipment is necessary at the BMU level for successful management activities. The problems with the LVFCP appear to emanate from the mid-level entities and the lack of clear authority that these entities possess over BMU committees, and the expropriation of tax collection duties (either directly or through tenders) that these entities possess. The BMU guidelines state that "[t]he BMU Committee shall in the performance of its functions consult and cooperate with local governments, relevant agencies of central government and lead agencies", it goes on to say that "for effective participation of various stakeholders in fisheries co-management, each party must understand its own role, that of others and the relationship and links between them" (LVFO 2005). Guidelines provide no further clarification of who local level entities should cooperate with.

Results

The following data reveal two components that cause delegitimization of BMUs as discussed above: unclear judicial authority and financial insufficiency. In 2005, participants of the Regional Workshop on Legal and Operational Framework for BMUs in East Africa noted that fisheries officers viewed the newly established BMUs as "enemies" and that one of the challenges to the success of BMUs was the failure to welcome them by government and local leaders (ILEG 2005). Perceptions of BMUs being illegitimate organizations by higher level government persisted in 2009–2010, the time of this study. The fishing communities are often viewed as being ill equipped and unable to manage the fishery effectively (pers. comm. with fishery officers; regional news paper articles). Indicative of research conducted by Ribot et al. (2006) perceptions if legitimacy like those previously stated, often lead to distrust of, and insufficient granting of authority to, local level authority by the central level government. Over 55 percent of respondents in this study indicated that illegal fishing takes place at their BMU and frame surveys (LVFO 2010b data (unpublished)), conducted by the fisheries departments and LVFO, indicate high levels of illegal gear (a quarter of all nets) are present around the lake.

Authority Real and Perceived

Lake Victoria's fishery is high-value and there is incentive for other entities, such as the police, army, or other political authorities, to gain from the fishing industry. Three attributes of conflict between BMU committee authority and higher-level authority exist, thus reducing the BMU's legitimacy: (1) there are no clear or consistent lines of authority by police or army force; (2) there exists an inconsistent or corrupt relationship between BMU and fisheries officers, and (3) there exist inconsistent lines of authority between village committee and BMU (Tanzania). These attributes delegitimize the BMU in relation to the fishermen, reducing effective fishery management activities.

No clear and consistent lines: None of the policies guiding the LVFCP address the participation of police or army personnel (see example of organizational structures of fisheries management, Uganda, Tanzania, Kenya in Bwathondi et al. 2001). Inconsistent lines of authority in the LVFCP have created a failure to execute enforcement at the local level. The BMUs were given authority, but intervention by police and army personnel confuse and delegitimize that authority. Where confusion of authority exists between higher-level authorities, punishment imposed by these entities sets precedent over lower-level authorities such as BMU committee members. Such actions, both procedurally and authoritatively inconsistent, decrease the authority and legitimacy of BMU committee members, as a BMUs punishment will likely never be stricter than that of a higher-authority. A BMU committee member in Tanzania described one instance where the BMU committee arrested illegal fishers and took them to the police, but "the police took a bribe and released them. We now don't have any punishment at our BMU because we have been disempowered by the police" (T17-CM-01; Qu3). In Kenya, a boat owner said that "police are not entitled to prosecute fisheries offenses and thus most of them take bribes" (K23-BO-01, Qu3). Others agree, stating that the police are corrupt and take bribes instead of prosecution and that the police do not consult the BMUs (K30-CM-01, Qu2). In Uganda too, a committee member expressed that the police take bribes and "suck the morale from the BMU" (U04-CM-01, Qu3).

Inconsistent and corrupt relationships: District fisheries officers are also able to arrest offenders (or as discussed above, BMUs have to consult with DOF to institute punishment). There are numerous instances from BMU committee interviews where corruption from higher political levels, mostly department of fisheries officers, was cited. One of the most common complaints among committee members was that actions taken from higher political levels delegitimized their efforts to control illegal fishing, collect taxes, or have any influence over fishing activities in general. "We are disempowered... last year in March, we caught 28 illegal gears and took them to fisheries officer, the day after, the gear was being sold to other fishermen by the department" (T17-CM-01; Qu3). Another committee member said that "it is discouraging to arrest illegal fishers because they are released by fisheries officers after taking bribes; illegal fishers are not prosecuted due to corruption at the department of fisheries" (T11-CM-03, Qu3).

Inconsistent lines of authority: In addition to unclear lines of authority between higher political levels and BMUs, there was confusion as to who has authority at the local level, specifically in Tanzania where village committees exist to govern local level social and political issues. Often, newly established BMU committees and village authorities conflict in Tanzania, while an incident of each were recorded in Uganda and Kenya. Village committees appear politically stronger in Tanzania than Uganda and Kenya most likely due to legislative powers granted them by the Local Government (District) Authorities Act of 1982 and other acts that outline a well designed, formal system for community involvement (Lawyers' Environmental Action Team 2012). This formal recognition gives village committees authority through electing local members as officials who then comprise a village assembly (forming the village committee) which govern local level social and political affairs including planning, finance, economic, social services, security and some resource management (e.g., forest protection and water resources) (Lawyers' Environmental Action Team 2012; Commonwealth Local Government Forum 2012a). Though Uganda has similar efforts to involve the village level, it appears that the village level is "consulted" by the next higher political level, the parish or ward, and not a strong political entity itself (see Ministry of Local Government 2003; Commonwealth Local Government Forum 2012b), and Kenya has no constitutional provisions for local government and no recognition of village level political entities outlined in its structure of government (see Commonwealth Local Government Forum 2012c). Formal recognition of the local community, by the central level government, may be the difference between increased conflict between village committees and BMU committees in Tanzania but not in Uganda and Kenya.

In numerous interviews in Tanzania, both boat owners and committee members stated that village authority overrides BMU committee authority. Boat owners and committee members also stated that efforts by BMU committee members to punish fishers for illegal activity were often negated by village chiefs or elders, especially those village authorities who were involved in the fishery (e.g., owned illegal nets). Village committees also often own lake-front land, which is used for BMU activities. Land ownership becomes contentious, taxes or rent must be paid by the BMU committee to the village committee, or the village committee supersedes tax collection and collects fishery-related monies themselves (T34-CM-02, Qu3). Other respondents—both committee members and boat owners—simply explained that "the village government doesn't want to cooperate with the BMU committee on [the village government's] premises" (T32-CM-01, Qu3), there is "conflict between local [village] leadership and [the] BMU" (T17-BO-01, Qu3), and "the village committee criticizes the BMU and disempowers the BMU" (T40-CM-01, Qu3).

Specific lines of authority need to be clearly determined so that BMU authority is not undermined, whether accidentally or on purpose. Whether it is during patrol or judiciary action, any activity or situation that reduces the legitimacy of the BMU undermines the purpose of the LVFCP. In the case of Lake Victoria, legitimacy is being challenged because the government often undermines BMU efforts. Ninety three percent of all respondents in the study believe that the Lake Victoria fishery is in trouble, and of those respondents, 76 percent indicated that the reason was because of illegal fishing activities. Yet, participants in the fishery are willing and want to have an effective management program—they see the benefits of legal fishing and sustainability, but are unable to control illegal fishing because of the circumstances (e.g., delegitimization and a lack of resources from the government). Indeed, 75 percent of fishermen indicated that to protect the fishery, stronger enforcement measures, including increasing law enforcement activities, patrols, punishment, and preventing the sale of illegal gear, are necessary. Of the total number of respondents, 92 percent believe that the fishing rules (preventing illegal fishing gear and practices) are important, and 95 percent believe that the fishery can be protected with the established fishing rules, if they are enforced and followed.

Financial Systems, Real and Perceived

The BMUs receive income generated from the fishery at the local level (e.g., licensing, fish tax, fines). This source of revenue is, theoretically, an indicator of legitimacy, demonstrating that the government trusts the BMUs to conduct their operations, and in that the BMU can be trusted with their own revenue generation, legitimacy is created by those that expect services—the fishers. Though the BMUs on Lake Victoria have been designed to fund their own operations through the collection of taxes and fines, they are largely underfunded and perceived to be so by both BMU committee members and boat owners. Three factors result in, and support, this view 1) with poor fish catches, there are less taxes to collect, 2) taxes that *are* collected, are done so by higher-level government representatives, or 3) taxes are collected by the BMUs who then are required to hand over all taxes to higher level government, and then receive in return a percentage of those taxes.

Government, at district, regional, and sub-county levels, are identified by the central governments of Uganda, Kenya and Tanzania as the level of government "collecting the revenues necessary to ensure sustainable local government, and to reinvest in fisheries development." (Ugandan Ministry of Agriculture Animal Industry and Fisheries 2004). In the interview-surveys, however, there was no mention of financial assistance from higher levels of government, other than that money which is collected by a government representative (tender) at the BMU and circulated back to the BMU as a percentage of total collection (money from the local level redistributed by higher level government). While this tender-collected money is "contribut[ing] to the revenue required by *local government* to fuel the development and administration processes" (Ugandan Ministry of Agriculture Animal Industry and Fisheries 2004) (emphasis added) 61 percent of the respondents in this study (34 percent BMU committee members), stated that the money that is obtained by the BMU committees is insufficient to run BMU operations, including enforcement activities. Furthermore, the portion of the taxes returned to the BMU is often only 25 percent of the total revenue collected at the beach-level. In effect, this creates a negative cycle: The BMU committee is supposed to, in part, collect taxes and fines at the beach level; from the monies collected, the BMU conducts patrols with purchased patrol equipment (fuel, boat, engine) in which the BMU committee can further collect fines. However, a large percentage (or all, in some cases) of the money the tender collects is not returned, thus reducing the efficacy of the BMU's ability to patrol and collect fines.

The Department of Fisheries, which is understaffed and has limited equipment, is funded, in part, by taxes and fines collected at the BMUs. Corruption by higher level political entities (e.g., the Department of Fisheries) was reported by 21 percent of the BMU committee members and boat owners. Out of 25 regions studied, only four regions reported no corruption from higher level government. Five regions (one in Tanzania, and two in Kenya and Uganda each) reported non-existent fishery management activities, such as patrols, meetings, or any other management activity, and high levels of bribery and corruption, by officers and politicians.

A BMU committee member at a Tanzanian beach said that "the [district] government tenders a person to collect taxes from fisheries activities and he [the tender] keeps half and gives half to the government. The BMU used to collect taxes and were planting trees and educating orphans, but that has stopped" (T04-CM-02; Qu3). At another BMU, the committee chairman again explained that there is no collection of taxes because the "district council does not return the percentage that it is supposed to return". This member explained that the BMU is required to, at the minimum every month, collect and provide to the government, 177,000 TSh to receive 5 percent back. If the BMU provides the government with less than 177,000 TSh they will not receive their share, however, in this case, the committee member also said that "whether we make the 177,000 or not" we never get our percentage (T15-CM-01; Qu3). Another committee member concluded: "No taxes are being collected for fishing activity by the BMU [the councillor (through an agent) does the collection]. Our BMUs only source of revenue is a monthly fee for members. The police and department of fisheries officers conduct patrols to collect bribes; the government has let us down" (T15-CM-01; Qu3). The fact that the central government has a hand in collecting and dispersing the fish catch taxes and fines reduces the local level ownership of the fishery, and lessens the notion of responsibility by the fishing communities. Intended or not, political delegitimization of the BMUs occurs.

The Balance between Unsupported BMUs and Over-supported BMUs

Legitimacy lies in the balance, in part, of fisher's perceptions of the BMUs relationship with the higher level government. Authority of appropriate actors within the LVFCP must be carefully identified so that clear lines of authority are present and consistent punishment from those same actors exists. When BMUs are under-supported through either lack of financial resources or unclear lines of authority their legitimacy is undermined. Undermined authority results from two reasons: first, procedural fairness is absent if a BMU committee cannot conduct critical operational functions from lack of financial resources. Second, authority is absent when higher level government (e.g., police or DOF officers) inappropriately interfere with BMU authority, confusing the line of authority.

Discussion

Though legal authority has been granted to the BMUs, this authority is often not recognized by the resource users (fishermen) for reasons emanating from weaknesses observed in the LVFCP's higher level authority, where: a) weak judicial action is taken by higher level officers as seen when fisheries officers undermine BMU efforts by releasing offenders with little or no punishment; b) corruption by higher level officers and others in authority, demonstrated when fisheries officers are corrupt and take bribes or sell confiscated illegal gear; c) unclear lines of authority, where police, army, department of fisheries officials, or village authority confuse fishery management actions reserved for BMU committee members; and, d) circumventing tax and fine action of the BMU by higher level authority tenders who collect money at the local level, undermining the BMU committee's authority and reducing their capacity to administer consistent enforcement procedures, this leads to insufficient financial resources to the BMUs which creates the inability to conduct consistent enforcement procedures.

The governments of Lake Victoria have provided some judicial authority to the BMUs, but the actions of higher political authority have subsequently taken that same authority away. Higher level government actions, and unclear lines of authority, have undermined the authority of the BMUs; no matter what stated authority the BMU committee members have, that authority is often not realized. Revisiting Ribot et al. (2006), who explain that "effective decentralization requires the construction of accountable institutions at all levels of government and a secure domain of autonomous decision making at the local level".

The central argument of legitimacy comes from the answer of a Tanzanian BMU committee member who stated that "the district council should give the revenue collection activities of the beach to the BMU [committee]" and that fisheries management "patrol[s] should be conducted jointly with the department [DOF] since they have better equipment" (T11-CM-03; Qu3). For the local level to have legitimacy, there appears the need for balance between autonomy—granted to the BMUs by higher level government as a demonstration that the government values and trusts the BMUs—and government support—a demonstration that the government is in a partnership with the BMUs to effectively manage the fishery.

The result of giving BMUs power of tax and revenue collection is the perceived legitimacy of the BMUs by the fishers, because it is also perceived that the government realizes the BMU as a legitimate entity. Further, with joint patrols between the BMU and higher level authorities (e.g., police or DOF officers), demonstrates the legitimacy of the BMU through the partnership, as the DOF provides protection of the BMU committee and allows effective execution of rules when rule-breakers are caught.

Disallowing the BMU committee to collect taxes also demonstrates a lack of faith in the BMU committee. The inability to collect their own taxes and fines means that the BMU does not have enough financial resources to then conduct patrols and enforcement activities; therefore, the BMU committee is unable to conduct consistent enforcement activities, reducing the committee's legitimacy by the fishers. And, there is no promise that a more effective BMU is delegated more money if a tender is in place, as the amount of money collected by higher level authorities is often less than that which is returned to the BMUs. This is a case of a lack of transparency in the tax system. Allowing BMUs to collect their own taxes creates incentive for the BMUs to patrol and enforce more effectively because taxes and fines would be collected as the BMUs become more effective.

In the case of Lake Victoria, lake-level revenue collection and control must be in the hands of the BMUs and patrols, arrests, and judicial activities, executed by the BMU must be supported by higher level government in the form of appropriate punishment through laws, security of BMU patrols, and equipment and fuel in partnership with the appropriate higher political authority.

Conclusion

Where government was once viewed as an obstacle for its unpopular, coercive, top-down fishery management programs, today the *lack* of appropriate government engagement is the obstacle. Insufficient financial resources and unclear judicial authority have delegitimized the local level management arm of co-management on Lake Victoria.

Legitimacy is a critical component of co-management. When resource users do not believe that consistent and transparent socio-political natural resource management institutions exist, poor resource harvest behavior may not change and the resource will be compromised.

A transparent system will reduce incidences of corruption at the mid-political levels. A clear tax structure must be created, with local level taxes being collected at the local level and higher level taxes being collected at the appropriate levels (e.g., taxation of exporters through national-level exportation tax structures). Additionally, clear responsibility structure must be created—if BMU committees are in charge of fishery management then it must be made clear what authority others (e.g., police, army) have, so that their involvement does not undermine BMU efforts, but rather, compliments it.

BMUs need to control revenue collection, but patrols should be conducted in conjunction with the appropriate government entities (police or DOF officers). Both actions demonstrate that government is invested in protecting the fishery, provides legitimacy and support to the BMUs, and allows effective execution of rules when it comes to arresting offenders.

The government must share in the responsibility as well as the management of the resource. The fishery resource should be viewed as a tool for the development of the local level users around the lake and can serve, if managed appropriately, as a sustainable source for economic growth. Weakness at the top will surely create weakness at the local level. With the loss of legitimacy comes the failure at all levels to successfully manage the resource.

While this chapter investigates factors that weaken the co-management institution on Lake Victoria, it must be acknowledged that co-management indeed, has had an overall positive benefit to Lake Victoria's fishing communities, national level development, and to some extent, reducing over-harvest. Results, observed by the author and well documented by others (see Odongkara et al. 2009; Ogwang' et al. 2009), demonstrate that the Lake Victoria fishery contributes to each of the partner country's GDP; income and revenue generation at local and regional levels; provides employment and produces foreign exchange (Odongkara et al. 2009); increases fisheries infrastructure and development; alleviates poverty among fishing community members; includes the disenfranchised (Ogwang' et al. 2009), and has indeed, prevented the collapse of Lake Victoria's Nile perch fishery (Lawrence, in progress).

References

Abdul-Aziz, B. 2009. Fishermen, govt differ on Lake Victoria's fish stocks. The Guardian. Mwanza, Tanzania. April 21.

Agrawal, A. and C.C. Gibson. Introduction: the role of community in natural resource conservation. pp. 1–31. *In:* A. Agrawal and C.C. Gibson [eds.]. 2001. Communities and the Environment: Ethnicity, Gender, and the State in Community-based Conservation. Rutgers University Press, New Jersey.

Butunyi, C. 2009. Sh20Bn fish export market faces threat. Daily Nation. Nairobi, Kenya. April 14.

Bwathondi, P.O.J., R. Ogutu-Ohwayo and J. Ogari. 2001. Lake Victoria Fisheries Management Plan (LVFMP). I. G. Cowx and K. Crean. Lake Victoria Fisheries Research Project (LVFRP)—Phase II, LVFRP Technical Document No. 16. LVFRP/TECH/01/16: Socio-economic Data Working Group of the Lake Victoria Fisheries Research Project, Jinja. pp. 34–55.

Commonwealth Local Government Forum. 2012a. The local government system in Tanzania. Retrieved February 1, 2012, from http://www.clgf.org.uk/userfiles/1/files/Tanzania%20 local%20government%20profile%202011-12.pdf.

Commonwealth Local Government Forum. 2012b. The local government system in Uganda. p. 232. Retrieved February 1, 2012, from http://www.clgf.org.uk/userfiles/1/files/ Uganda%20local%20government%20profile%202011-12.pdf.

Commonwealth Local Government Forum. 2012c. The local government system in Kenya. pp. 93–94. Retrieved February 1, 2012, from http://www.clgf.org.uk/userfiles/1/files/ Kenya%20local%20government%20profile%202011-12.pdf.

DFR. 2003. Guidelines for Beach Management Units in Uganda. U. Department of Fisheries Resources, Ministry of Agriculture Animal Industry and Fisheries. p. 17.

Ebong, I., M. Lwanga and J. Scullion. 2004. Beach Management Units and Integrated Lake Management. International Workshop on Community Participation in Fisheries Management on Lake Victoria, Kisumu, Kenya.

Gaden, M., O.C. Mkumbo, T. Lawrence and C. Goddard. Top-down and bottom-up approaches in the management of the Laurentian Great Lakes and Lake Victoria fisheries: A comparison of two shared water bodies. pp. 364–390. *In:* V.I. Grover and G. Krantzberg [eds.]. 2012. Great Lakes Great Responsibilities: Lessons in Participatory Governance. Science Publishers, Enfield, New Hampshire.

Hecky, R. 2003. What does the future hold for the Great Lakes of Africa? African Journal of Aquatic Science 28(1): iii–vi.

ILEG. 2005. Regional workshop on legal and operational framework for beach management units in East Africa. Institute for Law and Environmental Governance and the Environmental Law Institute. Imperial Hotel, Kisumu, Kenya. p. 17.

Jentoft, S. 1989. Delegating government responsibility to fishermen's organizations. Marine Policy 13(2): 137–154.

Jentoft, S. Co-mangement—the way forward. pp. 1–14. *In:* D.C. Wilson, J.R. Nielsen and P. Degnbol [eds.]. 2003. The Fisheries Co-management Experience: Accomplishments, challenges and prospects. Kluwer Academic Publishers, Guelph. E.K. Jul-Larsen, J. Overa, R.R. Nielsen, J. Zwieten and P.A.M. Can. 2003. Management, co-management or no management? Major dilemmas in southern African freshwater fisheries. F.F.T. Paper. FAO, Rome.

Kaimowitz, D. and J.C. Ribot. 2002. Services and infrastructure versus natural resource management: building a base for democratic decentralization. Conference on Decentralization and the Environment, Bellagio, Italy, World Resources Institute. Washington, DC.

Kasozi, E. 2008. Panic as Nile perch stocks dwindle. Daily Monitor. Kampala, Uganda.

Kasozi, J. 2011. Uganda: New Technology to Harvest Mukene Unveiled. The New Vision. Kampala, Uganda, AllAfrica.com. 5 January, 2011.

Kirema-Mukasa, C.T. 2011. Personal email cooreespondence on legal authority of BMUs. T. Lawrence. Ann Arbor, MI.

Larson, A.M. and F. Soto. 2008. Decentralization of Natural Resource Governance Regimes. Annual Review of Environment and Resources 33: 213–239.

Lawrence, T. and C. Watkins. 2012. It takes more than a village: The challenges of co-management in Uganda's fishery and forestry sectors. International Journal of Sustainable Development & World Ecology 19(2): 144–154.

Lawrence, T. (in progress). Preventing the collapse of a fishery: Successes of co-management on Lake Victoria, East Africa. School of Natural Resources and Environment. Ann Arbor, Michigan, University of Michigan. Ph.D.—Resource Policy and Behavior.

Lawyers' Environmental Action Team. 2012. District Local Government Authorities. Retrieved February 1, 2012.

Leech, N.L., A.J. Onwuegbuzi et al. 2011. Writing publishable mixed research articles: Guidelines for emerging scholars in the health sciences and beyond. International Journal of Multiple Research Approaches 5: 7–24.

Lemos, M.C. and A. Agrawal. 2006. Environmental Governance. Annu. Rev. Environ. Resour. 31: 297–325.

LVFO. 2001. The convention for the establishment of the Lake Victoria Fisheries Organization, Lake Victoria Fisheries Organization in conjunction with the International Union for Conservation of Nature.

LVFO. 2005. Guidelines for Beach Management Units (BMUs) on Lake Victoria. V. Ogwang', M. Medard, E. Kilosa, J.I. Nyeko and A. Bakunda (eds.). Lake Victoria Fisheries Organization, East African Community. Jinja, Uganda. pp. 12–15.

LVFO. 2010a. Regional status report on Lake Victoria bi-ennial frame surveys between 2000 and 2010. Kenya, Tanzania and Uganda. Lake Victoria Fisheries Organization and East African Community, Jinja, Uganda.

LVFO. (2010b data (unpublished)). Table: Comparison of Lake Victoria Fisheries Frame Survey 2000, 2002, 2004, 2006, 2008, and 2010 results for the three partner states, Kenya, Tanzania, and Uganda. Lake Victoria Fisheries Organization, Jinja, Uganda,

LVFO. 2011a. About the Lake Victoria Fisheries Organization. Retrieved December 4, 2011, from http://www.lvfo.org/index.php?option=com_content&view=article&id=50&Itemid=56.

LVFO. 2011b. Organs and Institutions of the LVFO. Retrieved December 4, 2011, from http://www.lvfo.org/index.php?option=com_content&view=article&id=52&Itemid=58.

LVFO. 2011c. Beach Management Units: building co-management in East Africa." Retrieved December 7, 2011, from http://www.lvfo.org/index.php?option=com_content&view=article&id=53&Itemid=59.

Ministry of Local Government, R. o. U. 2003. Harmonsied participatory planning guide for parishes/wards. M. o. L. Government. Entebbe, Uganda, Republic of Uganda.

Mkumbo, O.C., B. Marshall L. Muhoozi, T. Munyaho, P. Nzungi and B. Msuku. 2009. The commercial fisheries of Lake Victoria: How do we manage the Nile perch stocks? 139th Meeting of the American Fisheries Society. American Fisheries Society. Nashville, Tennessee.

Moore, M. [ed.]. 1997. Death Without Taxes: Democracy, State Capacity, and Aid Dependence in the Fourth World. Draft. (Later published *In:* G. White and M. Robinson (eds.). 1998. Towards a Democratic Development State. Oxford University Press, Oxford).

Muhoozi, L. 2008. Development Plan for Implementation under FMP 2: The Lake Victoria Fisheries Management Plan: 2009–2014. Regional Stakeholders Workshop. Colline Hotel, Mukono, Uganda.

Naluwairo, R. 2005. A review of the national fisheries policy and the proposed fisheries legislation. Promoting common property rights in fisheries management in Uganda, Advocates Coalition for Development and Environment (ACODE): ACODE Policy Briefing Paper No. 8.

Nielsen, J.R. 2003. An Analytical Framework for Studying: Compliance and Legitimacy in Fisheries Management. Marine Policy 27: 425–432.

Njiru, M., J. Kazungu, C.C. Ngugi, J. Gichuki and L. Muhoozi. 2008. An overview of the current status of Lake Victoria fishery: Opportunities, challenges and management strategies. Lakes & Reservoirs: Research and Management 13(1): 1–12.

Nunan, F. 2006. Empowerment and Institutions: Managing Fisheries in Uganda. World Development 34(7): 1316–1332.

Odongkara, K.O., R.O. Abila and J. Luomba. 2009. The contribution of Lake Victoria fisheries to national economies. African Journal of Tropical Hydrobiology and Fisheries 12: 47–51.

Ogwang', V.O., J.I. Nyeko and R. Mbilinyi. 2009. Implementing co-management of Lake Victoria's fisheries: Achievements and challenges. African Journal of Tropical Hydrobiology and Fisheries 12: 52–58.

Ostrom, E. 1999. Self-governance and forest resources. Occasional Paper No. 20. Center for International Forestry Research. Jakarta, Indonesia.

Ostrom, E. 2009. A general framework for analyzing sustainability of social-ecological systems. Science 325: 419–422.

Pinkerton, E. Introduction: attaining better fisheries management through co-management—prospects, problems, and propositions. pp. 3–34. *In:* E. Pinkerton [ed.]. 1989. Co-operative management of local fisheries: new directions for improved management & community development. University of British Columbia Press, Vancouver, BC.

Pomeroy, R.S. and W.J. Williams. 1994. Fisheries co-management and small-scale fisheries: A policy brief. ICLARM. Manila, Philippines.

Pomeroy, R.S. and F. Berkes. 1997. Two to Tango: The Role of Government in Fisheries Co-Management. Marine Policy 21(5): 465–480.

Republic of Kenya. 2008. National Oceans and Fisheries Policy 2008, Ministry of Fisheries Development. p. 25.

Ribot, J.C. 2002a. African decentralization: Local actors, powers and accountability. Democracy, Governance and Human Rights. Paper No. 8. UNRSD and IDRC, Geneva. p. 43.

Ribot, J.C. 2002b. Democratic Decentralization of Natural Resources: Institutionalizing Popular Participation. World Resources Institute, Washington D.C.

Ribot, J.C., A. Agrawal and A. Larson. 2006. Recentralizing While Decentralizing: How National Governments Reappropriate Forest Resources. World Development 34(11): 1864–1886.
Schneider, A. 2003. Decentralization: conceptualization and measurement. Studies in Comparative International Development. 38(3): 32–56.
Sen, S. and J.R. Nielsen. 1996. Fisheries Co-Management: A Comparative Analysis. Marine Policy 20(5): 405–418.
Therkildsen, O. 1993. Legitimacy, local governments and natural resource management in Sub-Saharan Africa. p. 76. *In*: H.S. Marcussen [ed.]. Institutional Issues in Natural Resources Management: With Special Reference to Arid and Semi-Arid Areas in Africa. Occasional Paper No. 9. International Development Studies, Roskilde University, Denmark.
Tyler, T.R. 1990. Why people obey the law. Yale University Press. New Haven.
Ugandan Ministry of Agriculture Animal Industry and Fisheries. 2004. The National Fisheries Policy. U. Ministry of Agriculture Animal Industry and Fisheries, Department of Fisheries Resources. pp. 16–18.
United Republic of Tanzania. 1977. The Constitution of the United Republic of Tanzania. N. Government. Government of Tanzania, Duty to safeguard public property, Act No. 15 of 1984, Article 6.

10

Barriers and Opportunities to Adaptive Co-Management of the Tisza River Basin in Hungary

Béla Borsos,[1,*] *Jan Sendzimir,*[2] *Peter Balogh*[3] and *László Koncsos*[4]

Introduction

Rivers and other surface water courses are indispensable for the welfare of human societies. They offer a number of ecosystem services essential for life itself as well as for meeting the needs of higher, organized civilizations. Large rivers are complex natural systems consisting of both living and non-living components. Human societies have striven for millennia to modify rivers and better exploit their resources, mainly in the form of water for drinking, sanitation or irrigation. Starting several centuries ago, the industrial revolution ushered in an era of explosive expansion

[1]Institute of Geography, Pécs University of Sciences, Pécs, Hungary; Mailing address: H-1016 Budapest, Tigris u 45/b.
Email: dioliget@t-online.hu
[2]International Institute of Applied Systems Analysis, Laxenburg, Austria.
[3]Alliance for the Living Tisza, Nagykörü, Hungary.
[4]Department of Sanitary and Environmental Engineering, Budapest University of Technology and Economy, Budapest.
*Corresponding author

of technological power that completely reshaped these natural systems. Large rivers globally have become so heavily modified (Gleick 2003) that their natural cycles and functions are mostly dominated by human infrastructure. This infrastructure also isolated society from water, shaping social perception of high water flows radically through centuries of living fearfully behind dykes.

For the first time in human history, technical river management has changed the human relationship with nature. Previously, human enterprise and social hierarchy developed following the possibilities provided by technology within the boundary conditions set by the natural environment. Now technology is being used to custom-tailor the natural environment to human needs and whims—triggered by social and economic pressures (Borsos 2000).

Historically, a distinction was made between high water and floods. In Hungarian, different terms distinguish natural (*"áradás"*) from harmful (*"árvíz"*) flooding of a river. While natural high water was recognized as a benefit, fertilizing and replenishing land along the river floodplain, flooding has increasingly gained a negative reputation from damages amplified by artificial hydromorphological alterations (Koncsos 2011). Dykes may restrain high waters up to a limit. But when they burst, infrastructure and lives are at much greater risk. Larger, faster, more powerful flood waves then flush the landscape with much greater impact and less time to react. During the last few centuries, flooding is no longer seen as an opportunity but as a disaster. It is time to ask: is the flooding of a river a disaster or an opportunity?

Current water and land use practices in the Tisza valley originate from an industrial river management paradigm originally intended to create more land for human agricultural use and habitation. This paradigm ignored the capacity of high waters to replenish floodplains and store water in landscapes prone to drought. It was founded on a narrow view that any water on the floodplain was too much (Sendzimir and Flachner 2007). At a time when urban expansion and war provisioning demanded more flood-intolerant grains, floods and marshland were recast as detrimental to agricultural production potential of fertile soil. However, the ultimate impacts of the hydromorphological alterations carried out in order to overcome "too much water" impaired ecosystem functions without increasing flood security. Floods continue to rise in height and impact, despite continued expensive infrastructure enlargement and repair. The customary techno-fix solution of building dams no longer seems to be a feasible option for water professionals, considering the tremendous social resistance against the Danube barrage system in Hungary some 20 yr ago (Borsos 1991). Such options no longer can be justified as an economic stimulus as the Tisza valley is considered to be an economically backward area characterized by high unemployment rates and low income levels, prone to flood risk, habitual waterlogging and systematic drought at the same time (Sendzimir et al. 2007).

Short Overview: Conventional River Management

A number of rising trends justify the revision of the currently prevailing concept of flood control: ever more frequent recurrences of extreme weather events, rising flood crest levels, the emergence of novel land use and nature conservation approaches, strengthening ecological and risk based attitudes, and the increasing recognition that statistical methods for flood control have strong limitations (Somlyódi 2002). Hardly more than a 100 yr after completion, the conventional system fails to adequately manage repeated incidents of flooding, droughts and excess surface water.[1] Theoretically, there are two inherently different ways out of this urgent dilemma: a) continue to apply the conventional water construction logic and build various technical structures to mitigate the risks, or b) to exploit the natural processes and dynamics of the river, letting them shape the landscape and then adapting to the changes. The latter approach would mean a completely new strategy to flood control and river management. Three conditions mandate change to a new management strategy:

- "the reserves of the existing operational system are depleted and the system is not able to fulfill its function any more;
- new and unforeseen threats emerge;
- the set of values which provided the basis for the old strategy is changed or becomes obsolete" (Koncsos 2011).

We review the historical development of river management policies in the Tisza basin to show how the emergence of these conditions justifies a transition from a conventional to a more experimental river management regime.

Historical Roots of Conventional Management

Historically, flood control strategies were shaped by socio-economic needs. However, actual and substantial action was always taken only when a serious disaster occurred. Up to the Modern Age and the Ottoman conquest, river management in Hungary seemed to work well. In the Medieval era, famine was quite rare compared to other parts of Europe (Zimányi 1976) and even in the 16th century one of the most significant export commodities,

[1]The term "excess surface water" is used in this context to designate stagnating water on arable land.

beef, was produced in great numbers along the Tisza valley (Benda 1983). The major flood control works in the 19th century were triggered by quite different reasons: road and railway construction, prosperity of the corn trade and the interests of land speculators. These pressures to enlarge grain fields were abetted by the need for grain to feed horse cavalry and standing armies, as well as a shift of ownership rights from peasantry to landlords and of economy from subsistence farming to cash crop production.

Conventional Wisdom and Past Actions

Large-scale river engineering works on the Tisza—called the Vásárhelyi Plan—in the second half of the 19th century aimed to promote grain transport by reducing the length of the river by shortcutting meandering bends and to promote grain production by draining the floodplain with canals and separating them from the river channel with earthen embankments—dykes—which prevented the water from entering the large areas previously inundated periodically. As a result, the river gradient was increased and the travel time of the water shortened. Later on it turned out that water trapped on the flood plain has to be drained artificially: a draining canal system was still needed as an auxiliary measure.

The hydrodynamic processes triggered by these alterations resulted in siltation of the floodway between the dykes, incision of the low stage river bed, draining of the floodplain of groundwater in times of low water and water stagnation on open fields in times of high water, intensive rainfall and/or snow melt. These factors—reinforced by other interdependent changes in the basin upstream such as the increasing amount of paved surfaces, reduced vegetation cover and strong water erosion—gave rise to ever growing flood crests. The habitual reaction was to raise the height of the dykes (Fig. 10.1). Over time dykes grew by additional layers as onion-like structures which reached the limits of their structural strength by the end of the 20th century. Further elevation of the dykes would increase the risk of dyke failure when hydrostatic pressure burst through the resistance of the underlying earthen material. The inherent faults of the system have been analyzed extensively elsewhere (Borsos et al. 2010). Additionally, it was also recognized that the mathematical models used to predict design flood levels were flawed as they could only make forecasts on past experience but are unable to take into account expected—or unexpected—future processes (Koncsos et al. 2000). One of these newly recognized unexpected and unpredictable factors is the local impact of climate change, which will definitely make historical data obsolete (Nováky 2000). Another factor that increases uncertainty is unpredictability of practices and policies, especially those affecting forestry and agriculture, in the upstream Tisza basin, which

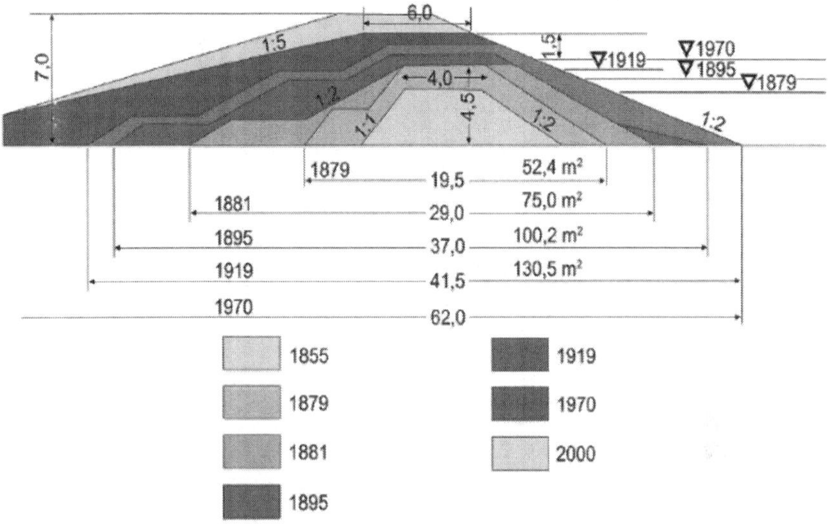

Figure 10.1 Rising of the Tisza dykes over time (Schweitzer 2001).

belongs to the national territory of other countries. One of them, Ukraine, is not even a Member State of the EU, and hence, is not completely beyond the control of the Hungarian water administration.

Thus, all three triggers of a new strategy are present: the current flood control works have reached their structural limits, unforeseen events from climate change or neighboring nations seem increasingly likely, and the perception of risk, threats, safety and prevention have changed. Also, new considerations such as sustainable development, ecological approaches, nature conservation and environmental protection all play a role in defining the development trajectory of modern society. This constellation of factors mandates that society explore new strategies for river management.

Alternative Solutions

Despite the confinement of the natural (active) floodplain to 5–10 percent of its former area, the geomorphology of the Tisza valley overall did not change much compared to the pre-engineered period before the industrial revolution. Figure 10.2 below shows a section of the Hungarian reach of the river on a schematic diagram indicating the lower elevations of the former floodplain and the high banks, which can still be clearly distinguished by the naked eye. The difference in elevation between the parts formerly inundated regularly by the river and the parts considered at low risk of floods is more than a meter. Infrastructure still follows more or less the former distinction, and most settlements have been and are still being built on high banks,

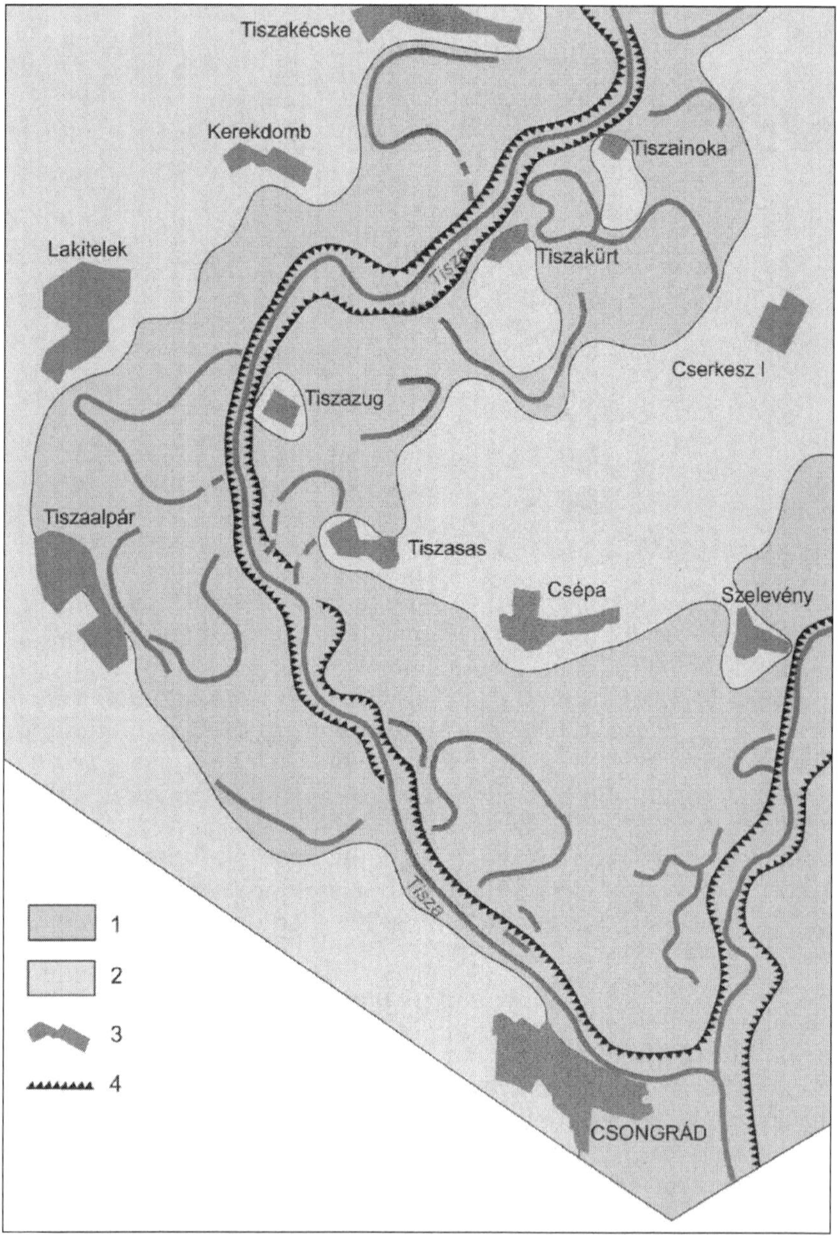

Figure 10.2 Geomorphology of the Tiszakécske-Csongrád river section. Legend: 1—low flood plain 2—high flood plain 3—settlements 4—embankments (Schweitzer 2009).

relatively safe from floods. Another important observation in Fig. 10.2 is that former river branches—now cut off from the main river channel and the floodway by dykes—can be clearly distinguished on the floodplain as deeper depressions on the flat land. Needless to say, water stagnation, the upwelling of water that lingers on the surface, is most severe on these parts (Schweitzer 2009) and causes millions of euros in lost agricultural production each year (Sendzimir and Flachner 2007).

Near the turning of the Millennium the search for river management alternatives began in earnest, prompted by a challenge and an opportunity. The challenge was the prospect of significantly higher costs from dealing with the structural problems associated with further increases of the dyke elevation and mass. The opportunity presented itself as an easily accessible potential to lower flood risk by diverting flood waters into deep floodplain areas (polders) with relatively cheap and easy modification of the water regime and the surrounding human infrastructure. The inevitable consequences of the challenge forced a much more serious consideration of the opportunities, once a series of flood crises in 1997, 1999 and 2000 undermined the conventional regime's reputation for security.

Two distinct and to some extent related design schemes emerged as alternatives to the conventional management strategy. Both alternatives are designed to overcome the flood problem by discharging surplus water at times of floods onto arable land on the former natural floodplain, taking advantage of the lower lying deep floodplain areas. One is a mild extension of conventional thinking that frames such challenges as problems of water quantity. Thus, diverting floodwater onto the floodplain simply relieves the challenge of too much water in the river channel. This alternative was conceived by the water administration professionals and is called the 'further improvement' or 'advancement' of the Vásárhelyi Plan. The design scheme is abbreviated from its Hungarian name as VTT (Váradi et al. 2003).

However, a number of authors (Balogh 2002; Bokartisz 2002; Molnár 2002; Borsos et al. 2010) see the first alternative as a failure to sustainably address the flood problem and to fully realize the potential services derived from floodplains functionally reconnected to the river channel. They mostly maintain that the expensive and energy/resource intensive solutions the VTT offers would not be appropriate for the long term. The research conducted in other alternatives has crystallized recently into another different concept that was dubbed by the promoters "integrated land development" or in shorthand ILD. The latter is a truly co-managing approach, because it demands the cooperation of managers and actors in both the water and agricultural sectors, with implications for industry, business and tourism. Below, we provide a short overview of the two competing concepts with their pros and cons.

Concept 1: Advancement of the Vásárhelyi Plan (VTT)

The new approach taken by the Hungarian water administration was to set up a *flood reduction and mitigation system consisting of engineering structures and reservoirs dedicated to the controlled discharge and eventual return of floods into the river as necessary (or transferring surpluses onto areas with water shortages).*[2]

The name of the new program was selected in remembrance of the original river engineering concept envisaged by the short-lived but influential water engineer Pál Vásárhelyi in the 19th century. The selection of the name tells a tale. The problems that emerged from the original Vásárhelyi plan have ongoing effects on the dynamics and functions of the Tisza valley up to this date. The first and main result of the Vásárhelyi plan—which was implemented poorly and incompletely even within the theoretical framework of the technocratic approach of the time—was that it has become a structural trap: the structures themselves increased the risk of floods.

The "advanced" VTT proudly boasts of a shift in attitudes, even a paradigm shift, because this is the first time that the conventional regime entertained the idea of allowing water to leave the river channel to be stored on the floodplain. Previously, water on the floodplain was a sign of catastrophe. And indeed, the focus was shifted from defense (and a military-like organization) to regulation, control and prevention, with the potential for a long-term sustainable solution based on ecological considerations. The most important change in the approach was the idea of retaining water on, not draining it from, the deeper floodplains immediately adjacent to the active river channel. The technical solution, however, did not really reflect the much quoted paradigm shift. The published program still gives all priority to one key objective: to enhance flood security in the Tisza valley. It ignores as irrelevant the implementation of an integrated land management and development practice, which challenge the supremacy of their underlying paradigm of risk, danger and exposure to floods.

There are three major segments in the advanced VTT program, of which only one is a relatively new idea. The other two are business-as-usual methods:

1. Improvement of the water carrying capacity in the high water stage river channel on the Tisza (in other words: clear the floodway)

[2]Act No LXVII of 2004 on the Improved Vásárhelyi Plan

2. Construction of a flood detention emergency reservoir system with a total storage capacity of 1.5 billion m^3 (10–12 reservoirs)
3. Development of existing flood control works and structures.

Later the concept was broadened to involve infrastructure development in the settlements concerned (excess water drainage in the built up areas, sewage system, waste water treatment plants, replacement and construction of by-roads, bicycle paths) and implementation of husbandry methods driven by natural conditions (landscape management). Yet the actual solutions treat only the symptoms. For instance, as part of the flood control measures, the bank protection works at the bottleneck in Kisar were reinforced, but nothing was implemented to overcome the bottleneck itself.

Cost cuts and funding difficulties resulted in shrinking and fragmenting the original concept and its implementation. First, it seems very unlikely that all originally-planned 11 reservoirs will be constructed. As a consequence, the reservoirs, which are thought to make up a complex system by the time they are all completed, are now prone to weird features that will reduce their functionality as described below. The first structure to be inaugurated was the Cigánd reservoir in the Bodrogköz region in 2008. The second structure, the Tiszaroff reservoir, was completed in 2009 with the expectation that it will be used once in every 30 or 40 yr. The "emergency reservoir" was obviously not designed in accordance with the ILD concept, much rather in the spirit of "Man conquering Nature". Additionally, both structures are poorly designed, in that they do not follow the natural depressions of the floodplain. Construction is underway at the third site.

The VTT has structural flaws that mainly result from a conservative engineering approach and the same institutional and legal barriers that are analyzed later on in this chapter. Poor design features are reflected by the following aspects:

- the functional features of the landscape such as depressions, former brooklets and high banks are not exploited for their potential to augment ecosystem services;
- the river floodway already lies higher than the floodplain itself because of decades of siltation;
- the design is subject to rigid artificial and legal constraints. For instance, a 60-meter protective zone must be kept as a buffer between public roads and areas that flood. As a result, a section of the new dykes had to be built on the areas best used for flood storage: the lowest elevations;
- inlet structures are oversized and with high threshold elevations, so they can only be opened at very high water stages;

- reservoirs (polders) are rigidly defined as structures dedicated for flood control only and hence, barriers to multi-purpose uses, such as fisheries, forestry or agricultural production;
- the system's operation is a self-contradictory paradox: during the flood of 2010, water was discharged into the Tiszaroff reservoir to skim the peak flows, but regional water authorities upstream pumped excess surface water into the river at the same time to drain open fields from stagnating water.

Concept 2: Integrated Land Development

As opposed to the VTT concept, the ILD avoids the considerable resource use and investment costs of massive river defense engineering by suggesting a structurally-different, complex and—to some extent—revolutionary solution. The theoretical background is based on spatial planning and development (Borsos and Bánvölgyi 2003). Key features of the concept include:

- The conventional wisdom of flood control engineering strives to drain floods from the plains or to discharge them into costly man made reservoirs in designated areas. This sustains the myth of water as solely a source of danger in the public mind, increasing the risk of compensation payments. It also drastically shrinks biodiversity at both the patch and landscape levels, reducing a complex mosaic of habitats to a monoculture and eliminating most ecosystem services. ILD policy, instead, retains water on the original natural floodplain in relatively large areas that are naturally shaped to absorb the water surpluses from high flood stages in the river. Control structures are to be opened in the dykes in several places, wherever relief information suggests that it can be done relatively easily and effectively, as described below.
- As opposed to the VTT, whose structural design builds flood waters higher before they spill onto the floodplain with devastating force, the ILD starts drawing off water from the river channel as soon as possible, when the water level in the main channel has risen above the middle stage level. This gives ample time to fill up the lower lying areas "gently", i.e., backing slowly up the slope often in the direction opposite to river flow (Borsos and Bánvölgyi 2008). In this way, water managers can foment the pulsing of water on the floodplain landscape that drives both productivity and biodiversity (Junk et al. 1989). This avoids sudden flash flooding on fertile land and keeps water cover on the fields very low.
- The ILD's "revolutionary" structure anticipates a dramatic shift in land use patterns and social perception of water. It is not the river which is customized and "trained" to meet perceived social and economic

needs but land use patterns and methods are custom-tailored to adapt to more natural river dynamics.

- Infrastructure is designed such that its operation allows managers to adapt to pulsing river dynamics. Current river management structures are static engineering works that are difficult to adapt to changing conditions because they simply resist and guide water into the river channel until water volumes exceed their capacity. Such excesses usually are too powerful to be productively exploited.
- Therefore, the goal of ILD is not to adapt to any given situation but to set up a system with sufficient internal structural flexibility to be adapted to future challenges such as drought or deteriorated water regimes, extreme weather events and demanding temperature conditions expected as local manifestations of global climate change.

ILD: A Comparative Analysis of Barriers and Opportunities

The International Arena

Throughout the world, development's implementation is regulated by legal provisions. The European Union tries to reinforce its integration by legalistic means, and is renowned for regulatory legislation on almost any issue, though their implementation often depends on what each member state negotiates with Brussels. This is so with regard to water management and agriculture, yet the two pictures are quite different. While the Water Framework Directive (WFD)[3] provides for wider opportunities to broaden river management concepts, the lack of integration across different pieces of legislation prevents consistent thinking: there are distinct pieces of legislation for flood, drought, and various aspects of water quality (Borsos et al. 2010). Agriculture, on the other hand, is regulated mainly in terms of production and markets; no special directive or regulation exists for land use practices or production technology, though options are possible through agri-environmental schemes. Rural development has become an issue only very recently, with repeated efforts to reform the Common Agricultural Policy (CAP) over the past decade (Kajner 2010).

Financing of development schemes—whether at national or community level—also have some serious difficulties. Tendering and aid systems are bureaucratic constructions, reacting to perceived needs only in a retroactive and follow up manner. The lobbying power of various business interests is reinforced by a positive feedback loop that bends policy considerations to their objectives which then further enhances their economic and political

[3]Directive 2000/60/EC of the European Parliament and of the Council of 23 October 2000 establishing a framework for Community action in the field of water policy *Official Journal L 327 , 22/12/2000 P. 0001 - 0073.*

reach. Funding schemes are narrowly focused on regional and sectoral economic development in ways that disregard natural boundaries like river basins. In the case of the Tisza river basin, for instance, Ukraine and Serbia, which are not Member States of the EU, are ineligible for conventional funding schemes of the European Community, so their development is haphazardly linked with other river basin nations in the EU. The tendering system is usually tailored to specific programs, projects and investments. Nobody cares how a completed project will earn revenues for maintenance and operation in the long term or how a piece of infrastructure will fit into the overall picture. In the case of the Cigánd reservoir, funding was only available for investment to build rather than operate the reservoir. Why invest in adaptively managing a structure, when it is unlikely to produce income for maintenance costs? Also, creation of jobs is promoted, but little emphasis is given to financing that helps stabilize the livelihoods for non-mobile residents. As opposed to the mega-farms promoted under the VTT, the ILD concept foresees that adaptively managing floodplains will exploit floods to produce products (fish, fruit, nuts, hunting, tourism) that sustain basin residents (Sendzimir and Flachner 2007).

Barriers

Structural, institutional, legal, financial, mental, and psychological barriers and value preferences currently work both individually and in concert to prevent the ILD concept's implementation. For instance, the prevailing conventional paradigm is reflected in institutional arrangements that are reinforced by the legal background and the financing mechanisms. General elections in 2010 rearranged the institutional landscape in Hungary fundamentally, and many sectors that impact ILD were united into one national super-ministry (Ministry of Rural Development). However, this potential to integrate cross-sectoral policies has not been realized, and the track record of the new ministry and the new government continues to sustain a path dependence on conventional management and development strategies.

The organizational chart of the ministry, the personal changes, concepts and practical experiences vary in terms of flexibility and novelty of approaches. For instance, the ILD concept assumes that river management operates as an integrated whole that coordinates land use practices, spatial development, infrastructure management, traffic, transport and other sectors of society. Instead, within the water sector alone, water administration in Hungary is simply a national reflection of how fragmented water management is internationally. There are distinct organizations in place that separately address flood control, excess surface water, municipal water supply, sewage works and agricultural land use.

These various organizations and authorities act in an uncoordinated manner that sometimes results in ironic and paradoxical situations. For example, sewage is discharged into inland drainage canals, or stagnating water is pumped into the river when there is a risk of flood.

Agriculture in Europe is driven almost entirely by subsidies. Community aid schemes and national top-ups both tend to disregard geographic differences, soil structure, ecological conditions or social aspects. They merely focus on a largely agribusiness-dominated market and the organizations thereof. As with any other bureaucratic intervention, they can only react to events that already have taken place and anticipate future events poorly (see for instance Lee 2011). Buzzwords directing such measures include "competition", "market", "competitiveness", "sales" and in the case of the current Hungarian agricultural policy: "national private ownership". Rural development is caught in the trap of luring foreign capital and investments, enchanted by the promise of "jobs". At the same time, creating jobs is impaired by counterproductive regulations (Kajner 2010).

Barriers to Flexible Land Use Practices

Research (Ripka 2005) has revealed that many of the obstacles hindering the feasibility of the ILD approach on the large scale go back to how property rights were shaped by privatization in the 1990s. For historical reasons the current ownership pattern of arable land parcels in Hungary has a very complex and unreasonably shaped structure. After lengthy and heated political debates, land ownership was restored after decades of Communist rule by organizing a so-called compensation process. Eligible former owners were entitled to marketable vouchers issued by the government in return for their confiscated former assets. The vouchers then could be used to acquire various properties and assets. As the system was the outcome of a political skirmish, the ill-designed and poorly implemented compensation scheme was not a political success at all, and its aftermath is still a burdensome heritage. Mainly, because the very concept of restitution was flawed and poorly designed and available land was scarce, the vouchers were very difficult to use properly but easy to manipulate and speculate with. The non-transparent and suspicious way this compensation process was implemented eroded the kind of trust needed for landowners to unite to forge a development path that integrates ecology and economy across boundaries like property lines, as demanded by concepts like the ILD.

In addition, social and economic factors contributed to a disproportionate structure of proprietorship and tenancy. Land classification is still based on an outdated and severely biased valuation system that is supposed to reflect the fertility of the soil and its suitability for crop production (Ripka 2005). This vague and poorly established, rigid categorization is called the

Golden Crown system and dates back to 1875, originally used to formally establish a cadastre or land registry. In the new restitution process, vouchers were distributed on the basis of this hypothetical valuation system. Many of the original beneficiaries were elderly people not interested in assuming the responsibility of farming and husbandry. Others received very few compensation vouchers, not sufficient to buy any reasonable amount of land. Prices were volatile and unrealistic, whilst land for compensation was not available in sufficient amounts. The restitution process resulted in an extremely fragmented and ill-arranged structure of cultivated plots and parcels. New owners were typically given the land acquired not in a single parcel but in the form of many, scattered plots frequently at some distance from each other.

As a result, the land privatization process created some larger parcels where several proprietors "possessed" a single piece of land without knowing which physical part of the administrative parcel belongs to them. Such a situation is called the *undivided common*. At the beginning of the 21st century, 1.5 million hectares were still listed as plots of undivided common. In addition, the number of absentee owners, who are not at all or only remotely related to agriculture, has grown substantially. For political reasons and fear from foreign encroachment, land ownership was restricted to natural persons and to the government. Business entities and foreign individuals are still banned from owning land. As a result, land use and title became separated, business organizations and large farmers mostly farm the land in the form of lease. Besides being ill-proportioned, current land use patterns are therefore strongly bipolar in character: poorly capitalized domestic farm structure dominates a large number of fragmented, tiny farms, while a mere 1.6 percent of all the agricultural enterprises with considerable means farm 75 percent of all arable land (Ripka 2005). Consolidation of the fragmented structure has not taken place since the restitution process and is one of the largest barriers to any reasonable transaction concerning land ownership or tenure.

Unlike in many other countries, a rigid delineation of various types of farming in Hungary hinders adapting land use and/or farming practices to the specific local context. Each piece of land is classified in one of these types and farmers are obliged to farm in accordance with that type with good stewardship. The *type of cultivation* is a concept used for the identification of the mode of utilization of agricultural land carried out systematically over the years. Arable land is typified in a limited number of categories: plough land (that is, anything under row crops), vineyards, orchards, gardens, meadow-land (hay making), pasture field (grazing), reed plots or reed beds, forest, forestation (plantations) and fish ponds.

As a result, this classification scheme hobbles attempts to experiment, innovate and adapt land use practices to address the challenges of shifting water dynamics. For example, the obligation to put plough land under row crops, which do not tolerate extended soil wetness, excludes the possibility to retain water on it for any substantial period of time. In particular, during early spring and summer, when water is abundant, arable land ought to be tilled and sown, which is impossible when it is under water. Therefore, land used for water retention needs to be converted preferably to meadow-land or grazing land, woodlots or plantation, and perhaps forest. Changing the type of cultivation on a piece of land, although not impossible, has its own difficulties. The owner or user is obliged to report conversion to the land registry authority. Any conversion of protected land is subject to the endorsement of the nature conservation authority. Special permits are required for forests, vineyards or orchards. You also need permits to withdraw land from cultivation or to put it to agricultural use again. Failure to report or to obtain the necessary license entails a land protection fine. Changing the physical boundaries of a parcel is also a very bureaucratic procedure called *parceling*.

One of the main reasons why such strict rules apply to land use is that agricultural subsidies and aids under the CAP system of the EU depend on the cultivation type of land. Currently, row crops or the heavy machinery dedicated to row crops are heavily subsidized, and therefore farmers and organizations are not interested in converting their lands into any other type of farming.

Barriers to Flexible Water Use

Current water regulation is also obsolete in its failure to meet the changing needs of human and natural communities in the floodplain. The Water Management Act lists various methods of use and utilization of water as a natural element for human purposes, and provides for the mitigation or elimination of damages caused by water through prevention, protection and control. Water management is a complex activity where the national government, regions, local governments, water authorities, water management associations and the water users (whether natural or legal persons) participate jointly. The legal structure built on this fragmented foundation reflects all the dichotomies that separate the various sub-sectors. For example, the VTT Act in 2004 was the first attempt to bridge sectors by coupling water usage (utilization) and water control (water damage mitigation or elimination).

Legislation is partitioned according to various types of water use. Depending on the classification of the water body or the water course, and the land covered by it, there are at least three to four players in water management operations: the government, the municipalities, the water boards and the land users themselves. Agricultural water use plays a special, integrating role in ILD, and, ironically enough, it is regulated in a separate piece of legislation in the form of water usage for business purposes.

Legislation to defend against damages caused by water clearly reflects this segmented approach. The entire system is geared up solely to discharge and drain any excess water, be it within the floodway (floods) or in the original floodplain ("excess surface water"). The building, development, maintenance, and operation of control works and activities are defined as the joint and several liability and obligation of the government, municipalities and other stakeholders. Thus, while funding and organization is quite different for excess surface water and floods, the legislation draws them under the same hat.

All rivers are assigned a design flood level from time to time by the minister in charge to which their respective flood control plans have to be developed. Longitudinal sections, crest of the dams/weirs, highest and design flood levels, the course of the river, the floodway and many other details of the flood control works need to be defined in these plans. However, there is no option for controlled discharge of high water levels, a key water steering mechanism under ILD.

On the contrary, stagnating water from the fields has to be drained, whenever it occurs, and even actively pumped over the crest of the dykes. For both floods and stagnating water, there are Grade I, Grade II and Grade III alertness levels specified with the respective associated actions to be carried out. At cross-purposes with the needs of ILD, inundation of the flood retention reservoirs of the VTT shall only be allowed when all three grades have been depleted and unusually high water provokes the announcement of an emergency flood incident. In other words, even if you had an appropriately sized reservoir under the VTT with a low inlet bottom sill to fill it up with mid-stage water, you could only fill it up legally once there was an extreme risk of bursting of the dykes.

In the same spirit, Grade I excess water emergency alertness has to be announced when the canal network is to be put to use. Grade II involves the active transfer of water from the drainage system into the active water course in two shifts, while Grade III means that all pumps are working at least with 75 percent of their capacity and emergency storage has to be contemplated. Emergency storage—that is, effective retention of water in the landscape, our main goal—is allowed only when an extraordinary emergency situation is announced, because excess surface water threatens residential areas, industrial areas or transport infrastructure, as specified by law. Again, this provision directly blocks the kind of flexible management envisioned for ILD.

A very interesting problem—a truly "interdepartmental one"—emerged during the flood of 2010 in Nagykörű, Hungary. As an outcome from a former project, three local inhabitants received proper training in professional fishing with traditional, old style methods, a part of the efforts to restore diverse land use patterns and mitigate the pressing unemployment and poverty. When the water left the riverbed and inundated the fields under crop in the floodway, they paddled out to practice their newly acquired skills. The officials of the business company exercising fishing rights along the whole section of the Tisza alerted the police, and poor fellows were caught in the red: as it happened to turn out, the law does not make any distinction between professional large scale fishing in the river—leased to big businesses—and subsistence fishing with traditional tools on the land—though temporarily inundated. A lawsuit followed and the fishermen were accused of poaching.

The Circumlocution Office:[4] A Labyrinth of Laws and Regulations

The aforementioned provisions are only a small piece of all the relevant laws and regulations governing complex water management projects. This dense legal network reflects decades of meticulous and sophisticated bureaucratic development, but it only entangles water users in unnecessary obligations that block any initiatives to adapt.

When, for instance, the Middle Tisza Water Management Directorate, the regional body in charge of the river and engineering works intended to dredge a drainage canal within its respective authority and dump the mud on both sides of the canal on agricultural land, a preliminary assessment had to be made first. The study conducted by an independent consultant firm was submitted to the relevant licensing authority. The relevant licensing authority happened to be the Middle Tisza Regional Environmental, Nature Conservation and Water Management Inspectorate. This body belongs to the same ministry and the same region. The two regional organizations—one in charge of water management operations and the other in charge of licensing them—are seated in the very same building.

The resulting decision runs on nine densely packed pages in 12 copies, each addressed to different stakeholders. Six different authorities and expert authorities had to be involved, three cognizant local municipalities informed and some 100 references made. In addition to the detailed description of all operations of how, when and where they have to be conducted, the decision referred to the applicable legislation on Environmental Impact Assessment, noise pollution, general rules of public administration procedure and services, protection of the environment, cultural heritage protection impact

[4]The Circumlocution Office is a place of endless confusion. Forms need to be filled in to request permission to fill in more forms. The term circumlocution describes roundabout or indirect speech, or the use of many words where a few would do (Charles Dickens: Little Dorrit).

assessment, the technical records maintained on routed line installations in outer areas, detailed rules for the excavation of archaeological sites and financial remuneration of discoverers of archaeological sites or findings, and the administration fee for official environmental, nature conservation and water management procedures.

The decision granted the permit subject to the preparation of a cultural heritage protection assessment study and solicited the endorsement of the following partner authorities: Northern Great Plain Regional Institute of the National Public Health and Medical Officer Service (NPHMOS), the Plant Protection and Soil Protection Directorate of the County Agricultural Technical Services Agency, the Cultural Heritage Protection Office and the town clerks of three communities in the neighborhood. It was also established that the project due to its size and nature does not require the completion of an environmental impact assessment. Thus, extensive bureaucracy actually provided too many details and the most critical goals were lost.

Although there seem to be insufficient funding and resources available for land consolidation, land registry offices are very diligent and active. While costs of consolidation are passed on to the owners, resources available for land registration and administration are sometimes used quite inefficiently. In another example, a document informed the owners of a parcel held in undivided common by 62 people that the cognizant land registry office has the right to amend the base map and the associated territorial data any time when there is an error detected in land surveying, mapping or areal calculations. Apparently such a failure was made, because given this opportunity, the bureau duly amended the type of cultivation, quality classification and Golden Crown[5] of the said parcel by Table 10.1 (Changes are highlighted in bold).

It can be clearly seen, that the 62 owners had to be informed, the administrative procedure conducted and the registration amended because of a three-hundredth part Golden Crown was mistakenly assessed. One tends to think that if there are resources available to correct such minor mistakes sparing neither trouble nor expense, why was it impossible to consolidate undivided commons on 1.5 million hectares for 16 yr.

Opportunities

Socio-political perception of barriers and opportunities in society to adaptive river co-management along the Hungarian section of the Tisza River were discussed in detail by Sendzimir et al. 2007. Research

[5]Golden Crown is an outdated land valuation system based on soil fertility and profitability. Golden Crowns are calculated per hectare.

Table 10.1 Decision delivered by the district Land Registry Office.

Before the changes						After the changes				
Top. No	Sub parcel	Cultivation type	Quality class	Area (ha)	Golden Crown	Sub parcel	Cultivation type	Quality class	Area (ha)	Golden Crown
087/6	a	plough land	3	0.7954	20.04	a	plough land	3	0.7954	20.04
			4	31.2933	**654.03**			4	31.2933	**653.89**
			5	131.6971	2054.**47**			5	131.6971	2054.**57**
			6	32.3290	336.2**2**			6	32.3290	336.2**3**
	b	ditch	-	0.5126	0	b	ditch	-	0.5126	0
	c	forest	3	0.9130	9.50	c	forest	3	0.9130	9.50
Total				197.5404	3074.2**6**				197.5404	3074.2**3**

applying system dynamics modeling tools explored the barriers and bridges to transformation of the current river management regime and to the development of capacity for participatory science to expand the range of perspectives that inform, monitor, and revise learning, policy, and the practice of river management. It was found that transformation may arise less from convergence of separate, parallel elements than from a chain or cascade of thresholds where each threshold passed triggers the accumulation toward the next. Adaptive river management regime and conventional management scheme are conceptualized as operating within separate stability domains, where each regime remains in its domain so long as its technologies, institutions, and paradigms reinforce each other. The current conventional system must be pushed against the resistance of all these reinforcing processes until acceptance of a new set of reinforcing factors marks the passing of a threshold into a new stability domain. For these reasons, the transition from one management regime to another is non-linear and unpredictable.

Thresholds started to be passed with the emergence of local and regional actors with alternative ideas and programs challenging the dominance of national decision makers, their methods and conceptual foundation: the conventional river management paradigm that river flood pulses and flows over the landscape must be prevented. Champions of change come from two distinct walks of life. National water authorities, which—caught between crumbling support at home and uncertainty from climate change, EU regulation and international investment from abroad—began to show unprecedented flexibility in considering the future development of the Hungarian Tisza reach. Also, a "shadow network" of scientists and local activists both in and outside of government has slowly grown in Hungary around a distributed set of dialogues to understand the management trap of the river and how it is perpetuated by the links that reinforce the current river management regime. This loose and informal alliance provides alterative visions and methods, supported with field experiments, to those that dominate the national agenda. Findings of the research are summarized in Table 10.2.

Social Factors: Psychology and Value Preferences

Stakeholder perception of the management problems in the Tisza region can be seen both as an obstacle and an opportunity. While a few hundred years ago a flood was considered a benefit and a blessing for the people living along the river, in that its pulses drove their flourishing fisheries, orchards and forests. However, it has been seen as a menace for more than a century, and it continues to be so. This impression is sustained and reinforced by stakeholder groups with a vested interest in maintaining the expensive

Table 10.2 Summary of major barriers and opportunities to transformation of the Hungarian reaches of the Tisza River Basin (HTRB); not in order of importance or priority.

Barriers arising from the current system	**Bridges/opportunities to an alternative system**
Basis: Long established paradigm "Protect landscape from the river"	**Basis**: Emerging new paradigm: "Live in harmony with the river"
1: Centuries living behind dykes entrenches a defensive mentality based on fear of water	1: More diverse view of options shared by a wider portion of society, including river engineers that we have to learn to live with a naturally flowing river
2: Sunk costs of massive infrastructure investment over two centuries inhibit any suggestion to modify or remove that infrastructure.	2: Increasing engagement of leaders and concerned citizens in considering and deciding on alternative management ideas
3: Momentum of national and globally driven development based on conventional paradigms leading to intensification of agricultural practices	3: Shadow network spanning entire river basin functions to generate new visions that influence regional debate
4: Lack of access to critical new information due to technological sophistication and due to lack of integration across domains of inquiry, i.e., disciplines, practice, training and governance, government jurisdictions	4: Flood, water stagnation, drought, and political crises shift political climate such that a window for alternative solutions appears in public debate
5: Concentrated lobbying power of prominent actors, i.e., individuals and organizations with ample financial and political capital, overwhelms dispersed and disorganized local farmers and activists.	5: Awareness-raising of importance of local culture, markets, regional brands, etc., and individual responsibility in decisions of where and how one lives
6: System self-organizes around reward loops of subsidies and paybacks, which reinforces a tight elite network such that all funding is funneled into their political machines within the present agricultural regime,	6: CAP reform (2nd pillar) promises new system to subsidize ecological farming and land use change and management practices that boost environmental services
7: Despite recent changes in political structure, present institutional setup does not permit implementation of complex, integrated programs that still follow administrative and sectoral lines	7: Innovative traditional and novel ideas show promise to concretely address drought and flood volume management as indicated by integrated basin computer models and pilot projects in western Europe and Hungary
8: Failure to reach and keep consensus. Some particular interests and alliances of the subgroups take overall precedence over solutions agreed upon by all parties	8: Legacy of knowledge and experience in extensive land uses and cultivation practices that provide sustenance in a periodically inundated floodplain

Table 10.2 contd....

Table 10.2 contd....

Barriers arising from the current system	Bridges/opportunities to an alternative system
Basis: Long established paradigm "Protect landscape from the river"	**Basis**: Emerging new paradigm: "Live in harmony with the river"
9: The diversity of views, knowledge, and terminology in all the separate formal and informal networks stall initiatives.	9: Information is disseminated by civic groups and individuals, facilitated by the shadow network.
10: Inertia of passive attitudes by local stakeholders sustained in the absence of leadership to build trust and understanding and motivate action across the region.	10: EU policies (Natura 2000, WFD, CAP reform) create a supporting reference framework with which to examine and modify river management policy.
11: Huge investment of financial, political, and social resources needed to provide convincing evidence of benefits of alternative river management strategies.	
12: Loss of natural capital, e.g., biodiversity, seed bank, and human capital, e.g., skills, local knowledge, due to death, termination of local practices, and regional emigration increase initial investments needed to re-establish the functional basis of a sustainable social-ecological systems.	

Source: Adapted from Sendzimir et al. 2007

engineering structures that keep the floodplain disconnected from the river channel. As a result, their participation in the regional discourse on water management always follows the same theme: convince the population that "there is no other way" to deal with the issue. However, appropriate learning and training efforts, accompanied with legislative and social organization changes have introduced and kept alternative ideas in the regional dialogue for decades and could eventually turn this perception upside down again. It will take successful pilot projects to provide convincing demonstrations.

Most people living along the river these days are not aware of the pre-engineered past. They do not have sufficient knowledge on what the river landscape looked like and how the former water steering ("*fok*" in Hungarian) system operated in historical times. Experimental pilot projects are extremely important to demonstrate how efficiently such systems could lower flood danger while restoring ecosystem functions. Also, after centuries of Imperial and then Communist rule, local autonomy—social, economic, political and financial—is unknown to today's villagers. They expect direction from the central government, which, in turn, ensures that they receive their social benefit payments. However, a vertically integrated structure of local economy, education and social cohesion would have the potential to turn this perception back as well.

Serious changes are needed in terms of attitudes to land ownership. Historically, land was prized as the most fundamental form of rural capital, almost the very root of existence, something that is not subject to purchase or sale. During the Communist period ownership was nationalized and generations grew up longing for their own individual plots. During the post-Communist restitution process, political will provided rights but imposed only poorly designed obligations. Today's land-owners do not feel responsible for the community on which territory their land is situated. All land is owned by legal entities, the institution of the commons is for the past.

As the post-communist era unfolds, the socio-political discourse has diversified as well. Attitudes differ as to whether to measure progress in narrow, economic terms, e.g., standard of living, or more broadly, e.g., quality of life. And if the latter, then what gives life higher quality? People of the Tisza want to be part of consumer society even if their financial situation does not allow purchasing a good quality of life with a high standard of living. In their pursuit of the material security of the modern, increasingly urban, society, many rural people adopt its consumer-based paradigms, e.g., commoditization for mass consumption sets the standard for quality. Given this view, good quality of food is defined separately by each link in the modern product chain: producers, shippers, commerce, retailers and consumers, respectively. Shelf life is considered to be more important by many than high nutritional value. For producers, high yield varieties needing high input are more precious than traditional local varieties with a diversity of taste and resistance to local diseases.

The upshot of adopting a consumerist valuation system is that most rural people may live in the river countryside but can no longer recognize or even imagine the ecosystem services that a restored, fully functional, landscape can deliver. The landscape has been reshaped and operated for maximum production and transport to mass markets, and only with recent flood crises has the common social perception begun to recognize how this industrial arrangement requires constant struggle and external input to sustain itself as is. Much experimentation with pilot projects will be needed to enrich the regional discourse with the possibilities of living more cheaply in a functional landscape where humans are adapted to the operation of complex systems. The discourse can also be enhanced through the intervention of knowledge brokering instruments to consider the complexities inherent in such alternative visions of land use. Earlier research demonstrated that a gaming tool can allow stakeholders, especially farmers and water officials, to explore different land use options whose success hinges on the interdependence of various factors, and to reconcile interests and to build on synergies (Stefanska et al. 2011).

The Land of Endless Possibilities

Provided the necessary institutional, social and legal changes are made, physical opportunities exist to implement innovative large-scale adaptive river management practices in the Tisza valley in conjunction with integrated land development. Paradoxically, the social and economic challenges of the region could constitute opportunities for restoring landscape functionality at meaningful scales. The socially backward situation of the eastern part of the country drives the political agenda toward regional re-development, and the lack of heavy infrastructure (heavy industries, large urban areas) makes larger-scale changes in land use possible. The relatively untouched river landscape also retains hints of what a functional landscape mosaic looked like in the remnants of the former flood plain structures, impressions, river channel sections, ox-bow lakes and the like that can still be easily tracked and measured by GIS and remote sensing tools.

Computer modeling (Koncsos 2006) of the geomorphologic features in the river valley demonstrated that controlled discharge of water into the deep floodplain can be accomplished relatively easily. A series of potential reservoir sites were identified which could be used to store water at low water levels and with minimum disturbance of the ecological conditions, provided that filling of the deep floodplain is started at an early stage by cutting through the existing embankments deep enough. Modeling results suggest that this network of reservoir sites could accommodate most of the uncertainty in flood dynamics anticipated under climate change for the next 40 yr. Since water in these reservoir sites would be pretty shallow (about one meter), no extensive infrastructure, engineering works or investments would be needed to execute the project (Koncsos 2006). Another potential is seen in the existing drainage canal system, which could be adjusted by accurate engineering to gravitationally lead the water to and from the fields (Balogh 2002). All these measures can be integrated into a comprehensive scheme of spatial and urban development when some fundamental principles of system theory are taken into account (Borsos 2009). In addition, such "disturbances" as periodic flooding can actually be exploited to enhance agricultural diversity as attested to by the extensive variety of fruits that existed when the Tisza flooded freely (Sendzimir and Flachner 2007).

At the international level, there are also emerging possibilities to implement experimental schemes in line with the principles discussed in this chapter. The Tisza Group within the International Commission for the Protection of the Danube River (ICPDR) was commissioned to develop an Integrated Tisza River Basin Management Plan as required by the Water Framework Directive of the EU and to begin implementation thereof by testing of new approaches on wetland and floodplain management through community-based demonstration projects (Borsos et al. 2010). The outcomes, findings and lessons from these projects are being reviewed to give ministry officials in Hungary fresh insights into river management issues. Both ICPDR and the Carpathian convention support active collaboration of the riparian countries.

Threats

Finally, we briefly consider some unpredictable threats, which are difficult to reckon with but have strong influence on strategies and the implementation of any design scheme. One interesting factor to be taken into account is the perception of truth in the context of concepts and paradigms actively kept in the public's attention by vested interests. The convoluted public debate over global climate change and the ozone hole furnish strong evidence that in spite of sound scientific facts and an emerging scientific consensus, political and business interests can block any progress for a considerable amount of time, and, even after the respective conventions were adopted, there are still some who question the very existence of anthropogenic climate change (for instance, Lomborg 2007).

Turning Away from Truth

The eminent climatologist Stephen Schneider pointed out in his book *The Patient from Hell* that during the development of the IPCC Working Group II report of 2001 good science presented at the session was manipulated until it satisfied all of the national representatives present. The words used to express the consequences of global heating were blurred until they were acceptable to representatives of all oil-producing nations, who saw their national interests threatened by the scientific truth. This incident underlines the importance of the chasm between reality, facts and science on one side and beliefs, manipulation, policy and politics on the other. Any measure or programs is only as good as much of it is implemented. According to Lovelock (2009) global climate fans divert interest entirely from real observation towards model scenarios and focus on carbon dioxide emission reduction instead of adaptation strategies. With integrated land management and development in the valleys of large low stage rivers, adaptation to more extreme weather and climate may be better suited if designed well and with the natural conditions taken as a priority.

Solidarity or Sovereignty?

The Tisza River Basin Analysis of 2007 of the ICPDR (Sheperd and Csagoly 2007) emphasized the importance of *solidarity*: the belief, that one region should not pass on water management problems to another. Albeit this is a very honorable proposition, the fact is that human communities are usually organized in groups and societies that distinguish themselves from other groups and societies by their degree of relative independence. Nation states exist on the basis of *national sovereignty*, the notion that each nation state was an independent actor in the international political arena and had unalienable rights on its own. In situations where ecological and physical geographic conditions of an area require a different type of division and segmentation than those defined by national boundaries such a notion may prove to be harmful.

Integrating the landscapes and functions of the entire Tisza river basin will not be easy. The basin is fragmented by the arbitrary political borders of five different nations with varied attitudes towards cooperation and collaboration. To name one example, upstream countries will not invest to retain floodwater on their land, which is shed as fast as possible on the increasingly vulnerable downstream countries. Building up the dykes as high as possible in one country has an effect of aggravating flood risk situations in another. The same applies to regions. However, innovations in governance of a national government may provide living examples of institutions and technologies that show how to avoid the emergence of such situations. Such governance innovations are the practical outcomes sought by participatory science processes, such as adaptive co-management. Such processes cannot be learned in academia and require repeated experimentation to see how they can effectively improve governance in a dynamic and complex world. Groups like the Hungarian Shadow Network have actively fomented co-management experiments for more than a decade, and the continuing barriers to regime transition mandate that such experimentation must continue.

Conclusions

Development problems of the Tisza-valley are marked by the vectors of water management, land use, resource exploitation and social tensions. A key factor in the region is the river itself. Due to new threats, increased risks and uncertainty, the current flood control and river management practice does not seem to provide a satisfactory solution. Adaptive co-management approaches could be used to explore the water retention potentials on large areas within the floodplain. Based on the experiences obtained on the design and implementation of the VTT scheme it became clear that a novel and more integrated approach was needed. Key features of such a change are as follows (Koncsos 2011):

- It is not possible to renew and adapt the current flood control strategy to the new threats only and exclusively by optimizing the water related technical measures.
- The potential to further raise dykes has been depleted. This solution tackled the problem in an isolated and unilateral manner.
- A broad consensus of stakeholders needs to be achieved if society wants to pursue experiments in alternative land use at landscape and regional scales. Technical measures related to flood control have to be adapted to the findings of these experiments.
- The land should be withdrawn from intensive agricultural exploitation and adapted to the prevailing agro-ecological conditions.
- The first strategic priority is to set up a system of interests where water retention and land use are integrated such that periodic flooding following natural river dynamics causes no danger or damages.
- A nationwide risk management plan is required which determines the *actual* risks by comprehensively listing and incorporating the full range of threats and potential opportunities. Alternatives need to take account of ecological considerations and potential uses. The resulting plan should regard risks at the level of the Carpathian basin and reflect on the existence of politically divided river basins as well as on the impact of interventions in foreign countries.

- Sites naturally suited for water retention and storage which may be seen as feasible options from the agro-ecological perspective should be identified.
- Several studies maintain that approximately two billion cubic meters of storage capacity was available within the deep floodplain.
- Shallow reservoirs in the deep floodplain need to be inundated in a natural way. Operational plans should be as plain and least artificial as possible. Inlet and outlet structures should ensure discharge of water at the lowest possible water levels so that they could be operated in the early stages of the floods with maximum efficiency. They need to be located at natural discharge and drainage points.
- Discharge and free drainage of minor floods should not be restricted. Due to this natural operating mode alone, flood levels could be reduced by 1.5 to 2.5 meters along the main flood control line of the Tisza, thus diminishing the risk of damaging floods substantially.
- The new and adaptive river management concept should be built on the concurrent personal interest of individuals, farmers and communities. They could contribute to energy independence, eco-tourism, fisheries, organic food production, promotion of special Hungarian land races, thus assisting in revitalizing local economy. Estimates suggest that the potential benefits obtained solely by planting and harvesting energy crops on the wetland thus created could reach the magnitude of flood risk mitigation costs. A part of the revenues could be used to set up insurance funds.
- The general public and local stakeholders need to be made aware of the financial benefits of such a program, including a sound consciousness and knowledge about the causes of the existing risks, whilst maintaining the fundamental principles of social solidarity.
- The strategy imposes a number of tasks on the academia primarily to set up integrated flood control and water management design schemes adapted to land use patterns and land use strategies adapted to water regimes and river dynamics.

Acknowledgements

Part of the original research discussed in this chapter was conducted by the generous support of the UNDP/GEF through the project entitled "Integrated Land Development (ILD) Program to Improve Land Use and Water Management Efficiency in the Tisza Basin" listed under Service Contract No IC/WD/384-HU. The project idea was conceived by the late Zsuzsanna Flachner, long term and ardent supporter of the alternative river management approaches in the Tisza basin. The authors are deeply indebted to her and would like to pay tribute to her memory with this chapter. Also, we are extremely grateful to Dr. Lóránt Fehér for revealing the legal background of water management and land use. Gergő Nagy was instrumental in preparing some of the ILD case study material.

References

Balogh, P. 2002. Basics and Method of Floodplain Management on the Middle-Tisza Valley. VATI Kht. Budapest.

Benda, K. [chief ed.]. 1983. Magyarország történeti kronológiája, II. kötet. Akadémiai Kiadó, Budapest.

[BME] Budapest Technical University. 2006. Flood control of the Tisza river in the Carpathian Basin. NKFP research project final report.

Bokartisz. 2002. Floodplain landscape management: concept and concrete steps at the Bodrogköz area. The 'Last Straw' program's ecological historical foundation and the synthesis of the environmental economics analysis. Karcsa, Hungary.

Borsos, B. 1991: Socio-political aspects of the Bős-Nagymaros dam system. International Water Power & Dam Construction 43: (5) 57–61.

Borsos, B.P. 2000. Három folyó között. A bodrogközi gazdálkodás alkalmazkodása a természeti viszonyokhoz a folyószabályozási munkák idején (1840–1910) Akadémiai Kiadó, Budapest. Hungary.

Borsos, B. 2009. Systems theory and ecological settlement design: a pilot project in rural Hungary, Hungarian Studies 23(2): 175–194.

Borsos, B. and T. Bánvölgyi. 2003. Átfogó területfejlesztési elképzelések lehetőségei a Kárpát-medence egyes területein az ökológiai adottságok figyelembe vételével. pp. 101–110. *In*: I. Fodor [ed.]. A fenntartható fejlődés környezetvédelmi összefüggései a Kárpát medencében. Bookmaster Kft. KVM, MTA RKKK and MTA PAB, Pécs, Hungary. (in English: Borsos, B. and T. Bánvölgyi. Possibilities for a comprehensive spatial development scheme in certain areas of the Carpathian basin with a view of taking into account local ecological characteristics. Proceedings of the Conference on Sustainable Development Challenges in the Carpathian Basin. Pécs. (in press)

Borsos, B., Z. Flachner, G. Nagy, L. Fehér and P. Balogh. 2010. ILD Manual: A practical guide to integrated land management methods intended to improve land use and water management efficiency in the Tisza basin. ICPDR/ UNDP/GEF Integrated River Basin Management in the Tisza Project IC/WD/384-HU, Budapest.

Gleick, P. 2003. Global Freshwater Resources: Soft-Path Solutions for the 21st Century. Science 302: 1524–1528.

Kajner, P. Végkiárusítás előtt. pp. 237–279. *In*: L. Lányi and G. Farkas [eds.]. 2010. Miért fenntarthatatlan, ami fenntartható? L'Harmattan, Budapest, Hungary.

Koncsos, L. 2006. A Tisza árvízi szabályozása a Kárpát-medencében. Magyar Természetvédők Szövetsége. Budapest, Hungary.

Koncsos, L. Árvízvédelem és stratégia. pp. 207–232. *In:* L. Somlyódy [ed.]. 2011. Magyarország vízgazdálkodása: helyzetkép és stratégiai feladatok, Hungarian Academy of Sciences, Budapest, Hungary.

Koncsos, L., J. Reimann and I. Vágás. Matematikai-statisztikai módszerek árvízvédelmi feladatok elemzéséhez. *In*: L. Somlyódy [ed.]. 2000. A hazai vízgazdálkodás stratégiai kérdései. MTA, Budapest, Hungary.

Junk, W. J., P.B. Bayley and R.E. Sparks. 1989. The flood pulse concept in river-floodplain systems. Canadian Special Publication of Fisheries and Aquatic Sciences 106: 110–127.

Lee, M. 2011. Can the Common Agricultural Policy save England's green and pleasant land? The Ecologist online: http://www.theecologist.org/how_to_make_a_difference/wildlife/1087551/can_the_common_agricultural_policy_save_englands_green_and_pleasant_land.html

Lomborg, B. 2007. Cool it: The Skeptical Environmentalist's Guide to Global Warming, Knopf Publishing Group, Cambridge, New York, US.

Lovelock, J. 2009. The Vanishing Face of Gaia. Penguin Books, London, UK.

Molnár, G. A folyó, mint ökológiai rendszer. pp. 182–190. *In*: L. Buka and I. Gyarmathy [eds.]. 2002. Élőtáj 2. Válogatott írások természetről, térről, teremtésről. Dél-Nyírség Bihari Tájvédelmi és Kulturális Értékőrző Egyesület, Debrecen, Hungary.

Nováky, B. 2000. Az éghajlatváltozás vízgazdálkodási hatásai. A hazai vízgazdálkodás stratégiai kérdései. Manuscript. MTA Stratégiai Kutatások Programja, Budapest, Hungary.

Ripka, J. 2005. Birtokszerkezet, nemzeti birtokrendezési stratégia és a földügyi igazgatás. Edited version of a conference lecture presented at the MFTTT assembly organised in Győr, Hungary 30 June–2 July 2005.

Schweitzer, F. 2001. A magyarországi folyószabályozások geomorfológiai vonatkozásai. Földr. Ért. L: 63–72.

Schweitzer, F. 2009. Strategy or disaster. Flood prevention related issues and actions in the Tisza river basin. Hungarian Geographical Bulletin 58(1-4): 3–17.

Sendzimir, J. and Z. Flachner. Exploiting Ecological Disturbance. pp. 211–228. *In*: S.J. Scherr and J.A. McNeely [eds.]. 2007. Farming with Nature: the science and nature of ecoagriculture, Island Press, Washington DC, US.

Sendzimir, J., P. Magnuszewski, Z. Flachner, P. Balogh, G. Molnar, A. Sarvari and Z. Nagy 2007. Assessing the resilience of a river management regime: informal learning in a shadow network in the Tisza River Basin. Ecology and Society 13(1): 11. [online] URL: http://www.ecologyandsociety.org/vol13/iss1/art11/

Sheperd, K. and P. Csagoly [eds.]. 2007. ICPDR: Tisza River Basin Analysis 2007. Summary Report. A Call for Action, Vienna International Centre, Vienna, Austria.

Somlyódi, L. [ed.]. 2002. A hazai vízgazdálkodás stratégiai kérdései. Magyar Tudományos Akadémia, Budapest. Hungary.

Stefanska, J., P. Magnuszewski, J. Sendzimir, P. Romaniuk, T. Taillieu, A. Dubel, Zs. Flachner and P. Balogh. 2011: A Gaming Exercise to Explore Problem-Solving versus Relational Activities for River Floodplain Management Environmental Policy and Governance. Env. Pol. Gov. Published online in Wiley Online Library (wileyonlinelibrary.com) DOI: 10.1002/eet.586.

Váradi, J. et al. 2001. A Vásárhelyi Terv Továbbfejlesztése koncepció-terv. KöViM Budapest. Hungary. Online: http://www.vizugy.hu/vasarhelyi/index.html

Váradi, J., L. Szlávik, I. Orlóczi, I. Kertai and I. Nagy. 2003. Az árvízvédelem és a tájgazdálkodás jövője a Tisza mentén (a Vásárhelyi Terv Továbbfejlesztése) Vízügyi Közlemények Különszám V.

Zimányi, V. 1976. Magyarország az európai gazdaságban. Akadémiai Kiadó, Budapest. Hungary.

11

Co-Management of Water Resources in a High Density Environment as a Stimulus for Sustainable Regional Development

Cheryl de Boer[a,*] and *Hans Bressers*[b]

Introduction

In addition to the concerns associated with the protection of delta areas from the sea, the Netherlands is in particular affected by the increasing frequency of high and low water levels as a result of climate change. This awareness has warranted a drastic change of approach in water, land and nature management towards a strategy that uses nature's resilience to provide for both human and natural environmental needs. Most notably in the parts of the Netherlands which are above sea level, water management is often practiced through co-management. In various projects seen here, many goals apart from water quality and quantity management are being integrated. Water retention, nature, recreation, economic diversification,

Twente Centre for Studies in Technology and Sustainable Development–CSTM, University of Twente, PO Box 217, 7500AE Enschede, The Netherlands.
[a]Email: c.deboer@utwente.nl
[b]Email: hans.bressers@utwente.nl
*Corresponding author

agriculture and cultural history are often included which necessitates the involvement of each of the responsible organizations and NGO's. This arrangement of relationships often creates the need for boundary spanning between actors, as well as rules and resources to cope with the complexity and dynamics of the process.

This chapter explores actions focused on planning for multi-functionality, increasing space for rivers and the connection of natural areas. These efforts are at the heart of new initiatives in the Dutch rural areas to meet European and National habitat and water quality and quantity goals (De Boer and Bressers 2010). The subject of co-management in this chapter is the restoration of the Regge River valley. The Regge is a 52 kilometre Dutch tributary river located in the eastern part of the country. Recreation, agriculture, nature and flood management are good partners under the case study of the Regge River Restoration Project. Project managers apply adaptive strategies and seek governance regimes to provide stimuli for implementing their innovative developments. The Regge River Restoration Project involves the integration of many sub-cases that are located at various places along the river. The particular sub-case, the Diepenheim Upper Regge project, was comprised of a remarkable variety of strategies used by practitioners from the various governmental (and non-governmental) organizations that had worked together. The chapter concludes with a systematic inventory of boundary spanning strategies that were found for the most part in this case of co-management, as well as in the rest of the Regge Renaturalization projects. Contexts are known to vary across time and place, however these lessons can provide worthwhile insights for consideration for those who work in similar projects or study their development.

Co-management Challenges: Water Management in High Density Environments

In the Netherlands water management has typically implied working against nature to ensure "progress" for mankind. A remarkable paradigm shift has however taken place in the last one or two decades. Several European countries, including the Netherlands, have experienced floods and high waters in large rivers. Although further improving dikes and embankments has been the standard first response, there in now a reconsideration of the basic underlying principles of water management. Instead of only containing rivers, the new paradigm seeks to maximize opportunities to make natural processes an ally in stabilizing water levels and preventing floods. In the Netherlands this new paradigm is accompanied by slogans such as "space for rivers", "living with water" and "building with nature". The predicted further increase of irregular rainfalls caused by climate change as well as the emphasis of the European Water Framework directive on respecting ecology and natural river basins have contributed to this paradigm shift.

Working *with*, rather than *against* nature to support human interests comes however at a price, which is especially relevant in densely populated countries such as the Netherlands. This way of working with water almost invariably costs a great deal of space. In fact, part of the reason behind the creation of "unnatural" interventions in the water system in the past was precisely the "rationalization" of the use of space. Working *with* nature also poses new challenges in the field of spatial planning.

Spatial planners are not necessarily unfamiliar with these kinds of challenges. Many see the integration of various spatial claims into productive "neighbourships" and multiple uses of the same area as their core business. Water managers only increase their administrative burden if they wait until the last minute and then try to integrate their plans alongside the previously included interests and purposes. Ideally, they seek to have the water system as the guiding framework, with water rules and policies providing the necessary administrative support. Of course, the reality of implementation processes is more complicated and powers can be sufficiently balanced to result in complicated processes within and around each project with which the new innovative paradigm is to be realized. In the Netherlands the authority to manage the regional water systems is in the hand of the waterboards. Waterboards in the Netherlands are not agencies but separate regional governments which report to democratically elected councils and governing boards, just as with municipalities and provinces. What separates them is that they have one specific task: to manage the water system for flood prevention, drought prevention and water quality while keeping it suitable for meeting human and ecological needs.

The eastern part of the Netherlands belongs mainly to the Rhine East sub river basin. The water basin approach supported by the European Water Framework Directive attempts to coordinate all of the major development in this area, adding to the challenge of connecting the actor networks to the measures being taken (Van Leussen 2011). Within this area the transboundary Vecht River provides most of the surface area drainage before it flows into the IJssel River in the Rhine delta, just before it enters into the IJssel Lake. The Regge River is a main tributary of the Vecht and drains into most of the Dutch region of Twente. In the past this river has been re-engineered and regulated alongside many others in the area in order to provide faster drainage of farm land. Consequently, a water system has been created that not only lacks natural qualities but is greatly degraded in its buffering capacity. Further climate change will lead to both more droughts and increasingly heavy rainfall, thus human development has further intensified the resulting flood risks (Schaap 2010). A broader view on the functions of water including its role in nature has gained wider

Figure 11.1 Rhine basin (Source: UNEP DEWA/GRID)—case study area indicated by hexagon.

Color image of this figure appears in the color plate section at the end of the book.

acknowledgment. Consequently, river restoration projects have commenced that are for a large part are "undoing the—recent—past" of channelizing projects.

With respect to the many river restoration efforts occurring in the Netherlands, the Regge restoration case is quite typical in a number of aspects. Multiple purposes and consequently many levels and arms of government and private organizations are involved. The project is not only a water management process, as it includes nature development aspects as well as many other policies and societal goals. Furthermore the Regge restoration is embedded in water buffering policies and plans to connect ecological corridors on a larger scale. On the other hand the Regge project can be seen as a collection of numerous smaller scale projects. Inevitably projects of the size and ambition of the Regge River restoration are "complex", moreover they are also dynamic (De Boer 2012). The period through which they are implemented is sufficiently long to allow ample room to negotiate and develop relationships, however it is also long enough to modify the possible context and is often beneficial to the actors involved. As such, analysis of the processes needs to take into consideration that not only the process, but also its contexts evolve and are made to evolve. Relevant policies in the Regge case underwent substantial changes within the project period. The Regge River renaturalization is characterized here as multi-sectoral and long term and is thus a "complex and dynamic" implementation process.

The complexity of project implementation and the various strategies of coping with the resulting level of unpredictability in the system are quite common in Dutch water and nature restoration projects. On the basis of the many case studies that have been completed (e.g., Projectteam Evaluatie NBW 2006) the case appears to deviate from about half of the others in that it is progressing in a reasonably successful manner. Many projects have a tendency to run into unforeseen obstacles during the implementation phase. Projects also specify results in the form of rivalries in resource use that are solved and in terms of natural and manmade resources' values being improved (Knoepfel and Nahrath 2005). From this perspective the Regge restoration can as well be seen as a success. To understand this relative success in dealing with the multiple policy context that urges co-management, we will first briefly explain our theoretical framework for analyzing multi-stakeholder interaction implementation processes and also situations of co-management. Thereafter we will illustrate it through the Diepenheim project, one of the Regge River restoration projects.

Sketch of Theoretical Framework for Analyzing Co-Management

Water and nature restoration projects are analyzed in this chapter as multi-policy implementation in complex and dynamic social interaction processes. When one studies the renaturalization processes these (inter) actions of the actors involved form the main portion of the story and are highly relevant from a policy science perspective. Given that the processes operate in a complex and dynamic, and thus unpredictable and uncertain environment, so-called linear project management is a recipe for failure. To be able to succeed in integrating multiple legitimate and desired uses, multiple actors' consent, sectoral policy schemes, funding rules, time frames and scale issues, members of project teams need to be skilled "boundary spanners" (Williams 2002) and able to see, use and sometimes create "windows of opportunity".

Consequently the narratives of the (inter)actions are highly informative in terms of what strategies can be used to achieve good results under various contexts. For that reason, the characteristics of the actors in these implementation processes are a vital part of the process. As core actor characteristics the motivations, cognitions (interpretations of reality held to be true) and resources (providing actors with capacity and mutual power) are seen as the ultimate shapers of inter-actor relationships (Bressers 2004). These characteristics are influenced on the one hand by governance and specific case context characteristics and on the other hand actors in the process apply strategies to make the most of these contexts. In the governance context even wider and more general contexts can also be influential. In Fig. 11.2 we give an overview of the various elements of the theoretical framework used, the Contextual Interaction Theory (De Boer and Bressers 2011).

The characteristics of the actors are thus influenced by factors from various layers of context. One of the layers of context is the structural context which involves the elements of governance and the relevant property and use rights. Previous research showed that the extent (completeness) and especially the coherence—which together make up the degree of integration of this context—are very relevant for guarding the sustainability of river basins (Bressers and Kuks 2006).

We have identified river renaturalization processes as being complex (multiple sectors and scales) and dynamic (long time horizon) processes, striving for improvement rather than protection of what is already there. The actors involved did not take the process setting for granted. Rather they tried to influence not only the course of the process but also its setting over the longer period. This was done by using externally oriented strategies that often take the form of "boundary spanning", spanning scales, times

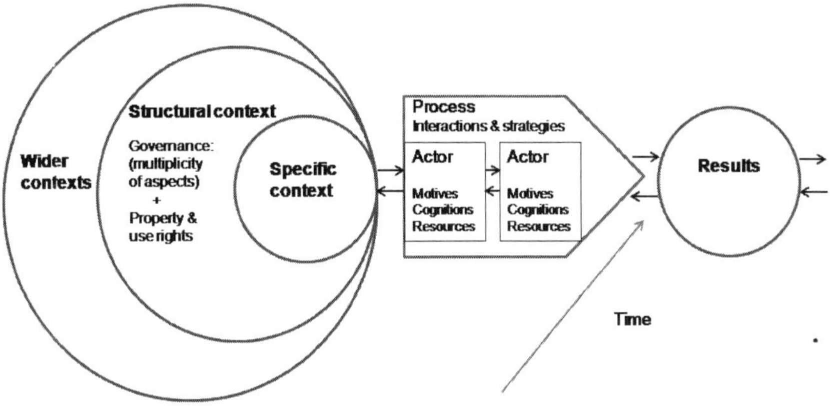

Figure 11.2 Contextual Interaction Theory model overview.

(Source: De Boer and Bressers 2011).

and sectors (Bressers and Lulofs 2010). These adaptive strategies can be reactive, but also responsive and even proactive. The external strategies can directly aim at influencing the motivations, cognitions or resources of the other actors involved. This can also be accomplished indirectly via the actor constellation or the operational rules of the institutional arena in which the process takes place.

To be able to do this requires that the actor organizations have a certain capacity for "receptivity" (Jeffrey and Seaton 2004). Receptivity is used by us to indicate the ability to combine new information with existing cognitions, to recognize new goals as matching existing motivations or the values behind them and to recognize the opportunities of new resources or combinations with existing resources to optimize their capacity and power. To increase their capacity for receptivity organizations can use internal strategies that do not influence other actors in the process but mainly their own organization.

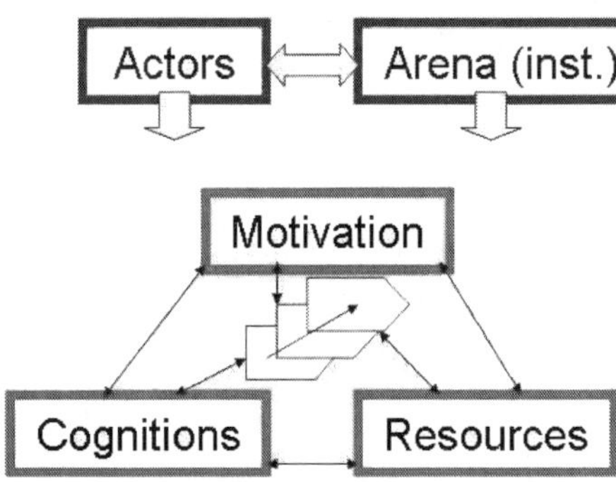

Figure 11.3 Intervention points for external strategies.

(Source: De Boer and Bressers 2011).

The use of various internal as well as external strategies as an adaptive response to complex and dynamic contexts cannot occur without a structural context that allows or even stimulates one to do so. This is not self-evident because often governance aspects such as policies and rules are more control than empowerment focused. Thus, especially with dynamic and change oriented projects, the flexibility of the governance context is important alongside the influence of the extent and coherence. Lastly the degree of change that is striven for also matters and this is referred to as the intensity of the governance context (De Boer and Bressers 2012).

In the case description and analysis below we will concentrate on the interaction and strategies used for co-management of the water resources. In the discussion we will also refer back to the role of a sufficiently stimulating governance context. As mentioned, a detailed description of a real life process is not only interesting as a story as such, but is also important for achieving a deeper insight and understanding of the more general context for such processes and the options practitioners have therein to develop the most successful case. By beginning with the detailed description of an existing case, the analysis and insights are most clearly elaborated. The case study area that is presented to illustrate the main points in this chapter consists of the upper origins of the Regge River near the town of Diepenheim.

The Estates of Diepenheim Project

Several streams in addition to the main river body contribute to the Regge in its upper regions. Due to a shipping canal that interrupts the river, some streams that once belonged to the catchment area now flow directly into this canal. However there are others that are led together under the canal and then reappear as the Regge River just north of it.

A lot of the projects occurring in the area surrounding the town of Diepenheim take the form of "area developments". These are officially recognized yet fully voluntary processes which are intended to improve the overall character of an area. The success of these processes is only conceived to be possible when carried out in a form of co-management including several public and private stakeholders. The Municipality of Diepenheim tries to fit the Regge projects and the area development together where possible into a joint process. One of the issues experienced in this process of integration is related to the ownership of the lands. Estate lands make up one third of the municipality and as well account for half of the project lands. These estates generally consist of either a mansion or castle and a large property surrounding them which are made up of a mixture of farmland, woods and open garden space. They are often hundreds of years old and were generally built by noblemen. Nowadays many of the families have had to sell off parts of their land and many are owned by foundations rather than private persons. The estates are however still managed under an attitude of "sovereignty and traditional ownership values". This is both a conservation oriented mentality (that has successfully preserved the estates and their natural beauty) but also a conservative mentality that makes it difficult to develop the trade-offs that often are involved in such complex projects. Additionally, since many estate managers follow the guidance of the largest estate manager, Mr. Schimmelpenninck, there is some decreased flexibility in working with the different estates. Mr. Schimmelpenninck has been successful in using his strong bargaining position in order to encourage cooperation to develop based on supporting his and the surrounding estate's own interests.

A number of these quite beautiful estates and castles are situated in the wooded areas in and around the town of Diepenheim. Such estates however, often have problems with their water systems. Canals dry out, woods and nature areas suffer from drought and agricultural fields can suffer from both excess and insufficient moisture. Following an inventory that was made of the problems experienced by each estate, various interventions have been developed in order to address them. This collection of work is referred to as the "Estates of Diepenheim" project.

The castles of Weldam, Warmelo and Nijenhuis and the "houses" of Westerflier and Diepenheim are working together with the Regge restoration project to address some of their own water related issues. The rivers belonging to the Diepenheim Mill brook, the Leide brook and the upper stretches of the Regge are being naturalized as part of this project. Between the houses of Westerflier and Warmelo a stretch of Regge of 700 metres is set for restoration. This was made possible due to a voluntary land exchange which made the required land available for the Waterboard to use. The potential to restore migration opportunities of various fish and crawfish in the area drew additional interest for the creation of an ecological pathway. These additional features were made possible by proactively choosing to exploit opportunities to improve the landscape value of the water whenever possible.

The estates located around the artistic village of Diepenheim were independently active in trying to "re-shuffle" the placement of agricultural activities on their estate and make the estate more natural in terms of ecology and landscape. They later solicited the help of the Waterboard to help improve upon the natural qualities of the Regge and the waters surrounding the castles. The Waterboard was happy to support the Estates because their actions aligned well with their general vision. The projects would address water concerns, add nature to the area and allow more space for water retention.

Another interesting activity that the Waterboard coupled its restoration efforts onto was the Regge Garden project in Diepenheim. The Waterboard deliberately waited approximately a year and a half to begin discussions with the Regge Garden project leaders after learning that this interesting initiative was developing. They believed that there was a possibility to develop a synergistic project between the two groups. The association of publicly accessible art, tourism and recreation were seen to be valuable aspects that would improve the resulting developments.

This Regge Garden project was initiated by Herman de Vries—a domestically well-known artist. Following the inclusion of the Waterboard he continued to be involved in the project as it developed. The final result included artwork in the project area along the Regge, and various kinds of gardens including butterfly, winter, marsh, and aroma gardens, as well as many walking paths.

Timing played an important role in this project. The Waterboard was the first to contact the local Municipality which includes Diepenheim with their idea for an integrated project. The Municipality had already been interested in the Garden of Diepenheim project for a long time. When the project started to materialize, the Waterboard and the Municipality contacted the artists and other stakeholders to discuss different options for the project. This process not only produced synergies which would help the Waterboard's interests to materialize, but also opened up additional venues for subsidies. Working together with the other parties (particularly the art community) made them eligible to receive sponsorship from the Mondriaan Foundation for which they otherwise would not have been eligible. The Waterboard refers to this deliberate combination of goals from different sectors to the projects as "schakelen" (coupling). They view it as a strategy that produces synergy ("added value of water") and enables the combination of various financial, legal, expertise, etc. resources to support the project. They are thus also willing to accept (where and when necessary) the added complexity that it brings to project organization and implementation.

In this case the initiative was not just a matter of coincidentally finding shared interests. Advice was given to the project team by another individual, not the artist, who had ties to the various organizations. He was a council member of the Waterboard, was the "rentmeester" (manager) of some of the estates and was also the director of the landscape architects consultancy Eelerwoude. Through his inclusion in the project he was able to provide advice in advance about what opportunities there may be to work together with different actors.

The estates are in need of new economy to support their future existence even though they strive to conserve their traditional situation as much as possible. From the perspective of the Municipality, the highly valuable

Figure 11.4 One of the estates: House of Diepenheim.
(Source: Tubantia).

nature areas, aesthetic landscapes and small rivers of Diepenheim have great capacity to support tourism. Nevertheless, difficulties have been experienced when the estates have been approached to participate in projects such as extending a bicycle path along a river. There is a difference in perspectives in terms of how to preserve and manage the lands for the future. The Waterboards often support recreation possibilities whereas the estates are generally concerned about preserving the traditional nature of their lands. One cycling path project was unsuccessful due to the estates withholding their lands for inclusion, despite the efforts that the Waterboard and the Municipality were making towards the development of a larger integrated cycle path system. This can lead to irritations at the Municipality since they have the responsibility of preparing land use decisions and feel frustrated by a refusal of the landowners to cooperate. An additional stumbling block is the differing views on the appropriate development of the land.

Although the estates are generally not development oriented in their long term approach, they are currently cooperating with a pilot project in another part of the Municipality. In this project, nature and landscape development and maintenance are set at the core of a new form of agriculture, sponsored by several governments. Two estate farmers want to switch to a more organic farming approach as an alternative to traditional farming. They have been mostly hindered in this by their current lack of capacity and knowledge to do so.

Another interesting aspect of the project is the number of the Waterboard's water goals which are linked to the nature development goals of the EHS (the National Ecological Network). In the last 10 yr the Province has sought out parcels of land in order to complete the linkage zones. In terms of ecology, the area along the Regge is ideal for the EHS due to the special biotopes found in the river. The surrounding woods and small scale landscapes also make the area a good candidate. In principle the estates benefit from being a part of these areas because they can choose whether or not they will participate in developing the various natural functions of the land. According to the Municipality the estates like being able to keep their options open for a variety of land uses. This aligns with the lack of coercion mechanisms present in "area development" restoration projects. This type of behaviour creates uncertainties for others in the planning progress. Since the project is considered to be part of the "area development" it has an official project status which is supported by EU funds. It is however still voluntary in nature because the government bodies will not use expropriation of lands as a tool to enforce cooperation.

The estates have very specific interests regarding which areas they choose to use for agriculture and nature. This is directly relevant for water issues, since the link between nature development and water courses is broken when considerations other than proximity to the water course are decisive to site nature development areas. Prior to the appearance of present day agriculture, economic exploitation of the estates was made through wood production; however, the estates are not well organized for the demands of modern large scale agriculture. In managing the estate lands they would for instance choose to change the designation of a plot of land to become naturalized as a response to whether or not they approve of the manner a certain farmer uses to manage his land. This causes problems for the Province since they would like to organize the EHS linkage quickly, and thus have specific interests in the lands adjacent to a continuous zone. The local government prefers to stay out of these negotiations because they recognize the ease with which conflicts arise. The EHS belongs to the Province's jurisdiction and thus the local government prefers that they take the lead.

The Waterboard sometimes requires that water levels be set higher in conjunction with the restructuring of the water system. There are a number of farmers who feel higher water levels negatively affect their ability to farm. This causes significant issues when these more traditional farmers are located near to other farmers who are interested in providing nature as an environmental service on their lands which is thought to require higher water levels. There is ongoing research on the effects of the water levels on farming, nature, fauna, etc. This is economically important because environmental services are supported by Blue and Green Services programs provided by the Waterboards and Provinces respectively, which are forms of "payment for ecosystem services". The higher water levels thus increase the profitability of some farmers that are generously compensated while reducing that of others. The benefits of the different water levels to the estates will also be an important result of the ongoing research.

The estates entered the project in this case to ask the Waterboard to help improve their water quality. The Waterboard took this opportunity to work with them to renaturalize areas as part of the overall Regge Restoration. Nevertheless it took years to prove to the estates that it would be in their best interest to cooperate as they continued to exhibit in this case their "wait and see" attitude. The original interest of the estates in raising the water tables was because the foundations of the castle could become unstable if they become too dry.

There are also examples which show promise for future collaborations. The estate of Westerflier has made an agreement with the Waterboard in which some land exchanges were successfully completed. There is also an example of a farmer in Diepenheimsbroek (north of Diepenheim) who was interested in developing large scale agriculture on his property. The

Figure 11.5 Upper Regge River: the very beginning.

(Source: Hans Bressers).

Waterboard and this farmer were able to agree on incorporating a number of high tech selective drainage solutions in order to solve the water level issues.

Locally, there are two groups involved that influence the process from a higher decision making level. The first is the Diepenheim Area Commission, and the second is a special committee on the exchange of lands that is responsible for pursuing the completion of landscape, water and nature goals of the area. In principle, legal instruments are available which force non-voluntary land use changes however they were avoided in this project. The reason for this is the historical background of the area. In the late 1990's a voluntary form of land re-allocation "Ruilverkaveling Administratief Karakter", was initiated. They had not proceeded very far in the process when at the beginning of the last decade, a new Reconstruction law was prepared that put the future of this tool at risk. This further halted actions of the local project as they were unsure of the future plans for this area. When the plans were finally developed, the Ruilverkaveling Administratief Karakter had indeed been removed as a tool that could be legally used. The local government and the farmers did not want to engage in a full-fledged non-voluntary land reconstruction project, because of experiences from other nearby areas like Rijssen and Haaksbergen, where land reconsolidation projects had lasted for 20–30 yr. The project teams involved in these processes had been working together for so long that they were able to celebrate when they had their 500th meeting. The actors in the Diepenheim process did not want that sort of planning and program and as such they chose to operate on a completely voluntary basis. This was also considered appropriate because the area has so many small-scale plots which they felt needed to be handled carefully so as not to destroy the landscape.

The real discussion was however not about the voluntary or involuntary nature but basically about whether or not the other parties that represent the landscape, nature, etc. should also be involved in the discussions regarding the exchange of land. This is important because legally all stakeholders must be involved in the process. This significantly increases the time required for the project. The alderman in this case made a strategic decision based on the recognition of the strong bargaining power of the farmers. The general opinion was that it seemed likely that the farmers would in the end be expected to alter their practices to conform to the new plans. This expectation reduced the enthusiasm with which the farmers would participate with an open mind, and increased the chances that they would take an approach designed to minimize the resulting restrictions on them. He solved this issue by conveying the situation to the farmers in a way that emphasized meeting the needs of the other groups. He suggested that the farmers should make their own proposal for how the different interests could be addressed. In such a way they needed to seriously consider how their proposal would be accepted by the other parties. By putting the ball in the farmer's court to make a plan with integrated solutions that would suit everyone's needs, they were able to get things moving and to get them involved in a constructive attempt to propose what was best. They predicted that otherwise they were not likely to participate openly and would continually be preparing for resistance to any proposed actions. The local NGOs agreed to this procedure since their experience in the past was that the farmers would continue to plague the process with complaints for a very long time (10 yr in one case) and would in the end eventually succeed. They too saw this as an opportunity to get the farmers active from the start in the joint process. A large amount of money was provided by the government to perform the land exchanges which significantly aided the process. This proved to be a good position to work from and overcame the previous concerns about the conflicts between nature, landscape, agriculture, etc. This voluntary method also had the benefit of getting things done more quickly because of the reduction in red tape. The risk of proceeding in this way is however that by compromising among various local goals you may not meet the requirements of the EU programs. In this case, the Netherlands would be required to pay some of the money back that was received for nature and biodiversity development. The risks associated with using involuntary measures can however also be significant and can include stalemates, increased court time and fees and reduced trust amongst the parties.

The Soil Exchange Commission has taken the same approach as that given to the farmers in the Diepenheim case. They have chosen to develop their own plans for land exchanges and until now they have been successful in getting general approval and consensus. The farmers are aware that they have to agree with the decisions of the commission and that in general the farmers are well represented by this. The estates have however remained a bit outside of this process as they generally do not exchange land except when it improves their position according to their own deliberately stated interests. There are examples of where the estates could have contributed to collective problem solving of this sort and they have chosen to abstain. In one case, they chose not to contribute to the construction of a bridge that would directly benefit one of their farmers. Sometimes the relationship between the estate owners and their farmers is not very supportive. It is also important to note that according to our interviewees the economic position of the estates, one of which has connections to the royal Dutch House of Orange, varies widely.

As mentioned earlier, one big disappointment occurred in this process where the estates did not want the proposed cycling paths to go through their properties. The Municipality wanted to create the possibility for continuous cycling all along the Regge. At that time cyclists had to travel partly along the road and instead the Municipality wanted to provide a route that followed along the river. The estate owner continued to refuse to participate because she was concerned about the effects of too many tourists and the negative impacts of the trash they would generate. Generally it was experienced by the Municipality of Hof van Twente, that the estates have limited interest in going beyond what they are required to do and are not supportive of collectively seeking what is possible to increase recreation and tourism. There are other areas however such as in the adjacent Achterhoek region where the estates are actively changing and are opening up their castles and increasing tourism (including restaurants and terraces). Some castles however feel that they can still survive with their traditional ways and choose to avoid going in this direction.

Another example of where these issues have arisen is the Kunstwerk Diepenheim ("Artwork of Diepenheim") project. This project supports both the Municipalities' and the Waterboard's goals (tourism and "experiencing water") and great efforts are aimed at further enlarging it by involving many new gardens and artwork. It is part of a 10 million Euro innovation project. The artists involved want to create a historical connection to the Huize Diepenheim and its gardens. As part of this they want to build a bridge and the Waterboard has offered to help fund it. The overall goal is to connect the various historical features in the area. The castle owner is concerned about the extra traffic and garbage that will result. It was made clear that this feeling was shared by Mr. Schimmelpenninck and thus backed informally by other estate owners. Additionally in the case of the project of Eelerwoude, Mr. Schimmelpenninck's influence was visibly exerted. The estate manager involved at the beginning with the Waterboard, allowed the negotiations with the farmers to be conducted almost exclusively through Mr. Schimmelpenninck. The actions taken in the preservation of the estate interests have caused issues in the development of the overall nature and water planning of the Waterboard and the Province. They play a strong role in many projects since they are such a large landowner.

Timelines are also a concern in terms of the land exchanges that take place. The Regge restoration project members had set a deadline of 2011 to complete all land exchanges with the estates in the realization of the water goals. At the end of 2018, the Province is committed to having the EHS (the National Ecological Network) completed. However the Municipality and the entire committee continue to strive to finish their project by the end of 2012 as they have already agreed to with the various stakeholders. They set a high priority on being action oriented and as such place emphasis on not becoming a "tea club".

There are still discussions taking place regarding the implementation methods and rates between the Municipality and the Province. The Municipality expects the Province to cooperate more in terms of providing the means to accomplish the tasks that have been asked of them. At the time of the interview (spring 2010) the provincial staff member that was interviewed was still under the assumption that the Province would continue to try and do whatever they could in the next few years despite the project not being on target to meet their goals. Money was not considered to be an issue for these investment projects due to the availability of reserve investment funds[1] which could be used even as budgets shrink. Later however, in view of the national budget cuts and low priority given to restoration by the new government that came into office in October 2010, the Province has decided to temporarily stop investing money in buying land for restoration purposes.

In the Netherlands, the European Nature 2000 regulation is translated into the Nature Protection Law and is often mentioned as a source of fragmentation and inflexibility from a regime level. One area protected under this legislation (a designated Habitat area) is located close to the Regge on the Estate of Weldam. This estate with beautiful gardens, has had big problems with increasing the size and intensity of their farms because of the Nature 2000 designation. The estate owner claimed that had he been aware of this consequence of the designation as a Habitat area he never would have suggested that this area become protected. The Municipality of Hof van Twente sees it as potentially disastrous when these protection laws are put into practice strictly in the manner in which they were designed by the EU parliament in Brussels. They are thus still negotiating with public administrators in the capital city of "The Hague" on this issue. The Hague is then required to work with Brussels to come to an agreement. The Borkelt is another habitat area in the vicinity. As a result of the designation there is a 4 kilometre area surrounding it that is restricted for development. A large scale agricultural area exists there and cannot be expanded despite it being officially designated as an "agricultural intensification" area under the provincial planning strategy.

In terms of the Water Framework Directive (WFD), there is an inherent risk that if the project remains voluntary that people would not feel the urgency of the nearing WFD timeline. Even as the timeline approaches, the Waterboard will not have the ability to use legal instruments to realize their goals. The Waterboard is at this point in time fortunate that there are other processes happening in the area like the art project that they can connect with.

[1]The reserve funds are actually known as "Essent money" which is a financial reserve that was previously earned by the Province through the selling of their shares in a large energy company.

Observations from the Theoretical Framework

Results: Dealing with Rival Uses

We begin by addressing the results of the project, both in terms of dealing with resource use rivalries and values of the resources involved. The term *"rivalry"* stems from natural resource management literature (Knoepfel and Nahrath 2005) and points to homogeneous overuse or conflicting heterogeneous uses of the resource that can threaten the sustainability of the resource. The drainage of the Regge and other creeks were once "improved" to serve agricultural interests and prevent the drying out of both castle moats and nature areas. As a result, the initial differences of opinion over the resource use were predominantly experienced over the level of the water table. The Diepenheim projects started out as an attempt to resolve these issues. New ambitions developed that attempted to add value to the natural and landscape resources and thus created new rivalries to be included and to be resolved by the projects. One such rivalry exists between the preservation of solitude and privacy associated with the (natural) areas of the estates and their sometimes centuries old ensemble and the wish to expand infrastructure to enable more recreation and tourism, e.g., the cycle path along the Regge. The estate values can sometimes collide with the construction of the EHS despite the fact that they need to be uninterrupted geographically to maintain their value. On the other hand, existing land uses are not often easily dismissed and estate owners prefer to determine for themselves where land for new nature would be most suitable for them. While restoration efforts generally require higher water tables, the actors who were satisfied with the previous situation, as many farmers were, see this as a new rivalry. New factual information is sought to show whether there really is a disadvantage or whether the preference for lower water tables is based on false assumptions of a specific case. It is also possible that high-tech solutions can be used to produce different water levels at small distances to overcome these rivalries. Finally, the remaining "new" rivalry that was uncovered was between the designation of a protected nature and its sometimes unforeseen consequences for agricultural practices and their growth.

Results: Adding Value

The identification of new rivalries that the Diepenheim Regge restoration projects create should not overshadow the new values provided to the natural resources. These enable the provision of more goods and services to numerous people and natural processes. From a general perspective, the new rivalries result as a consequence of reshuffling the old land use arrangement that primarily served agricultural interests into one that also serves the interests of other people and nature. It is also argued that in the old situation the rivalries were already present, but were just hidden because the challenging rival uses were only latent at the time. The extent of uses and users served was clearly enlarged by the project, perhaps not so much by new post-project use arrangements, but by the new realities created by the project itself. The project added value to nature and its corridors of passage, including those for fish. It added to the robustness of the water system, which is necessary for climate adaptation through adding retention capacity and buffering additional water to prevent droughts. Cultural historical values were also served, for instance by restoration of water mill creeks and basins, but also by preventing damage to the castles' foundations caused by dry moats. The value of the landscape and the "experience value of water" were served by the restoration itself, but also by joining forces wherever possible with projects such as the art and gardens of Diepenheim and by contributing to recreational infrastructure like marked walking and cycling paths. These positive characteristics and results have nevertheless taken a great deal of time.

Process Characteristics

This brings us to the process through which the project is designed and realized. What strikes one's attention is that partners have activly sought to engage each other in the form of a supportive set of actors from the very beginning. This was even seen to be true in the one instance, where the NGOs were left out of the design process and the farmers were given the first chance to design an integrated proposal. In fact, one can argue that this strategy was not really leaving the NGOs out, since they were aware that the alternative of the farmers dropping out of the interaction process and deferring to a "wait and see" attitude could greatly delay the process. In the instance of the Diepenheim gardens, the interaction with the citizens and their organizations was actively sought. The main actors seen throughout are the Waterboard, the Municipality, the estates, the farmers, and some citizen groups. In more of a background role, the Province, NGOs, and potential sponsors played a part. Apart from more informal contacts, an important platform is the area committee and its land exchange subcommittee.

Actor Characteristics

Relating to the conceptual framework used to guide this research, the actors have been analyzed through the specific motivations, cognitions and resources that impact their (inter)actions. While most of the motivations and resources are similar throughout the Regge restorations projects, here we highlight a few aspects of cognitions (frames of reference and information or interpretations of reality held to be true) that stand out.

The estate owners often work towards maintaining the continuity of their ways of life as they have been over hundreds of years. They hold a relatively dynamic vision of continuity: there has always been change occurring, however this takes place on a much longer time scale. There is also a cognizant desire to maintain control over these changes to ensure that they occur along the lines of what the estate considers as its core values. Their initiatives are often responses to changes that are induced from the outside, for example the inclusion of dry moats due to water drainage. The Municipality also sees itself as the governor of the area and finds it sometimes hard to see the estates' visions of reality as anything other than insensitivity to the recreational needs of the modern citizen. The Municipality thus views the estates as being negatively motivated regarding most of the proposed changes.

The Time Dimension

An interesting development has taken place over time with respect to how the Waterboard operates. Generally one can say that they have developed from a strictly engineering attitude towards managing water with an attitude in which "coupling", linking and producing synergies from various interests is regarded as their core business. This is what they call "contextual water management" (Kuks 2005, De Boer et al. 2011) and is not about foregoing their water goals, but about recognizing that working with rather than against other stakeholders avoids stalemates and produces more "value of water" to people. The Diepenheim projects were approached in a considerably advanced stage of this transformation of self-conceptualization. Looking towards the final stages, the farmers find themselves at crossroads. Those using traditional agricultural practices often consider themselves to be the modern entrepreneurs as a result of 50 yr of stimulation. There is also an increasing number that see this as a path without a future and are open to looking for alternatives. This is not just a matter of motives; it is also a frame of reference, which gives meaning to the information. An example can be seen in the previously discussed questions over the appropriate height of the water table. Generally, a lower water table has been considered better because of the heavy equipment that needs to be able to enter the field. Some farmers are more aware of and open to including the alternative argument that unseen, but real drought damage is in fact restricting yields more than damage from wet periods.

Institutional Arena: Rules of the Game

The project initiators could in principle choose to set the project up under various types of agreements or "rules of the game": voluntary on a case-by-case basis, a voluntary "area development" process or more legally specified forms of land reconsolidation. The last option was deliberately not chosen and only the first two were used. A setting that on the one hand appears to be "stronger" with more public authority (legal land reconsolidation) can be on the other hand perceived as risky and conflict prone. Preventing opposition is now regarded as superior to overcoming opposition. The nearby experiences where such processes under the traditional rules-of-the-game were implemented took up to 20 or 30 yr. Instead, in Diepenheim they tried to "dissolve" rivalries through a voluntary approach that would not evoke fear and preliminary anger and instead tried to create win-win package deals that would satisfy all stakeholders.

Applied Strategies

On the whole we see a wealth of strategies used by the actors in the Diepenheim cases to create a maximum likelihood of a positive setting for the institutional arena, actor constellation and their characteristics of motivations, cognitions and resources. Figure 11. 6 refers to the intervention points for such strategies and are presented in a summarized version below.

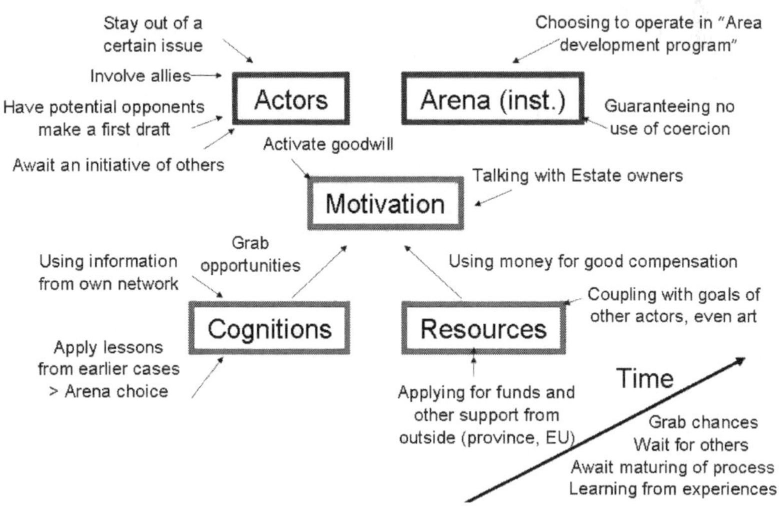

Figure 11.6 Overview of external strategies used in the Diepenheim cases.
(Source: De Boer and Bressers 2011).

Governance Regime Flexibility

In addition to this wealth of both proactive and reactive strategies, the Diepenheim cases also reveal some regime inflexibilities. The deadlines approaching for project completion could limit the use of certain strategies and techniques; the self-imposed 2011 deadline for land exchange for water management and the 2018 deadline for the completion of the EHS. There is however very little clarity about how these deadlines will be dealt with in practice. Another inflexibility is formed by the restrictions that are imposed by the presence of a Habitat protection designation in parts of the area. Here again, apart from the restrictions that are feared to reduce opportunities for work and development, the uncertainty about what extent these will really "bite" in practice is also a factor that hampers the progress of the process.

Strategies for Water Co-Management

An essential feature observed in all of the Regge projects is the coupling of several goals stemming from various policies and stakeholder's interests. This can in fact be very productive because competing claims for land use need not be mutually exclusive. Through the inclusion of these multiple policies, multiple arenas and actor constellations and multiple governance contexts also need to be joined together in what we can label as an "inter-regime" (De Boer and Bressers 2010). When the goals are similar, overlapping, mutually reinforcing or even unrelated, important synergies can be discovered. The nature NGO that was involved for instance accepted that the main project priorities are often related to water, landscape and recreation development, since they are confident that nature development will follow as a result of improvements in the other three. When the development of these synergistic projects also includes some level of trade off and compromise the resulting package as a whole can end up being better for everyone. Accomplishing this without falling into the traps that prevent achieving the best outcome is the key point in moving towards a successful project.

Not only in the Diepenheim case, but in all Regge projects the Waterboard applies the use of direct personal communication, which they regard as essential to preventing future problems. They promote the slogan of: "two days of drinking coffee in kitchens and living rooms is better than two years of dealing with legal consequences". This also reduces the risk of spending months in litigation and halting the project. It is thus felt that, using the most direct options for communication is the most productive strategy for cooperating with private landowners and inhabitants. The importance that the Waterboard places on this aspect was exhibited through its choice to have research performed about the people involved in one of the projects. The survey that was conducted was regarding how they experienced the program, the communication process and their level of participation.

With respect to institutional stakeholders a similar strategy was used to ensure everyone gathers at the table even when interests are perceived as being different. It can be that their positions are not really in opposition but that they are only dissimilar and so partnering had not been an obvious way forward. Creating the right atmosphere in which the actors do not begrudge gains for the others, and where the atmosphere is one where getting everyone's goals achieved to the greatest extent possible, are quite similar. Persistent communication and approaching each other as equals is the preferred method of undertaking these projects. It is believed that it works best when parties really attempt to do their best in helping to achieve each other's interests. This creates upwards spirals and in the end leads to higher rewards for all parties involved.

The development of a team atmosphere among all actors involved was one way in which they were able to accomplish these synergistic activities. When actors consider themselves to be and appear to be primarily members of the project team more so than representatives from their individual organizations it allows them to see the interests of the project as their own. This adds greatly to the likelihood of an optimal project design for all of the parties involved.

The determination of the actions or setting that leads to the development of this sort of "cooperative-game" situation and the avoidance of competition is an important task. In the Netherlands, there have been examples of projects which did develop into the sort of competition situation which was actively avoided in the Regge projects. In this way they still see the process as a sort of game but more so where one can only benefit or win at the expense or detriment of the other players (parties). Even when the game is played in a fair way, this preconception influences the likelihood of achieving a well-integrated arrangement or agreement. "Who is getting what and who gets the most" becomes the central question. The question that must be asked is "how do you get the people to adopt this other frame of mind which revolves around joint project development?" Integrated project teams are thought to be of key importance in the process. The art is in finding the right and most important players to make up the team. Various parties and their goals will always need to be met however it is most important to discuss them and work them out in the project team.

Repeatedly coming into contact and working with the same parties on various projects, was observed to ease cooperation in future projects. Participants already know each other, have developed trust between one another and they have also learned important information about each other and their organizations. Learning also occurs in the sense that they have learned what to do differently in their successive involvements with the project members. This makes it easier to find the right people to talk to and can also make certain processes less formal (and thus more efficient).

Some of the external strategies that are not only observed in the Diepenheim case, but also in the other subprojects of the Regge River restoration (see De Boer and Bressers 2011 for a full account) are presented in the following paragraphs. They are in fact not really separate, but clearly reinforce each other's efficacy.

Openness to Synergies with Other Policies' and Actors' Goals and Interests

The wealth of combinations of goals and interests that are observed in these projects is presented as a strategy as such. Openness to synergies is not only a way to make the most efficient use of public money from various sources, and of scarce space in a densely populated country, but is also a way to increase the likelihood of achieving *actor constellations* with supportive *characteristics (motivations, cognitions, resources)* for the progress of the process.

The Management of Relations

This strategy relates to the building of relationships and trust with other relevant actors before the project begins (*actor constellation, timing*). Often there is a choice of institutional arena, an option that exists because there are different legal and voluntary possibilities for framing, e.g., river restoration sub-projects. Sometimes it's better to refrain from institutional settings that provide legal coercion options, because they are hard to use and will only cause widespread resistance.

The choice for a voluntary approach (*institutional arena*) can thus also be seen as a strategy to improve the likelihood of development of sufficient trust and mutual commitment of the stakeholders involved.

Blurring the Boundaries of the Process Phases

In many examples we saw the involvement of some actors at the very beginning that would otherwise typically only appear at later phases of the process (*actor constellation, timing*). This occurred by asking landowners in the area and neighbouring citizens very early on in the process what their wishes for the development of the area were. What was very important in a number of the projects was the early involvement of Landscape Overijssel (or other nature organizations that would end up managing the project area). The traditional distinctions between the various phases of the process are deliberately blurred through this process. While this can increase complexity when done in an extreme manner, it can also prevent situations in which the later involvement of new actors blocks the process or provides other unpleasant surprises. One way to reduce the additional complexity is by dividing the project into smaller geographical sub-projects. This is exactly what we observed in the Regge restoration process (De Boer and Bressers 2011).

Surfing the Waves

The Waterboard also found that on a number of occasions it was not optimal to start a project on its own, but to wait and to latch onto an existing initiative or Area Development project/plan (*institutional arena, actor constellation*). Thus not the Waterboard, but the Municipality would be the main director of the process. This can have disadvantages under adverse conditions, but has mainly advantages for the Waterboard when the goals are in accordance with one another.

Seizing Opportunities When They Arise

There are also good examples of where the *timing was used advantageously*: opportunities that would support the broader restoration vision were taken as soon as they occurred. Actions that would enable the project to move forward with quick wins were taken in order to build momentum, leaving issues related to tougher areas for a later time when more resources are available.

Learning to Build Trust

Trust is also of key importance in the relationships between the members of project teams. Learning from past projects plays an important role: who to ask (or not), how to build trust, how to build informal contacts. Likewise, good cooperation can be presented as a positive example to support the development of relationships desired in the future. More generally, conceding on some issues can be used as a calculated risk to help to build a level of shared trust that will have returns later on.

Knowing Your Context

Proactive information gathering can result in acquiring information on municipal plans, which when received early enough can in turn enable cooperation on further studies that can be used to help inform decision makers. Getting acquainted with local knowledge can improve the projects as it is generally very useful to be aware of various types of opportunities. Chances to create goodwill in ways that can be included into the project without much difficulty (*cognitions*) are then less likely to be overlooked.

Seeking Alignment of the Characteristics of the other Actors

Through thoughtful and early communication it is possible to understand the *motivations* of the people involved and can make it possible to influence them. A rather interesting strategy practiced by the Waterboard related to this was to deal with the opposition of some neighbours, not by being responsive to their official complaints, but rather instead to the objections that they presumed to lay behind them. As such they adapted the controversial plans in such a way that the concerns of the inhabitants regarding their loss of view of the river were removed. Following the first negative court decision of their initial appeal, the proponents accepted the decision without pushing the case up to a higher court level. This was actually the only case in the Regge Renaturalization projects in which a legal objection was brought to court. This is considered as very successful since in the Netherlands court cases regarding land use changes are quite common. Actively investigating the interests of groups in the community is also done in order to increase resources in an innovative way through access to "community funding".

Direct Personal Communication

It is very important to have as much direct personal communication with stakeholders as possible. Directly communicating with farmers and neighbours is often the only way to overcome clashes of fundamentally different "readings of reality" (*cognitions*). Open consultation is also the key when dealing with institutional stakeholders. It is also important to be creative in a way that aims to be able to support each other's interests and thus creates an upward spiral which eventually results in the development of other valuable resources, such as trust. Consequently it is not just a matter of communicating, but also of being open and moreover trying to advance others' interests whenever they are or can be made sufficiently compatible with one's own.

Strengthening Position in Advance

Purchasing land in the time preceding project development in order to hold a private landowner *resource* position in the area is also often used as a strategy. Sometimes this is a matter of stepping into a "window of opportunity" at the right time such as when a farmer decides to quit farming and is willing to sell his land. In several ways this kind of resource can be put into use during later phases of the process; the land itself can be used for the project, although it can also be exchanged for other lands which are needed for the project. Buying land before a project has been developed is of course an investment, but it also has the benefit of avoiding both resistance and possible price pressures compared to buying the land when a project needs to be realized at a particular spot.

Conclusion

The Regge Restoration project is an intermingled and complex group of projects designed in a strategic way to overcome obstacles and maximize opportunities. The use of co-management and other strategies towards the improvement of the natural ecosystem and the social, cultural and economic environment requires the researcher to look past the typical project management phases to understand the dynamic and innovative processes at work. External influences from various policies and resources as well as internal histories, relationships and geographical contexts all contribute to the development of a successful project. The strategies highlighted here show how deliberate attempts to increase the owners of an integrated project can be used to take advantage of project momentum, trusting relationships and synergistic opportunities to overcome inflexible policy intrusions, lack of resources and historical segregations and conflicts between actors. Only in this way can such projects maximize their value for sustainable regional development.

References

Boer, C. de and H. Bressers. 2010. New Rurality: The Netherlands country screening, Report New Rurality project ANR-08-STRA-09, Enschede: University of Twente, pp. 1–137.

Boer, C. de and H. Bressers. 2010. "Inter-regime" Effects on Local Stream Restoration Projects, Advancing Sustainability in a Time of Crisis: International Society for Ecological Economics (ISEE) 11th Biennial Conference, August 22–25, Bremen and Oldenburg Germany.

Boer, C. de, H. Bressers and S. Kuks. 2011. 'Coordination of Policies and Governance: Regime Requirements in Dutch Freshwater Management', Policy Quarterly Volume 7, Issue 4.

Boer, C. de and H. Bressers. 2011. Complex and dynamic implementation processes: The renaturalization of the Dutch Regge River. University of Twente books. Enschede.

Boer, C. de. 2012. Contextual Water Management: A Study of Governance and Implementation Processes in Local Stream Restoration Projects. (thesis) Enschede: University of Twente. ISBN 978 90365 3427 7, pp. 1–265.

Boer, C. de and H. Bressers. 2012. New Strategies for Implementing Locally Integrated Stream Restoration Projects, in Environmental Practice, 14, pp. 26–34.
Bressers, H. Implementing sustainable development: how to know what works, where, when and how. pp. 284–318. *In:* W.M. Lafferty [ed.]. 2004. Governance for sustainable development: The challenge of adapting form to function. Edward Elgar, Cheltenham, Northampton MA.
Bressers, H. and S. Kuks. 2006. Water basin regimes in Europe and institutional conditions for their sustainability. pp. 235–268. *In:* V.I. Grover [ed.]. Water: Global common and global problems. Science Publishers. Enfield NH USA.
Bressers, H. and K. Lulofs [eds.]. 2010. Governance and complexity in water management: Creating cooperation through boundary spanning strategies. Edward Elgar, Cheltenham.
Jeffrey, P. and R.A.F. Seaton. 2004. A conceptual model of 'receptivity' applied to the design and deployment of water policy mechanisms. *In:* Environmental Sciences, 2003/2004, 1(3) : 277–300.
Knoepfel, P. and S. Nahrath. 2005. Sustainable management of natural resources: From traditional environmental protection policies towards institutional natural resources regimes (INRR). Idheap, Chavannes-Lausanne.
Kuks, S. 2005. 'Contextueel waterbeheer', presentation with PowerPoint for the IPO, the Association of Dutch Provinces, Ellecom, 16 June.
Leussen, W. van. 2011. Leven met water: Waterbeheer in Nederland volgens de stroomgebiedbenadering in een tijd van klimaatverandering. University of Twente. Enschede.
Projectteam Evaluatie Nationaal Bestuursakkoord Water. 2006. Evaluatie Nationaal Bestuursakkoord Water. LBOW, Rotterdam.
Schaap, S. 2010. Klimaat en overstroming: Een verleidelijk verband. Inaugural speech. Technical University of Delft. Delft.
Williams, P. 2002. The competent boundary spanner. *In:* Public Administration 80(1): 103–124.

12

Irrigation Water Co-Management in The Limarí River Basin, Chile

Alejandro León Stewart[a,*] and *Rodrigo Fuster Gómez*

Location of the Limarí River Basin and its Climate

The Limarí province—which comprises nearly the entire Limarí River basin (Fig. 12.1)—is one of three provinces in the region of Coquimbo. This part of the country is relevant for the purpose of this chapter because a) the region corresponds to the semi-arid realm, and hence, irrigation water is scarce; and b) the basin contains the so-called "Paloma System" that consists of three interconnected reservoirs allowing water storage for irrigation during 5 or 6(dry) yr, and its administration has been recently transferred from the state to private irrigators.

Rains are highly seasonal in this Mediterranean climate, concentrating 75 percent of precipitation in the southern hemisphere's winter (June, July, and August); summer rains do not occur, limiting the development of vegetation. The mean annual precipitation in this province varies from 100–200 mm with a high inter-annual variation: there is a high frequency of dry periods that can last 4 to 6 yr, and rainy periods of 1 or 2 yr (León, 2008). The Limarí province has a total surface area of 13,461 km^2 and an

Department of Environmental Sciences and Renewable Natural Resources, Universidad de Chile.
[a]Email: aleon@renare.uchile.cl
*Corresponding author

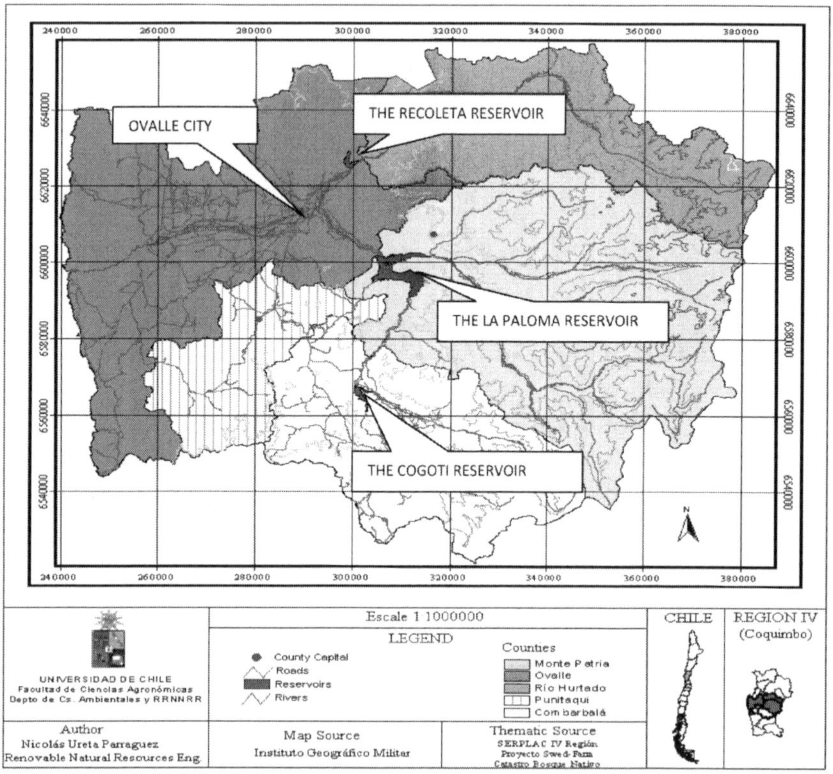

Figure 12.1 The Limarí Province and the Limarí River basin (adapted from Fuster 2006).
Color image of this figure appears in the color plate section at the end of the book.

irrigated portion that varies between 40,000–60,000 ha depending on the availability of irrigation water in the Paloma system.

Water Resources and Infrastructure

The "La Paloma system" has a storage capacity of about 1,000 hm^3 of water. The three inter-connected reservoirs can be characterized as follows: La Paloma reservoir, finalized in 1969, stores 750 hm^3 and covers an average of 3,000 ha. It is located 23 km east of Ovalle city at the confluence of the Grande and Huatulame rivers. Its tributary hydrographic basin is 6,253 km^2; the Cogotí reservoir, built in 1939, stores 150 hm^3 and is located at the confluence of Cogotí and Pama rivers 43 km to the south of the La Paloma reservoir. It covers 850 ha and its tributary hydrographic basin is 1,450 km^2; and the Recoleta reservoir, built in 1934 can store 100 hm^3 and is located on the Hurtado River, 14 km upstream the confluence of the Hurtado and Grande rivers. It covers 555 ha and its tributary hydrographic basin is 2,210 km^2.

National Irrigation Policy Highlights

In terms of public policy two key changes occurred in Chile during the last decades of the XX Century. First, the institutional modifications in the legal framework regarding economic activity; and second, a decrease in public investments in irrigation infrastructure. Both decisions were inspired by the application of a neoliberal economic model (as opposed to a centralized system with a strong presence of the State in place until 1973), and the utilization of the State's subsidiary role concept. Under this concept, the State should not have a relevant role in matters that could be accomplished by private parties. Some activities were deemed "reserved" for the State, such as the national defense, while others were to be transferred to the hands of private parties—production of agricultural and industrial goods, irrigation, and the provision of most services. Others—such as education and health care—could be shared by both. Eventually, in the case of irrigation water, as will be seen in the following pages, the Chilean State transferred the management of the Paloma system to private parties after a period of public management and another, transitional period, of water co-management.

Regarding water and irrigation, two landmark policies were passed in the 1980s: i) the *Water Code* (which corresponds to the Decree of Law N° 1,122), which has been widely cited as an exemplar case of the application of an extreme liberal economic model, passed in 1981 (MOP n.d.), and ii) Law N° 18,450 aimed at promoting the investment in on- and off-farm irrigation infrastructure, passed in 1985.

The 1981 Water Code is the main legal body regulating water management in Chile. It created a water market in accordance to the neo-liberal economic model and the subsequent market mechanisms to allocate resources. In its 5th article, the Code states that "water is a national good for public use", and article 6 indicates that the "right of exploitation" of water is given to private parties as "a real water right" and consists in the possession and use by the right holder. According to Bauer (2004) the most important features of the Code are 1) strengthening of the private property condition of water rights; 2) the fact that water rights are separated from the land ownership; 3) that there is freedom to determine and change water utilization by the right holder, in the sense that it can be decided privately if and how to use it; 4) that the rights are conceded perpetually by the Sate at no cost[1] for the right holder; 5) that the State has a reduced interference in water-use conflict resolution, and private negotiations and the courts are the way to solve them; and finally 6) that irrigation (and agriculture) is the dominant factor in this Code.

A right corresponds to a proportion of the quantity available in the water supply source such as a river, canal, or reservoir. Different types of rights are recognized by the Code. For example, according to frequency, rights can be permanent or temporary. Permanent rights enable "water use in the corresponding amount, except when the supply source does not have enough quantity to completely satisfy them. In that case the flow will be distributed in proportional parts among different users." This type of right allows the distribution of water from supply sources that have not been declared as 'exhausted' by the authority. Temporary rights enable "water use only in times when the flow has an excess amount and the permanent rights have been served" (MOP 1981).

In the Limarí basin water rights were originally applied as early as 1928, and they currently range between 1–1.4 l/s depending on the water source. To date, few market transactions have taken place here since water is scarce and highly valued. However, after the Code was passed in 1981, many smallholders and communal landowners decided to sell part of their few rights (Fuster 2006) "as a fast way to improve their living standard."[2] According to the authors' field experience in the basin, some of those sellers have probably, in not few cases, returned to peasantry after spending the sale's lump sum, contradicting the notion of Cristi and Vicuña (2001) who suggest that the ability to sell water rights improved their living standard on a permanent basis. These authors underscored this notion as one of the benefits of the water market.

The second relevant instrument is Law 18,450, which established incentives for on- and off-farm irrigation investment on improvements by the private sector. Through the application of this subsidy, the State originally funded 75 percent of the investment (during the 1990s this proportion changed, especially in the case of small farmers, reaching a 90 percent of funding) and thus helped increase the cropping area, allowing the planting of new crops such as table grapes, citrus, and avocados, and fostering private investment in cold-storage, transportation and other related services for agriculture. As a consequence, there has been a major change from traditional (like furrow) irrigation methods to more modern, technified ones such as drip irrigation in private lands. In 2007 these crops

[1]An authors' note: an amendment to the Code passed in 2005 as Law No. 20,017 prevents perpetual water rights by stating that in cases where the applicant does not use the right within a reasonable time, he/she would have to pay a non-use fee.

[2]Palqui Canal Manager, personal communication, January 12th, 2007.

covered a surface area of 17,677 ha, almost a 40 percent of the irrigated lands. Other relevant crops are fresh produce and fodder.

In this sense, the orientation of this policy was clear, shifting responsibility from State to private intervention, as is reflected by the *State of Chile Development Program, 1983–1989*, which states "the construction of irrigation infrastructure will be a responsibility of the private sector. The State will only fund major irrigation works (such as the reservoirs of the Paloma System) when the economic and social evaluations show that the social benefit is higher than the private one. However, management will be the users' responsibility" (República de Chile 1983–1989, cited by ODEPA 1994). As a consequence, no major or medium-size projects were constructed by the State between 1974–1989. After democracy was reestablished in 1990, large irrigation infrastructure projects were reborn. One exemplar case is the Puclaro reservoir, in the nearby basin of the Elqui River, whose management was immediately transferred to the private irrigators. However, an editorial from a local newspaper highlights the fact that the advancement in reservoir construction has been poor: since 1995 only four of them have been added to the irrigation infrastructure (i.e., Santa Juana, Corrales, El Bato, and Puclaro). The newest addition will be the Ancoa reservoir which will be filled up in September 2012. The 3,000 beneficiaries will have to fund a 30 percent of the total investment (=US$ 122 million). But, and according to previous experiences (e.g., that of La Paloma System) the farmers tend not to honor their signed agreements, and they have pressed over their local representatives to decrease or even avoid their payment (El Mercurio 2012).

Local Irrigation Initiatives in the Region of Coquimbo and the Province of Limarí

In 1914 the Ministry of Public Works started studies on the potential improvement of irrigation in the Elqui, Limarí, and Choapa basins (which altogether conform the region of Coquimbo). Different divisions of the ministry, such as the Irrigation Inspectorate, the Department of Roads, and the Hydraulic Inspectorate were the first institutions that promoted economic development in Coquimbo through the construction of several works that are the basis of the current regional economy.

In 1948 the "Urbanization Plan of the Province of Coquimbo" launched the plan to construct the Paloma reservoir in order to store water of the Grande River. Having started its operation in 1967 this reservoir completed the Paloma System and normalized irrigation in more than 85 percent of the approximately 50,000 ha of the irrigable land in the Limarí river basin.

The cost of the Paloma system was, in current value:

- *The Recoleta reservoir*: The dam had an estimated investment cost of US$ 17.9 million, including expropriations, technical inspection, and canals building. Other associated works had an estimated cost of US$ 21.7 million.
- The Cogotí reservoir was built at an estimated investment cost of US$ 23.5 million including both the dam and the canal network. Improvements made in 2003 had a cost of US$ 2 million (Manager of the Cogotí Reservoir, personal communication, August 2004).
- The Paloma reservoir was built between 1959 and 1967 (MOP 1978) with an approximated investment cost of US$ 117.8 million (MOP 1961).

Thus, the total investment in the three reservoirs and canals adds up to approximately US$ 183 million. However, according to local experts the real amount should be close to US$ 300 million in current currency.[3] In terms of the efficiency of the utilization of public funds these amounts compare positively to the annual revenues obtained just by the fresh fruit exporting business, which in 2007 summed up to almost US$ 35 million.

Water Users' Organizations (WUO)

Article 186 of the Water Code states that "if two or more parties draw water from the same river, reservoir or canal, or use the same facilities for pumping groundwater, they will be allowed to conform different types of Water Users Organizations (WUOs) with the purpose of drawing water from the source, or distributing it among right holders, constructing, exploiting, conserving, and improving all related infrastructure." Whenever water is drawn from natural water courses (i.e., a river), users organize a *Junta de Vigilancia* (JV). If the water source is an artificial canal, the organization is denominated a Canal Association (CA). The *Juntas* and the CAs are private non-profit, organizations, funded by their members.

The JVs and CAs are legal entities and both have the legal power to solve conflicts among users (Puig 1998). The government structure of both types of organizations is identical and is based on a general assembly, a board elected by the assembly (each member has as many votes as water rights he holds), and a president elected by the board who represents the entity either judicially and extra-judicially (Puig 1998). According to the Code, the board must also hire a manager in charge of the technical aspects of

[3]Jorge Romero, former DOH Director Region of Coquimbo, personal communication, February 18, 2004.

water distribution. Both the JVs and CAs are in charge of resolving conflicts among their members (Puig 1998).

In the Limarí river basin there are four JVs and five CAs, which consist of the Paloma System. Approximately 6,200 right holders are members of these organizations. Every year and once the rainy season is over, the DOH (Directorate of Public Works in Spanish) together with the organizations allocate the quantity of water that corresponds to each right holder according to the amount of stored water, the modeled projections of the snow pack, and the distribution of water rights (Alfaro and Honores 2001). The following criteria are considered for distribution during the irrigation season:

a. If the water storage is 1000 Hm^3, there is free distribution,
b. If the availability is more than 500 Hm^3, 320 Hm^3 (85 percent) are distributed, and
c. If the availability is less than 500 Hm^3, 50 percent of the availability is distributed.

Allocation and Distribution

The construction of the Paloma reservoir provided 240 Hm3 of additional water that could be distributed among the existing organizations and water right holders (Sandoval 2003) according to the original allotment of rights (Table 12.1). This additional amount of water increased irrigation security up to 85 percent.[4] With the existence of this reservoir, the irrigated surface

Table 12.1 Potential amount of water from the Paloma Reservoir that JVs and CAs can receive per annum.

ORGANIZATION	Maximum annual delivery (m^3)	% to be distributed
CA Recoleta Reservoir	64,966,728	27.07
CA Cogotí Reservoir	59,528,000	24.80
JV Grande and Limarí Rivers	58,651,200	24.44
CA Canal Camarico	25,280,000	10.53
JV Hurtado River	9,433,000	3.93
CA Canal Derivado Punitaqui	8,544,000	3.56
JV Huatulame River	6,310,008	2.63
CA Canal Palqui Maurat-Semita	4,164,800	1.74
JV Cogotí River	3,121,992	1.30
TOTAL	240,000,000	100,00

Source: CNR 2008.

[4]Cristián Vílches, DOH, manager of the Paloma Reservoir, personal communication, January 24, 2011.

with water directly drawn from rivers decreased due to the positive impact of the reservoirs (Table 12.2).

Thus, there is a direct benefit for the land located downstream, while another portion of land receives indirect benefits: when the reservoirs are working at full capacity, irrigators upstream of the reservoir may use up to 100 percent of the water from rivers that feed the reservoir.

Table 12.2 Irrigated area directly and indirectly benefited by the Paloma system (figures are in ha and correspond to the estimated acreage when the system operates at full capacity).

Below Reservoir (Direct benefit)

Recoleta Reservoir	14.831
Cogotí Reservoir	13.083
Camarico Canal	5.500
Limarí River	8.661
Huatulame River	953
Punitaqui Canal	866

Above Reservoir (Indirect benefit)

Río Grande	3.723
Río Cogotí	2.083
Río Hurtado	3.283
Total Surface under direct and indirect benefit	52.983

Source: CNR 2008.

Co-management of the System

During the 1990s and the first half of the 2000s efforts were made by the DOH to transfer the complete managerial responsibility of the Paloma System to private users. Not all beneficiary organizations, however, agreed with this proposal arguing that the operation and infrastructure maintenance costs were too high. Until then, these costs were covered by the State.[5] Different stakeholders acknowledged the operational cost issue as a source of conflict between the organizations.[6] Another reason that the organizations argued to reject the private management was the inexistence of a WUO aimed at monitoring the correct application of the law, and at ensuring water distribution according to the amount of rights of each holder (Peña 2004). This argument resulted in a combined public-private effort to create a new organization with this role. In 1986 a consumptive right of continuous and

[5]Manager of the Cogotí Reservoir, personal communication, January 2006; and Manager of the JV of the Huatulame River, personal communication, January 2006.

[6]President of the Grande River and Limarí River JVs, personal communication, January 2006.

eventual exercise for an amount of up to 1,200 million m^3/year from the Grande and Huatulame rivers was created for the DOH[7] (DOH 2006). The purpose of this right was to allow the DOH to make managerial decisions in accordance to the legislation. This water right was transferred entirely to private parties. Hence, in September 2008 an agreement was signed between the DOH and six WUOs (the JVs of the Grande, Limarí, Hurtado and Huatulame rivers, and the CAs Canal Semita-Palqui, Canal Camarico, and Canal Derivado Punitaqui) and in April 17, 2009 two complimentary agreements were signed with the other WUOs to complete the transfer. A new WUO grouping the aforementioned organizations, in charge of the administration of the system, was created. It was the *Asociación de Canalistas del Embalse Paloma* (ACEP), to which the water rights created in 1986 were transferred. The administrative steps were to first transfer ownership of the infrastructure, followed by the operation of the system, and finally the transference of the water right to the ACEP (CNR 2008).

Thus, late in 2008 the Council of Ministers of the Government of Chile finally transferred management of the entire system. In a first stage, the users only had to cover the operational and infrastructure maintenance costs. The DOH also supported the WUOs so, in practice, a combined public-private administration begun operating on July 1, 2008. The public component included the infrastructure management (of the dams themselves) and the simulation of future water availability through hydrological models. The DOH also provided technical support to the new administrative entity in order to recruit management and technical staff.

As of 2010, private parties have been on their own for the operation of the system, and each water right owner facing the obligation to provide around US$ 0.50 per l/s each growing season to cover operational costs.

Conclusion

After 30 yr, the law passed in the 1980s was completely applied. There is yet, however, little evidence to show whether private management has been more effective than public, or *vice versa*. Thus, the State has finalized a combined managerial arrangement (or co-management) after more than 30 yr. This process had different stages: i) initially, the State determined water distribution in accordance to water rights originally assigned, and planned the short and mid-term usage based on climatic and hydrological conditions; ii) a transition period of a combined public and private administration during which private parties were in charge of water distribution according

[7]These rights can only be used when the volume of affluents to the reservoir is above normal, and all consumptive water rights of permanent use are satisfied by the water supply of all water sources.

to their assigned legal water rights and the State was in charge of managing the reservoirs; and iii) as of 2010, the private sector holds the property and management of the infrastructure, provides short and mid-term planning, water distribution, and other management roles. No State funding is provided and little intervention power is kept by the State.

This process has not been trouble and conflict-free. Disagreements between different WUOs were common, as not all were capable of running the system and covering its costs. Secondly, there was no consensus as to the organization that would be in charge of running the system and distributing water according to the rights. And last but not least, the role of the State in the new arrangement is questioned, as in accordance to the law, the State should not have played a role since the 1980s nor covered the financial costs of the transition to private management. But this chapter in Chile's water co-management history is not finalized yet.

References

Alfaro, C.A. and C.A. Honores. 2001. Análisis de la disponibilidad del recurso hídrico superficial en cauces controlados de las cuencas de los ríos Elqui, Limarí y Choapa. Thesis. Universidad de La Serena, Facultad de Ingeniería, Departamento de Obras Civiles. www.sad.dga.cl/ipaczo/ipaczo/ipac/jsp? session=13504866L4XV5.231202&profile=cirh&source=~!biblioteca&view

Bauer, C.J. 2004. Canto de Sirenas. El derecho de aguas chileno como modelo para reformas internacionales, Bakeaz y Fundación Nueva Cultura del Agua, Bilbao.

CNR (Comisión Nacional de Riego), 2008. Traspasan administración de embalse La Paloma a regantes de Limarí. *In:* Dirección de Obras Hidráulicas DOH: Sistema Paloma. Documento de difusión. Dirección Regional de Obras Hidráulicas.

Cristi, O and S. Vicuña. 2001. Mercado de agua para irrigación: una aplicación al Sistema Paloma de la Cuenca del Limarí, Chile. World Bank-Netherlands Partnership Program (BNWPP). Trust Fund TF024014. Chile

El Mercurio. 2012. Editorial Lento avance de embalses para riego. Available at http://diario.elmercurio.com/2012/07/21/editorial/editoriales/noticias/68fde524-f72b-4eb0-a54a-278e54b3e5e7.htm

Fuster, R. 2006. Equidad en el acceso al agua en la agricultura. Evaluación del modelo chileno de gestión de las aguas: El caso de la cuenca del río Limarí. Master Thesis. Universidad Autónoma de Barcelona, España.

León, A. 2008. Políticas macroeconómica y de promoción del riego y cambio climático en la montaña semi-árida de Chile. Revista Pirineos 163: 111–133, ISSN 0373-2568.

MOP (Ministerio de Obras Públicas), 1961. Memoria Anual. Santiago de Chile.

MOP (Ministerio de Obras Públicas), 1978. Memoria Anual. Santiago de Chile.

MOP (Ministerio de Obras Públicas), 1981. Memoria Anual. Santiago de Chile.

MOP (Ministerio de Obras Públicas). n.d. The 1981 Water Code. http://www.dga.cl/otros/publicacioneslinea/archivos/codigodeaguas.pdf

ODEPA (Oficina de Planificación Agrícola), 1994. Políticas de desarrollo del riego en Chile: 1974–1993. Documento de Trabajo No. 2. www.ciren.ci/cgi-bin/cedoc/wxis?isisscript=plus.xis&mfn=010429&base=biblo

Peña, H. 2004. Sistema de asignación/reasignación. En Investigación sistémica sobre regímenes de gestión del agua, El caso de Chile. Global Water Partnership. www.eclac.cl/drni/proyectos/samtac/inch01304.pdf

Puig, A. 1998. El fortalecimiento de las organizaciones de usuarios para una gestión integrada de los recursos hídricos. Paper presented at the International Conference "Water and Sustainable Development." Paris. www.bvsde.paho.org/bvsacd/cd08/usvarious.pdf

Sandoval, J. 2003. El riego en Chile. Gobierno de Chile, Ministerio de Obras Públicas, Dirección de Obras Hidráulicas. Santiago, Chile.

13

Adaptive Co-Management and Learning

Developments in Coastal Management in the Netherlands from 1985 to 2010

Leon M. Hermans,[a,*] *Scott W. Cunningham* and *Jill H. Slinger*

Introduction

The complexity of managing our water and coastal systems is increasingly being fed by two simultaneous drivers. The first is the need to involve various actors, and accommodate their sometimes conflicting demands and interests. The second is the need to cope with uncertainty and change. Whereas the first challenge can be addressed by more interactive and participatory forms of management, such as co-management, the second challenge can be addressed by more adaptive forms of management. In combination, this leads to adaptive co-management (Berkes 2009). Adaptive co-management is a response both to the demands of the network society and to dealing with uncertainty and change. Adaptive co-management requires

Delft University of Technology, Faculty of Technology, Policy and Management, P.O. Box 5015, 2600 GA, Delft, The Netherlands.
[a]Email: L.M.Hermans@tudelft.nl
*Corresponding author

learning. Learning about changes in the conditions and environment that influence the system that is being managed, as well as learning about the possible ways to respond to these changes.

Learning and the use of feedback information is a key aspect of adaptation. However, the involvement of multiple actors in co-management processes poses additional demands and challenges to learning. Multiple actors need to engage in collaborative learning efforts and multiple actors need to decide on desirable ways to adapt to changing circumstances. How do actors engage in such learning efforts, and how do they decide on adaptations? Do actors learn from the impacts of their policy decisions? Is there evidence of deliberate learning efforts, and are decisions over time connected? In other words: how do actors learn in situations of co-management?

We address this question for coastal management in the Netherlands in the period from 1985 to 2010. Here, multiple actors have been involved in a mutual effort to maintain the coastline in the Netherlands. We will consider the coastal management case over a period of 25 yr, as policy change and learning often become visible only after a longer period. In this period, an initial decision was made in 1990 to maintain the Dutch coastline actively, through an annual program of sand nourishments along the shoreline, first on beaches and later also underwater through shoreface nourishments. In the first years after this decision, some six to eight million m^3 of sand was distributed annually along the coast of the Netherlands, at locations where erosion was most severe. The 1990 decision for a structural annual program of sand nourishments has culminated, provisionally, in a 2010 decision to place a mega-nourishment of 20 million m^3 of sand on the coast near a small village called Ter Heijde. This will create a semi-permanent artificial island in front of the coast. This island will erode in a period of 10 to 20 yr—or it may be replenished at some time in the future. Can we trace the logic that connects these decisions? Who were the actors involved, how did they make their decisions, and what role did learning play in the subsequent decisions?

To address these questions, this chapter first reviews key insights on adaptive co-management and learning, after which events in the case of coastal management in the Netherlands are reconstructed. This is done using concepts from the game theory and using the notion of critical assumptions as key elements to connect decision-making and learning. We then compare the learning needs as identified in this reconstruction with the manifested learning efforts, after which we conclude with observations on the role of learning in the adaptive co-management of the Dutch coastline over the last 25 yr.

Adaptive Co-Management and Learning

Co-management, Adaptive Management and Adaptive Co-Management

Co-management is rooted in collaborative management and in community-based approaches. According to Carlson and Berkes (2005), most views of co-management essentially see it as a power sharing arrangement between states and local users. Indeed, some authors describe such co-management processes in particular, discussing for instance arrangements between fishermen's organizations and government agencies (Beem 2007). This strand of co-management literature bears close resemblance to similar discourses on multi-stakeholder platforms in water management and to other forms of community-based approaches. However, there are also somewhat broader definitions of co-management, such as the one provided by the International Institute for Environment and Development (IEED) and the International Union for the Conservation of Nature (IUCN): "a partnership by which two or more relevant social actors collectively negotiate, agree upon, guarantee and implement a fair share of management functions, benefits and responsibilities for a particular territory, area or set of natural resources" (Borrini-Feyerabend et al. 2004). This description is broader, and no longer specifies the exact character of the social actors involved.

Here, we will build on the broader definitions of co-management, whereby we assume that, in many instances, local communities are represented by local government-bodies or by non-governmental organizations and associations. This is in contrast to a situation where local communities first need to be organized as part of a co-management effort. Such may be the case when water user associations or watershed councils are established specifically with the purpose of allowing users and communities to participate in the management of natural resources. In many instances, at least in countries with well-developed governance institutions, communities are represented by local governments such as towns and municipalities, or by business associations and interest groups. These representative bodies in turn receive inputs and signals from the local users or members, without many local users having the need or desire to become actively involved in co-management themselves. An active engagement in co-management provides influence, but also requires the investment of time in participation and in gaining the knowledge and experience necessary to act as a knowledgeable partner. Thus, the assumption that local users will want to become active partners in co-management arrangements, is not always warranted (Beem 2007).

In recent years, co-management is increasingly combined with adaptive management, as co-management is viewed as evolving and continuing over a longer period of time (Berkes 2009). A central notion of adaptive management is the need to learn and to support resilience, as a way of coping with the uncertainties inherent in the management of complex socio-ecological systems. In comparison to co-management, adaptive management focuses more on the resource managers and decision-makers, rather than the resource users and communities (Berkes 2009). In the merger into adaptive co-management, the scope also broadens and includes all participants, not just decision-makers and managers and not just local users versus state governments.

Given our emphasis on learning, we connect to the newly emerging adaptive co-management paradigm, which we interpret to be broader than the original co-management concept. Where co-management in the "strict" sense refers to collaborative arrangements between communities and states, we refer to collaborative management arrangements in the broader sense, as arrangements between various actors, be they government entities, private sector and non-governmental organizations and interest groups, or non-organized resource users. In essence, there is the notion of (one or more) central government agencies versus other actors that can be considered representative of a more decentralized and networked system of actors (cf. Berkes 1994 as cited by Plummer and Fitzgibbon 2004).

Learning in Adaptive Co-Management

As adaptive management and co-management discourses have connected in recent years, more attention is being paid to learning in relation to co-management. Reviews of literature on learning in relation to adaptive co-management indicate that the more influential schools of learning theories in this field include social learning, experiential learning, and organizational learning. These schools stress the influences of different aspects on learning processes.

Social learning has received quite some attention in relation to adaptive management and adaptive co-management. Often, it is linked to the work of Bandura, who pointed out that individuals learn from observing what others do (for instance Pahl-Wostl 2006). From this basis, social learning is then applied to refer to the social processes of learning: Learning by a collection of individuals and/or organization, or through participatory processes. Often the benefits are emphasized that such more open and inclusive processes can have in terms of learning, knowledge mobilization and knowledge generation. Reed et al. (2010) provide an overview of the various ways in which social learning is used in recent works. To clarify concepts they turn to work on social networks and participation, but also to the theories of Kolb and Argyris and Schön, to which we also turn.

Kolb stresses that "learning is the process whereby knowledge is created through the transformation of experience" (Kolb 1984). The cornerstone of the experiential learning theory is a learning cycle that includes the four elements of concrete experience, reflective observation, abstract conceptualization and active experimentation. Furthermore, according to Kolb learning involves activities of "grasping" and "transforming". Combined, this may result in four different types of knowledge. Kolb is famous for the identification of different learning styles, whereby different individuals typically use different ways of learning, shaped by psychological, educational and professional characteristics (Kolb 1984). We note that experience, either through an apprehension of concrete experience, or as generated to extend comprehension through active experimentation, is a key aspect in the learning cycle. Furthermore, we note, based on recent developments in experiential learning theory, such as the meta-cognitive experiential learning model shown in Fig. 13.1, that learners tend to learn about objects through linkages of monitoring and control (Kolb and Kolb 2009). A similar process may be expected to occur when learning about a co-management arrangement and/or a natural resources system.

Experience also plays a key role in the work by Argyris and Schön on organizational learning. Argyris, Schön and Kolb explicitly ground

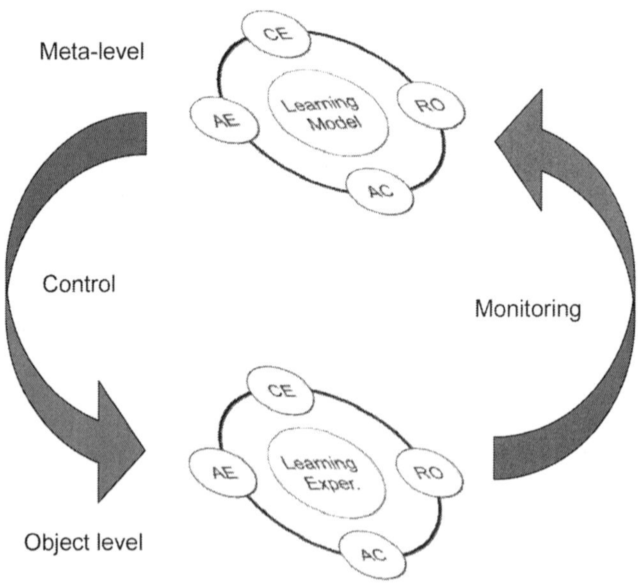

Figure 13.1 Meta-cognitive experiential learning model based on Nelson.
(Source: Kolb and Kolb 2009).

their work in the philosophy of John Dewey. According to Argyris and Schön(1996):

> *"In Deweyan inquiry ... doubt is construed as the experience of a "problematic situation," triggered by a mismatch between the expected results of action and the results actually achieved. Such a mismatch—a surprise, as we experience it—blocks the flow of spontaneous activity and gives rise to further action aimed at re-establishing that flow."*

Also for Argyris and Schön, learning is connected to knowledge. Key in their work is the representation of organizational knowledge through "theories of action" (1996):

> *"We define a theory of action in terms of a particular situation, S, a particular consequence, C, and an action strategy, A, for obtaining consequence C in situation S. The general form of a theory of action is: If you intend to produce consequence C in situation S, then do A. Two further elements enter into the general schema of a theory of action: the values attributed to C that make it seem desirable as an end-in-view and the underlying assumptions, or the model of the world, that make it plausible that action A will produce consequence C in situation S."*

Thus, for Argyris and Schön, learning requires a reflection on the underlying values and assumptions. By "testing" a theory of action against the responses experienced in a real-world action situation, organizations may hope to learn. Combining this with the "meta-cognitive model" discussed by Kolb and Kolb, monitoring of events and developments at the object level will need to be part of this learning effort (see also Borrini-Feyerabend et al. 2004: 'Learning-by-doing' through monitoring and evaluation").

Types of Learning

Publications of the past few years indicate an interest in learning about co-management arrangements and in facilitating such learning among the participants in co-management arrangements (e.g., Heylings and Bravo, 2007; Armitage et al. 2008). In these works, attention is given to the role of co-management arrangements in fostering the sharing and co-production of knowledge and learning through the social interactions that take place in co-management processes, which is then connected to the notion of "social learning" (e.g., Berkes 2009). In another set of publications, the focus is more on learning about co-management arrangements. Here, attention is given to learning about the process through which co-management works. Is everyone included, heard, are arrangements equitable, do they enhance legitimacy (Heylings and Bravo 2007)? Beyond such "process"—learning on the process of co-management, is the need to learn about the outcomes of the co-management processes themselves, in terms of the ecological and livelihood components (Plummer and Armitage 2007). Learning will not be easy, as highlighted by Armitage et al. (2008), who describe a learning paradox in relation to adaptive co-management. Complications arise because, among others, learning involves risks, learning is not value free, and there is little experience on which to build learning efforts (Armitage et al. 2008).

When talking about learning, one can distinguish different types of learning. There is a distinction to be made between learning about processes and about substantive outcomes of the processes (cf. Plummer and Armitage 2007; Koppenjan and Klijn 2004). Armitage et al. (2008) also make a distinction between what is called single-loop (instrumental) learning, double-loop (fundamental) learning and triple-loop learning. This distinction rests on the three types of learning discussed by Argyris and Schön (1996) in their work on organizational learning. Instrumental learning refers to learning on how to better realize intended objectives, and about the efficiency of the instruments employed. Double-loop learning involves a reflection on the objectives and underlying values: why do we want to achieve this? Is this really what we value most? Deutero-learning, in the terminology of Argyris and Schön (1996), refers to "learning how to learn." This is a type of process learning, whereas, in the case of natural resources management, single-loop learning often addresses substantive learning, about impacts on the resource base and the ecological system.

Ultimately, learning-how-to-learn should be striven after as the other forms of learning follow logically. However, studying more direct forms of single-loop learning is at least as important and interesting: How do participants learn about the developments in the physical system, related to the natural resources that are being managed in this collaborative fashion? Such simple substantive single loop learning will generally provide the basis on which higher-order learning becomes meaningful in the context of natural resources management. Generally, actors participate in a co-management process because they have an interest in the functions provided by the physical systems. They act on these systems, thereby altering the outcomes of these systems. We think it interesting to trace single-loop learning, based on information about the physical system, and to place it in a broader process of co-management.

Ultimately, higher-order learning is conditioned by lower-order learning. We do not think that these lower-order learning processes can be taken for granted as complete and without constraints or complications. Literature on evaluations, as an important instrument for learning, provides ample support for this caveat.

Why is Simple Substantive Learning Difficult in Co-Management Settings?

Literature on learning and evaluations in multi-actor environments indicates that there are several challenges that need to be addressed in these settings. They include: who should learn and who is responsible for joint learning processes? Who can claim success or who would receive blame for failure? Whose objectives are to be taken into account in looking for instrumental lessons? How to address the known difficulties in learning about complex environmental issues? We will shortly address these challenges in this section to explain why even lower-order learning processes are difficult. Because of these difficulties, lower-order learning processes are interesting fields of investigation: how do actors in practice respond to these challenges and dilemmas?

Who should Learn? Who is Responsible for Learning?

If different actors are involved in a co-management process, with different roles and responsibilities, it is not necessarily clear who should take the initiative or the lead to ensure learning occurs. Collecting and storing monitoring information is key to learning, but in co-management processes such monitoring information can be considered as a public good, as the demand for the knowledge produced by this information tends to be spread out across many actors—and across time (see the discussion of evaluations as a public good by Levine and Savedoff 2006). When monitoring information is being collected and stored, it may be relevant only for a specific group of actors at a specific point in time.

An obvious way to ensure that learning processes fit with co-management practices, would be to ensure that these learning processes are taken up jointly, by a group of actors. However, practice indicates that the fear of being held accountable for perceived failures prevents many actors from cooperating with, let alone initiating, a thorough independent and public learning effort (cf. Van der Meer and Edelenbos 2006). When it comes to learning, many policy-makers and decision-makers are more interested in "success stories" than in failures but unfortunately, the outcome of a learning process may go both ways (Levine and Savedoff 2006). Also, power differences and the structure of participation arrangements are likely to influence the opportunities actors have for learning (Armitage et al. 2007).

Who Receives Credit for Success? Who is to Blame for Failure?

Also, if different actors are involved in a co-management process, collaborating in policy development and implementation, it will be difficult to determine who is to be held responsible for what (Van der Meer and Edelenbos 2006; quoting Mayne 2003). And if everybody is responsible, who is to bear the consequences in the case of failure? Or, who may claim success? Also, it may mean that an actor who was part of a failed policy process risks being associated with that failure even if the particular actor was not responsible for this failure. Wildavsky (1987) describes a paradox of policy learning whereby small errors are easy to correct, but hard to spot. Large errors are the other way around: easy to spot but difficult to correct. For these reasons, these tasks are often specialized in a bureaucratic environment, with correspondingly different incentive structure. Designing a policy explicitly to go for the small or the big wins is therefore highly consequential.

Whose Values Count? Whose Objectives are to be Monitored?

Even more fundamental is the question of values and policy goals. Although many policies have explicitly stated policy objectives, these may not be the only possible policy outcomes of interests to those involved in co-management. This applies especially to learning in relation to natural resources management, where significant externalities and side-effects often occur (cf. Gysen et al. 2006). This poses a risk to learning and monitoring approaches that focus too narrowly on the (official) objectives of the sponsoring agency. Official policy objectives hopefully motivate a public agency that bears policy responsibility, but other actors may cooperate or hinder policy implementation for completely different reasons. On the other hand, agreeing on goals to be included in learning processes may be a difficult and time-consuming process (Armitage et al. 2008). Furthermore, including all possible values and objectives in evaluations may well result in an expansion of the learning agenda beyond the realms of practical feasibility. Thus, learning in multi-actor systems must pose the question: "Whose values are to count? (cf. Sanderson 2000)."

Single-actor Complications to Learning in Water Management

Finally, in natural resources management, the notion of system complexity is important. Often, the systems that are being managed, are insufficiently understood or known. Threshold effects and time-lags make it difficult to connect interventions to outcomes, and often, policy instruments may also be changed long before their actual impacts are visible (Sanderson 2000);

in these cases, it will be even more difficult to establish cause-and-effect relationships between policies and observed system changes.

Methodology

Case and Data Collection

Even the simplest form of substantive learning, in the form of instrumental or single-loop learning, is not self-evident in multi-actor settings such as co-management. Hence, we have studied if and to what extent such learning has occurred in the development, implementation and adjustment of coastal policy and coastal zone management in the Netherlands over the past 25 yr (1985–2010).

We have done so by using a combination of data sources. During the entire process, we have collaborated closely with a few key experts in Dutch coastal policy, who have acted as sounding board and key informants. This was done as part of a joint research project into "Games and Learning in Delta Infrastructures." Semi-open interviews were conducted with 21 respondents, who represented various roles and positions in Dutch coastal management process during the past 25 yr. Insights from these interviews were combined with information available from policy documents, research reports and meeting transcripts, as well as with information from analyzing physical system data that were available in national coastal monitoring databases for the period. The resulting reconstruction of the processes in Dutch coastal policy was discussed and fine-tuned in a one day validation workshop with independent key experts (whereby the participants of the workshop partially overlapped with the set of interview respondents).

Analysis Framework: Games and Assumptions in Different Rounds and Institutional Levels

As the co-management dimension was central in our investigation, we have sought to represent key events in decision-making through a game theory lens. Inspired by previous political scientists who have used game theory as a tool or lens (e.g., Scharpf 1997; Ostrom 2005), we have sought to model each game through the identification of the following elements:

- Who were the main actors involved?
- What were the rules of the game?
- What decision was at stake?

Furthermore, to gain insight in the strategies and the learning needs of the various actors in these games, we addressed the main elements that are known to influence actor behaviour in a network context (Hermans and Thissen 2008). Thus, for each main actor we have sought to address their:

- Objectives (what did they want to achieve, what was important for them?)
- Resources (what means did they have to further their objectives?)
- Perceptions (how did they think the "world worked" ? What was their perception of the state of the system and the "game"? And what causal relations did they assume?)

Looking into objectives and perceptions of actors allowed us to connect learning processes to the games as they evolved over time. For each game, we identified some of the critical assumptions that actors made, while playing those games. Critical assumptions are typically those assumptions that are very important in warranting a decision, but, at the same time, are quite uncertain (Walker et al. 2001; cf. Mason and Mitroff 1981). These are key elements for learning. Critical assumptions may well be proven wrong. And if proven wrong, this would be ground for reconsidering earlier decisions, and possibly adapting policies.

As a last step in our reconstruction we then compared the identified critical assumptions with the reported monitoring efforts during the studied period. What efforts were made to obtain information on impacts on the physical coastal system? Here, we looked at monitoring both in terms of annual monitoring programs as well as ad-hoc monitoring through more isolated research-projects or incidental data-collection activities.

In order to limit the research to a feasible scope, we made a selection of decisions to model, based on a more complete time-line. The selection was motivated by the desire to trace possible linkages between an initial policy decision that was made on the national level, down to particular local implementation choices years later. Rather than staying on one particular institutional level, for instance national level policy-making, we wanted to see if learning and adaptation processes would indeed connect national level decisions with experiences from local level implementation.

Coastal Management Processes in the Netherlands, 1985–2010

Rounds and Layers in Coastal Management in Netherlands from 1985 to 2010

Roughly, three main phases in Dutch coastal management can be distinguished, departing from the decision in 1990 at national level to establish a structural coastal management policy.

1. Pre-1990. Ad-hoc maintenance of sandy coasts and increasing awareness of severity of coastal erosion affecting the sandy coasts of the Netherlands. Development and discussion of national coastal policy to address coastal erosion. Ended with official establishment in 1990 of first official coastal policy. This policy required a dynamic preservation of the coastline of the Netherlands. This meant that coastal erosion would be combatted through the use of annual sand nourishments (initially mainly on-shore), in order to safeguard sustained safety and maintaining the existing functions of the dune area.
2. 1990–2002/3. Implementation and evaluation of the 1990 coastal policy on dynamic preservation. In the 1990s, experience had to be built with the implementation of the dynamic preservation policy for coastline management. In part, implementation was devolved to the regional and provincial level. Annual recurring questions here were: where, when, how to implement the sand nourishments? During this period, the practice of dynamic preservation was refined and extended to include also underwater sand nourishments in deeper water.
3. 2003–2010. The trends started in the early 2000s continued. The policy agenda for coast management in the Netherlands broadened considerably, with new (ad-hoc) issues emerging, such as the need to address a number of urgent weak links in the Dutch coastline. The focus in coastal management broadened from an initial focus on coastline management and erosion control to a focus on integrated coastal zone management, giving more weight to economic and nature interests, next to the priority to maintain safety along the Dutch coast. And, fairly recently, especially with the advice of the second national Delta Commission, the initial question that dominated the first round is back on the agenda: How to continue with sand nourishments on the long-term? One possible option was further explored in the form of a mega-nourishment, a so-called sand motor, for which construction began by the end of 2010. This would create a semi-permanent coast extension south of The Hague.

Establishing a National Coastal Management Policy for the Netherlands (1980–1990)

Staging the Game: Experts as the Actors that Set the Agenda

Engineers and scientists working for the Dutch Public Works Department *Rijkswaterstaat* and at universities and research institutes thought it necessary to develop a proper policy for coastline maintenance in the Netherlands. The context for this early period is formed by the official completion of the Deltaworks, which were started in response to the devastating floods. Despite the desire for an overarching national policy there was not a structurally assured budget for countering coastal erosion (Hillen and De Haan 1993).

Insight into the magnitude and extend of coastal erosion was growing, although initially, there was not yet clarity on the preferred technical solutions to counter erosion. The practice in the ad-hoc measures for coast maintenance was to restore dune areas using sand nourishments on dunes, often from the land-side (Roelse 2002: 19). In the 1980s scientists and engineers gradually accepted the idea that coastal erosion was best combatted using sand nourishments on beaches.

Rules and Play of the Game

In the absence of an existing policy or regulatory framework, the rules of the game were relatively open; the game occurred within the rules and procedures set by the existing constitutional democracy in the Netherlands, meaning that ultimately parliament would decide, based on proposals developed by the Minister for Public Works, Water Management and Transportation. The decisions to develop such a national level policy, were driven by increasing recognition of the need for coordination of ongoing local and regional efforts for coast preservation. The upside of the choice for a soft solution based on sand nourishments was that it also secured coastal engineers of the support of the nature conservation groups who were concerned with the dune area (Rijkswaterstaat 1989). The coastal experts had thus a strong lobby for their policy proposal to maintain the Dutch coastline through the use of sand nourishments. The idea was that this lobby would then create some public pressure on decision-makers to release budgets for coastline maintenance (De Haan 1995).

At the end of the process, early in 1990 there was an agreement on a preference for a moderate policy alternative for coastline preservation along a continuum of potential policy options. However, this consensus over a policy alternative was not accompanied by consensus over the financial implications and the allocation of the required budget. This apparent stalemate changed due to an external event—storms early in 1990. A five day storm that year was severe and attracted national attention (De Haan 1995).

Critical Assumptions

Looking at the game as outlined above, one can see that three types of actors were involved: experts, societal interest groups (especially nature conservation groups), and the decision-makers who had the decision-making authority on budget allocation at national level. The outcome of the game was a decision to maintain the coastline through the use of sand nourishments, for which a budget of 60 million Dutch guilders would be available annually. Underlying this decision are at least the following critical assumptions:

- One key assumption underlying the outcome of the 1990 game, was that maintaining the base coast line, through sand nourishments, would be both effective and efficient to halt coastal erosion.
- It was assumed 60 million Dutch guilders per year would be sufficient for this.
- Another assumption was that this would then mean that safety would be ensured.
- Nature groups were supportive because they assumed that sand nourishments would be better for nature than hard maintenance interventions.
- Finally, the decision was made under the condition of the measured scenario of 20 cm sea level rise per century. The estimated increase in costs associated with an expected scenario of 60 cm sea level rise seems not to have been taken into account.

Implementing the New National Policy: 1991–2001

In parallel with the preparation of national policy, a white paper, also a new Coastal Defence Bill (*Wet op de Waterkering*) was being drafted (Roelse 2002). Part of this new Bill was the establishment of Provincial Consultative Bodies for the Coast (*Provinciaal Overlegorgaan Kust*, POK). These new provincial consultative bodies had an official role in the new annual procedures in which the yearly distribution of sand along the coastline were to be established. These procedures mandated some room for interpretation and discussion.

These resultant discussions show a conflict of interest, at least in emphasis, between safety, recreation, and nature. The coastal policy was designed primarily to serve safety interests. *Rijkswaterstaat* had been made responsible for this Base Coast Line maintenance until the dunefoot. The regional water boards were responsible for securing safety through the maintenance of the coastal defence structures, such as dunes (see Fig. 13.2). Hoarding sand in the dune areas, as previously mandated by water boards, was no longer necessary or supported. Further pressure on water boards increased to allow more room for other functions in the dune areas. Ultimately equating Base Coast Line maintenance with long-term safety, as done on national level by *Rijkswaterstaat*, was not accepted for areas with narrow stretches of dunes by the local safety keepers, the water boards.

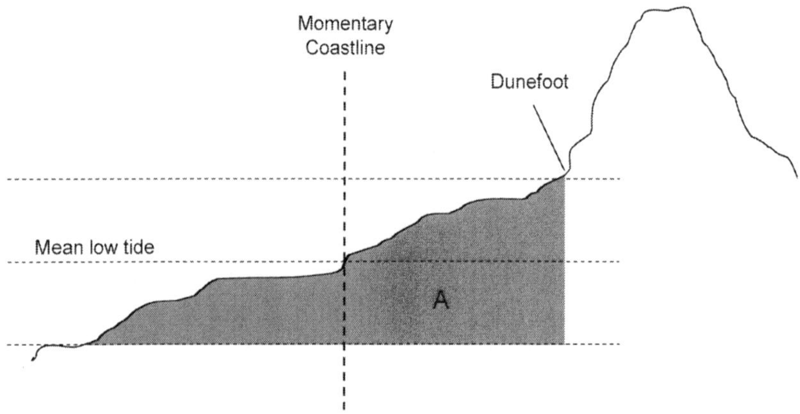

Figure 13.2 Calculation of coastline position.

(Source: Roelse 2002. Also contained in RIKZ, 1993 and other documents).

Furthermore finding funds for additional nourishments for recreational purposes was often difficult, especially for smaller municipalities. Other issues, but seemingly less important, were related to nature conservation concerns.

Critical Assumptions

In these annual games in the provincial consultative bodies, the actors involved based their reasoning on, inter alia, the following assumptions:

- Maintaining the base coast line primarily served safety interests
- Maintaining dunes for safety was easiest to organize by prohibiting other activities such as construction of buildings or paths and recreation by the general public
- Local economic development would be served by allowing more access to dune reserves
- Nature interests are best served by leaving nature undisturbed from human activity and intervention
- Recreation interests were best served by stability—i.e., keeping pavilions and coastline on fixed locations (no problem of moving pavilions, and importance of a beach width that would be neither too broad nor too narrow for tourists)
- The main alternative to sand nourishments would be hard measures like dikes and concrete water works.

Looking at these assumptions, it is possible to identify a scale on which to plot frequency and size of local sand nourishments, favoured by different interests: recreation was best served by frequent but small sand nourishment (annually), safety by somewhat less frequent and larger nourishment (once every 4 to 6 yr). Finally, nature interests would best be served by leaving certain areas undisturbed—even less frequent nourishments, if at all.

Sand-motor Game (2002–2010)

The Sand-motor and its Underlying Philosophy

In the second half of the 1990s, there was more attention paid to spatial quality concerns, next to safety, in the political decision making over coastal management. In the province of South-Holland, these processes led not only to plans to improve the coastline, but also to a pilot with a so-called sand-motor. This sand-motor, also known as sand-engine, basically is a large pile of sand that is used to create an artificial appendix to the coast. The sand-motor would gradually change shape and erode over a period of 20 yr or more, to nourish beaches and dunes further north. Eventually the sand-motor should be fully incorporated into the dunes and beaches, thus contributing to a broader and safer coast. The sand-motor is located near the municipality of Ter Heijde, in between the cities of The Hague and Rotterdam. It will support coastline management up to the beaches of The Hague to the north. Proponents of the sand-motor see it as an important innovation that they have labelled "Building with Nature" (http://www.sandengine.nl). However, creating a large artificial peninsula with a lifespan of at least 20 yr also reduces for instance the flexibility of the sand nourishment strategy that was heralded in the 2009 National Water Policy.

The process culminating in the sand motor featured some political wheeling and dealing across different institutional levels. The sand-motor pilot was championed by the province of South-Holland, which was looking for ways to expand it coast to meet needs for recreation and nature. Simultaneously, the idea of a sand motor had been raised within the Ministry of Transportation, Public Works and Water Management. The Ministry was looking for more innovative ways to further optimize the technology of sand nourishments for coastline preservation. Eventually, the Province and the Ministry became the key funding actors in a larger coalition of national and local government agencies, researchers and the dredging industry, who jointly supported a pilot project with a sand-motor.

Despite the coalition building process several parties raised serious doubts. Municipalities feared the effects that the sand-motor might have on (the access to) their beaches and port. At Rijkswaterstaat, some experts had questions about spending large sums of money on a pilot which was not designed and located at an optimal location from a safety perspective. In later stages, concerns were raised by the local drinking water company which operated a dune water filtration site near the sand-motor. Local inhabitants raised concerns over swimming safety, due to changing currents. None of these concerns proved insurmountable; all concerns were effectively resolved, sidelined or neutralized.

Critical Assumptions

The decision to construct a sand-motor was explicitly framed as a pilot project. Officially, according to the Environmental Impact Assessment report, there were four main objectives: safety, nature, recreation, and knowledge development and innovation. Reasoning from those objectives, one could say that at least the underlying assumptions were that the sand-motor, in its final design and location, would contribute to all these four objectives. Furthermore, the various alternatives for sand-motor design were evaluated also against the expected costs.

Monitoring and Learning Strategies

If we now look at this narrative account of decision-making events as games, it is clear that at each stage of decision making, the parties involved made certain assumptions. We highlighted certain critical assumptions for each of three games. These were the assumptions that were important for the outcome of the co-decision procedures, but they were also uncertain. Thus, these were typically the assumptions that could support learning, especially if they received attention in monitoring or research strategies. For this, these assumptions should be linked to indicators that could be measured and monitored in practice, to "test" assumptions against empirical data of impacts as they developed over time. Table 13.1 summarizes the critical assumptions and identifies some possible indicators for each of these assumptions.

Official monitoring and evaluation activities for these decisions were, and are still being, carried out by the government actors with formal responsibilities at the operational and policy levels. Especially *Rijkswaterstaat,* which maintained the JARKUS data system, and published annual "coastline books" with yearly results of monitoring. Also, Rijkswaterstaat and the Ministry undertook and/or commissioned specific evaluations of the adopted policies.

A comparison of the needs identified in our games and the official efforts for monitoring and evaluation shows that, for a long time, active collection of feedback information from the system was biased towards the information needs of the dominant actor in the game, *Rijkswaterstaat*. More specifically, the information needs of nature groups were long neglected. They have

Table 13.1 Critical assumptions and possible indicators.

Assumptions in Games	**Possible Indicators**
ROUND 1: Towards coastal policy 1990	
Maintaining the base coast line, through sand nourishments, would halt coastal erosion	Actual/momentary coast line with the base coast line as reference
	Volumes of sand used for nourishments
This would then mean that safety would be ensured	Calculated safety levels (with volume, width, height as variables?)
	Actual incidents, damage done
Sand nourishment to halt coastal erosion is an efficient way for coastal protection	Costs of sand nourishments (with estimated costs of alternatives as reference)
Sand nourishments would be better for nature than hard maintenance interventions	Biodiversity: species, population age groups, … (with no actual reference?). Dunes, beaches, but also sea (sand mining). Influenced by frequency and size of disturbance (sand nourishments), timing (seasons), exact location, sand quality (particle size)
ROUND 2: Provincial consultations on programmes for sand nourishments	
Maintaining the base coast line primarily served safety interests	As above, coastline and safety indicators
Maintaining dunes for safety was easiest to organize by prohibiting other activities	Safety indicators above. Costs for dune management water boards
Recreation interests were best served by stability—i.e., keeping pavilions and coastline on fixed locations, certain width of "dry beach"	Turn-over at beach pavilions, comparative, for varying dry beach width dynamics and correcting for other variables
Local economic development would be served by allowing more access to dune reserves	Local production, jobs—for varying access regimes (correcting for other variables)
Nature interests are best served by leaving nature undisturbed from human activity and intervention	See above for nature indicators (biodiversity)
The main alternative to sand nourishments would be "hard" measures like dikes and concrete water works	Cannot be tested, is (partially) a "mental construct"
ROUND 3: Sand-motor	
Sand-motor will help to support:	
Safety	As above, coastline and safety indicators
Nature	As above, nature indicators (biodiversity)
Recreation	Turn-over at local establishments, number of visitors before/after construction, additional recreational area created (target for province)
Knowledge and innovation	Among others: orders for similar sand-engines from elsewhere
	Longevity of effects (in all categories)

presumably for a longer time been relying more on theoretical assumptions and data from elsewhere, not backed by empirical data for the particular coastal system in the Netherlands. Likewise, data on recreational needs are limited, found only in two reports for specific locations, commissioned by provinces and published in 2009 (Decisio 2009; Ecorys 2009).

One of the consequences is that the knowledge on the non-safety aspects remains soft, whereas knowledge on safety aspects appears to be much more solid as it is backed by extensive monitoring records. This provides more room for interpretation when it comes to demands on these other "non-safety" interests. In this light, it is not so surprising that an evaluation conducted in 2002 found that there was a lack of clarity in how Rijkswaterstaat decided on trade-offs with recreational and nature interests in its decisions for annual nourishments (DHV 2005). But note that the accompanying recommendation was to develop more transparent decision procedures, not to collect additional monitoring data on the effects of past decisions.

Also, some critical assumptions have neither been tested, nor been probed. Probing could have been done for assumptions that were difficult to test, given the outcomes of decision making.[1] For instance, the assumption that maintaining dunes for safety is easiest to organize by prohibiting other activities, thus restricting access to dune areas and prohibiting construction of buildings or (cycling) paths. Testing this assumption is impossible, as long as restrictions are not being lifted. Still, probing could have been possible, by engaging in a dialogue to jointly explore the pros and cons. Instead probing has given way to advocacy, with parties arguing in favour and parties arguing against the validity of these assumptions.

As for the sand-motor, an extensive monitoring and research program accompanies the sand-motor. This program also intends to monitor impact on nature, in addition to impacts on coastal morphology and safety. Some recreational aspects are covered as high importance is given to monitoring the conditions for recreational safety, due to changing sea-currents that may affect the safety of people who swim or surf in the area. How other local and regional recreational impacts such as changes in visitors or spending, are being monitored is not yet clear. Although these recreational impacts were stated to be a key objective for the province, so far no specific efforts seem to have been taken on this aspect.

[1]We explicitly use the term "probing" here, in reference to the use of this term by Charles Lindblom (1990) Inquiry and Change, Yale University Press.

Learning and the Role of Monitoring Information in Coastal Management in the Netherlands

Adaptations in Coastal Management Since the 1980s

The developments in coastal management in the Netherlands have seen important changes and adaptations. The decision that was made in 1990, could be seen as an important change from the past. It marked a transition from predominantly hard solutions (groynes, dikes) in coastal engineering to soft solutions to combat erosion, as well as a transition from ad-hoc interventions to a structural annual intervention, signalling a more pro-active stance. This decision, driven for an important part by a coalition of experts, marked the start of a new paradigm in Dutch coastal policy. Despite adaptations and modifications in subsequent years, this paradigm itself has not been questioned, and still is accepted as the underlying basis for coastal policy in the Netherlands.

However, within this paradigm important shifts are visible. Annual discussions between actors in the provincial platforms highlight that additional interests in the functions provided by beach and dune areas were not always automatically served by sand nourishments for safety and coastline preservation. In particular sometimes recreational and nature interests posed conflicting demands, in terms of the frequency and size of sand nourishments. Although these demands have been accepted as legitimate, striking a balance and using the available budgets to cater to all these demands have proved recurrent bones of contention. It proved impossible to reconcile these differences in the provincial consultative bodies. Started in the early 1990s and given an official status in the Coastal Defence Act accepted early in 1996, these Provincial Consultative Bodies were again stripped of their formal role in the coastal management process in the Water Act that succeeded the Coastal Defence Act in 2007.

The soft paradigm for dynamic preservation of the coastline was initially accepted for its flexibility and its ability to respond to annual data on coastal erosion. Initial annual volumes for sand nourishments of 6–7 million m^3 were later expanded to 12 million m^3, and still later, debates were held over needs to further increase the annual volumes to 20 million m^3 in response to sea level rise. In 2011, a sand-motor was constructed, depositing 20 million m^3 of sand at a cost of some 70 million Euro. With these characteristics, the sand-motor is larger than the annual national nourishment program, both in size and budget, while covering only a relatively small stretch of coast. It still builds on the idea of "building with nature" in terms of using natural dynamics and sediment flows to maintain coastal safety. Although provisions have been made for certain adjustments if needed, once in place, such a mega-nourishment leaves less room for flexibility than the smaller annual nourishments; it is anticipated to suffice for a period of about 20 yr. Debates on the sand-motor have been quite heated at times, with experts disagreeing on expected impacts. The final design and location have been modified due to various constraints and considerations posed by actors both at local, provincial and national level. Although nominally the sand-motor is sponsored by a range of actors, in practice the province of South-Holland appears to have been the main driving force behind its realization, with effective support from key persons in the Ministry of Infrastructure and Environment—the successor of the previous Ministry of Public Works, Water Management and Transportation.

Did Learning Occur?

Although important changes can be observed, the question remains how these decisions were made, and what role was played by monitoring and other feedback information provided by the coastal system that was being managed.

In this case, it is clear that learning, based on monitoring information, did play an important role. Prior to the 1980s, data on the position of the Dutch coastline and nearshore environment were already being collected, in the form of so-called JARKUS data. These data played an important role in the initial decision made in 1990, and continued to play an important role in the implementation of this decision. Annual sand nourishment program as discussed in later years in the Provincial Consultative Bodies, were mainly informed by the calculations based on these data. Also, learning occurred in the community of coastal engineers and morphological scientists; they learned about the impacts of sand nourishments on coastal erosion, and they developed new techniques. Onshore nourishments were gradually replaced by underwater or shoreface nourishments, and currently the sand-motor is part of a learning effort to investigate the potential of replacing smaller annual nourishments with mega-nourishments. This learning has played a key role, mainly building on processes that were already engrained in the existing expert system in the 1980s: JARKUS data and morphological and geophysical expertise.

Concerns of newer actors, who became more active in coastal management in the 1990s, were acknowledged. However, no specific learning efforts were made, until very recently. From the early years, nature interest groups supported the policy choice for dynamic coastline preservation, based on the assumption that this would also be beneficial for nature. However, the actual ecological impacts of sand nourishments remained unclear and unknown. Although sand nourishments were practiced since 1990, only in 2009 an explicit provision was made (and funded) for research into the impacts of sand nourishments on ecology (Rijkswaterstaat et al. 2009). For the impacts of sand nourishments on recreation, no explicit monitoring efforts were being made. Or at least, they do not seem to play a role in the discussions on coastal management. Only in 2009, provinces commissioned some studies into these aspects. Coping with limited data, these studies focused only on a few particular beach locations, and did not yield results that would provide ground to contest prevailing knowledge (Decisio 2009; Ecorys 2009).

Although relatively late in the process, and more limited than learning on safety aspects, the coastal management process has opened up, which, in some ways, has given more room for experimentation and learning. The sand-motor is an example of this. This experiment was only possible by an active coupling of decisions and events by some of its political champions in the province of South-Holland. These champions were not the actors present in the 1990 decisions, but were new entrants into the coastline management arena. Only when these new provincial actors threw weight behind the experiment, was it possible for the dredging sector and certain experts to implement their experiment. At the same time, the experiment itself rests on several assumptions that are truly critical, highly uncertain and of key importance. Yet the project is considerable in terms of size and costs. Provisions for learning have been made for the sand-motor, but these provisions are most sound on the safety and morphological aspects, and less so for interests such as recreation.

All in all, it is clear that the actors involved in the co-management of the coastal zone in the Netherlands did learn, and did use monitoring information and feedback from the system. Nevertheless, there has been an uneven basis for learning, as not for all aspects and functions monitoring data were being collected and used. The collection of feedback information from the coastal system was skewed towards the State interests in safety and erosion control. This provided *Rijkswaterstaat*, as the main safeguard of these State interests, with an important leverage for their interests. The other parties involved mainly used the more incidental feedback signals they received from the community as indicators. Nature groups seem to have acted on signals from the public, municipalities and provinces have used signals they received from owners of beach pavilions, etc. Only recently, nature organizations have demanded more structural efforts for research and monitoring, to be provided by the State. This confirms that learning is not neutral, but a tool that can be employed in shaping and influencing co-management processes.

Monitoring Information and Learning as a Strategic Resource?

Although not all critical assumptions have been monitored with the same intensity, one cannot claim that actors deliberately have designed monitoring strategies to strengthen their position in subsequent games. In this sense, the neutral character of monitoring apparently has not been questioned by any of the actors involved.

Given that risk plays such an important part in coastal policy, the importance of perceptions is even more pronounced than in other policy making settings. Risk perceptions are influenced by crisis and visibility —hence the importance of the 1990 storm in changing the initial coastal policy game. With these risk perceptions, comes a feeling among experts of right and wrong risk perceptions. In line with theory (notably the political science work on the Advocacy Coalition Framework, see, e.g., Sabatier and Weible 2007), it is the opposing parties that are accused of having a "wrong" risk perception, based on misinformation and lacking a sound scientific underpinning.

If one then still wants to engage in a dialogue or a collaborative effort, this dialogue easily takes the form of education and information. Bringing science to local communities and informing local communities of the latest insights. Communication in the other direction is framed as hearing the concerns of local communities. Although these tendencies may have been reduced in recent years, they are still visible. For instance nature groups like to educate the general public, just as government agencies like to inform the public—especially if they have specific plans they want to realize such as a sand-motor.

Thus, the actors in the process do not question the neutral character of research and learning. And although they may be right that the findings of sound scientific research may be as neutral as can be, they do seem to overlook the fact that setting the agenda for research is a political process. The politics is not so much in the research itself, but in the choice of what is to be researched and what is not. For instance, in the absence of sound long-term monitoring data on ecological trends, the insights on morphological dynamics, supported by decades of data, simply carry more weight.

Conclusions

In this chapter we reviewed if and how actors learn in longer-term processes of co-management. For the case of coastal management in the Netherlands in the period 1985 to 2010, it is clear that actors did learn. They learned, in part, as a result of explicit strategies to collect and use monitoring information from the system that was being managed. However, there also appeared to be an uneven basis for learning, as monitoring data were not collected for all aspects and functions. Explicit learning efforts were skewed towards the State interests in safety and coastal erosion control. This has enabled learning on these aspects, but has also provided the actors with an interest in safety and coastal engineering with ammunition carrying considerable weight in policy debates. Only recently, has learning on other aspects such as nature and recreation been demanded and initiated. Learning thus occurs, but not as a neutral endeavour, but rather as a tool that can be employed in shaping and influencing co-management outcomes.

Looking back, it is possible to support more balanced learning efforts, by paying more attention to the critical assumptions that actors make in agreeing to decisions in a collaborative process. More explicit recognition of the diversity of assumptions is likely to broaden and balance learning efforts to more adequately reflect the interests and needs of all concerned. Still, blind spots are likely to persist. And, sometimes, it may be legitimate to focus learning predominantly on "one-dimensional" interests. For coastal management in the Netherlands, safety is the bottom line. It may well be warranted to focus learning efforts on this interest. However, the consequences of such a strong focus then need to be accepted: we can develop efficient and innovative ways to manage coastal safety, but we will remain impaired in managing our coastal zone to serve the multitude of functions valued by users in addition to safety.

Learning does occur and is effectively supported. This is the result of a political decision. Focusing monitoring and learning efforts on specific types of questions and issues not only provides knowledge and insights on these issues, but also carries weight in subsequent discussions and debates, whereas insights based on empirical data and analysis are more difficult to sideline than issues that remain uncertain and under-researched.

Acknowledgements

We would like to acknowledge the contributions of Ankie Bruens and Jan Mulder from Deltares, and Marja Menke from Arcadis, to the research reported here. Furthermore, we are indebted to the experts who participated in interviews and in the validation workshop. Our research was funded through a grant provided by the Next Generation Infrastructures Foundation.

References

Argyris, C. and D.A. Schön. 1996. Organizational Learning II. Theory, Method, and Practice. Addison-Wesley Publishing Company, Reading, Massachusetts (first published 1978).

Armitage, D., M. Marschke and R. Plummer. 2008. Adaptive co-management and the paradox of learning. Global Environmental Change 18(1): 86–98.

Beem, B. 2007. Co-management from the top? The roles of policy entrepreneurs and distributive conflict in developing co-management arrangements. Marine Policy 31(4): 540–549.

Berkes, F. 2009. Evolution of co-management: Role of knowledge generation, bridging organizations and social learning. Journal of Environmental Management 90(5): 1692–1702.

Borrini-Feyerabend, G., M. Pimbert, M. Taghi Farvar, A. Kothari and Y. Renard. 2004. SharingPower. Learning by doing in co-management of natural resources throughout the world. IIED and IUCN/ CEESP/ CMWG, Cenesta, Tehran.

Carlsson, L. and F. Berkes. 2005. Co-management: concepts and methodological implications. Journal of Environmental Management 75(1): 65–76.

Decisio. 2009. Toeristisch recreatief belang van zandsuppleties in Noord-Holland. Eindrapportage. Opgesteld in opdracht van de Provincie Noord-Holland. (in Dutch) Touristic recreational importance of sand nourishments in North-Holland. Consultant's report commission by Province North-Holland.

De Haan, T. 1991. Public Support, Keep it Awake. Paper for the IABSE Colloquium. International Association for Bridge and Structural Engineering. Nyborg, Denmark. pp. 279–286.

DHV. 2005. Beleidsevaluatie 'Dynamisch Handhaven'. November 2005. (in Dutch) Policy Evaluation 'Dynamic Preservation' Consultants' report commissioned by Ministry.

Ecorys. 2009. Pilot kosten en effecten strandafslag: uitwerking Brouwersdam en Bloemendaal. Opdrachtgever: Provincie Noord-Holland, Provincie Zuid-Holland, Provincie Zeeland. (in Dutch) Pilot costs and effects beach erosion: elaboration for Brouwersdam and Bloemendaal. Consultant's report jointly commissioned by three coastal provinces.

Gysen, J., H. Bruyninckx and K. Bachus. 2006. The modus narrandi: a methodology for evaluating effects of environmental policy. Evaluation 12(1): 95–118.

Hermans, L.M. and W.A.H. Thissen. 2009. Actor analysis methods and their use for public policy analysts. European Journal of Operational Research 196: 808–818.

Heylings, P. and M. Bravo. 2007. Evaluating governance: A process for understandinghow co-management is functioning, and why, in the Galapagos Marine Reserve. Ocean & Coastal Management 50 (3-4): 174–208.

Hillen, R. and Tj De Haan. 1993. Development and Implementation of the Coastal Defence Policy for the Netherlands. pp. 188–201. *In:* R. Hillen and H.J. Verhagen (Eds.). Coastlines of the Southern North Sea. Series Coastlines of the World. American Society of Civil Engineers, New York.

Kolb, D.A. 1984. Experiential Learning. Experience as the Source of Learning and Development. Prentice Hall, Inc., Englewood-Cliffs, New Jersey.

Kolb, A.Y. and D.A. Kolb. 2009. The Learning Way. Meta-cognitive Aspects of Experiential Learning, Simulation & Gaming 40(3): 297–327.

Koppenjan, J. and E.H. Klijn. 2004. Managing Uncertainties in Networks: A Network Approach to Problem-Solving and Decision Making. Routledge, London.

Levine, R. and W.D. Savedoff. The Evaluation Agenda. pp. 183–193. *In:* N. Birdsall [ed.]. 2006. Rescuing the World Bank. A CGD Working Group Report & Selected Essays. Center for Global Development, Brookings Institution Press, Washington, DC.

Lindblom, C.E. 1990. Inquiry and Change. The Troubled Attempt to Understand & Shape Society. Yale University Press, New Haven.

Mason, R.O. and I.I. Mitroff. 1981. Challenging strategic planning assumptions: Theory, cases and techniques. Wiley, New York.

Olsson, P., C. Folke and F. Berkes. 2004. Adaptive comanagement for building resilience in social-ecological systems. Environmental Management 34(1): 75–90.

Ostrom, E. 2005. Understanding Institutional Diversity. Princeton University Press, Princeton, New Jersey.

Pahl-Wostl, C. 2006. The importance of social learning in restoring the multifunctionality of rivers and floodplains. Ecology and Society 11(1): 10. [online] URL: http://www.ecologyandsociety.org/vol11/iss1/art10/.

Plummer, R. and J. Fitzgibbon. 2004. Co-Management of Natural Resources: A Proposed Framework. Environmental Management 33(6): 876–885.
Plummer, R. and D. Armitage. 2007. A resilience-based framework for evaluating adaptiveco-management: Linking ecology, economics and society in a complex world. Ecological Economics 61(1): 62–74.
Reed, M.S., A.C. Evely, G. Cundill, I. Fazey, J. Glass, A. Laing, J. Newig, B. Parrish, C. Prell, C. Raymond and L.C. Stringer. 2010. What is social learning? Ecology and Society 15(4): r1. [online] URL: http://www.ecologyandsociety.org/vol15/iss4/resp1/.
Rijkswaterstaat. 1989. Kustverdediging na 1990. Discussienota. 1 mei 1989 (in Dutch). Discussion paper coastal defence after 1990.
Rijkswaterstaat and Natuurbeschermingsorganisaties. 2009. Samenwerkingsovereenkomst zandsuppleties. 24 maart 2009. (in Dutch). Cooperation agreement sand nourishments.
Roelse, P. 2002. Water en zand in balans. Evaluatie zandsuppleties na 1990: een morfologische beschouwing.Ministerie van Verkeer en Waterstaat, Directoraat-Generaal Rijkswaterstaat. Rijksinstituut voor Kust en Zee / RIKZ. Report RIKZ 2002.003.108. (in Dutch). Water and sand in balance. Morphological evaluation of sand nourishments after 1990.
Sabatier, P.A. and C.M. Weible. The Advocacy Coalition Framework: Innovations and Clarifications. pp. 189–220. *In:* P.A. Sabatier [ed.]. 2007. Theories of the Policy Process Westview Press, Boulder.
Sanderson, I. 2000. Evaluation in complex policy systems, Evaluation 6(4): 433–454.
Scharpf, F.W. 1997. Games Real Actors Play. Actor-Centered Institutionalism in Policy Research. Westview Press, Boulder, Colorado.
Van der Meer, F.B.L. and J. Edelenbos. 2006. Evaluation of multi-actor policy processes: Accountability, learning and cooperation. Evaluation 12(2): 201–208.
Walker, W.E., S.A. Rahman and J. Cave. 2001. Adaptive policies, policy analysis, and policymaking. European Journal of Operational Research 128(2): 282–289.
Wildavsky, A. 1987. Speaking Truth to Power: The Art and Craft of Policy Analysis. Transaction Publishers, Piscatsaway, NJ.

14

The Roles of Knowledge in the Emergence of Co-Management Initiatives for Transboundary Groundwaters

The Case of the Génévois Aquifer

Martin Walter

Introduction

Underground water, or groundwater, accounts for some 95 percent of the global freshwater reserves and is one of the most critical sources of water for the future supply of cities, industries, and agriculture (UNECE 2007; Zekster and Everett 2004). Indeed, the use of these waters has increased steadily since the late 19th century, producing massive economic and welfare gains globally (Moench 2003; Morris et al. 2003). The intensive exploitation of groundwater has, however, also led to significant declines in the quantity and quality of the resource and poses significant political and economic threats (Llamas and Martinez-Santos 2005). Although groundwater problems have triggered the emergence of initiatives directed at the active management of groundwaters, internationally-coordinated

Northwestern University, Political Science Dept., Scott Hall, 601 University Place, Evanston, IL 60208, Institut d'Etudes Politiques de Paris, 199, boulevard Saint-Germain, 75007 Paris.
Email: martinwalter@u. northwestern. edu, martinwalter@sciences-po. org

policies for the management of shared groundwaters remain relatively scant, especially compared to those that have been formulated over the years for surface waters (Burchi et al. 2005; Delli Priscoli and Wolf 2009; Puri and Aureli 2005).

Explanations put forward to explain the lack of co-management arrangements argue that it may result either from the lack of knowledge and information about the resources (Salman 1999; Eckstein 2005; Eckstein and Eckstein 2003) or from the unwillingness and incapacity of political actors to translate available information into management policies (Matsumoto 2002; Jarvis et al. 2005; NATO and Darnault 2008). These explanations are rooted in more general theories about the emergence of internationally-coordinated policies that, it is argued, reflect the preferences of powerful States (Zeitoun and Warner 2006; Zeitoun and Allan 2008; Phillips et al. 2006), as well as bargaining processes between political stakeholders engaging within the constraints of particular institutional frameworks (Risse-Kappen 1995; Hochstetler 2002; Scott and Milman 2010). The literature suggests that the co-management of shared waters is the product of political struggles at different levels of governance; it claims that the co-management of transboundary groundwaters is the functional response to known problems according to the preferences and possibilities of the powerful. Despite the consensus that "power matters", existing theories advance multiple and at times contradictory interpretations of the role of knowledge in the emergence of co-management schemes for transboundary groundwaters.

This chapter examines the different roles of knowledge in the political process that led to the co-management of the Génévois Aquifer shared between Switzerland and France (see Fig. 14.1).[1] The Génévois Aquifer is the first documented case of formalized cooperation for the joint management of transboundary groundwaters and remains to this day one of the few successful cases of co-management of these resources. The case demonstrated that formal mechanisms can be used effectively to manage shared groundwaters, which has been instrumental to the development of international guidelines for the co-management of transboundary aquifers at the global level (Yamada 2004). It reviews the process of social recognition of groundwater problems, their entry into the political agenda, and the implementation of policies for their resolution. It emphasizes the impacts of scientific knowledge in the different stages of the policy process. The analysis focuses on often forgotten, but critical, constitutive dimensions

[1]The empirical materials that support the analysis were gathered during extensive field work in the Franco-Genevese region during Fall and Winter 2009-2010. It consists of original out-of-print documents and interviews with several political actors involved in the governance of the aquifer.

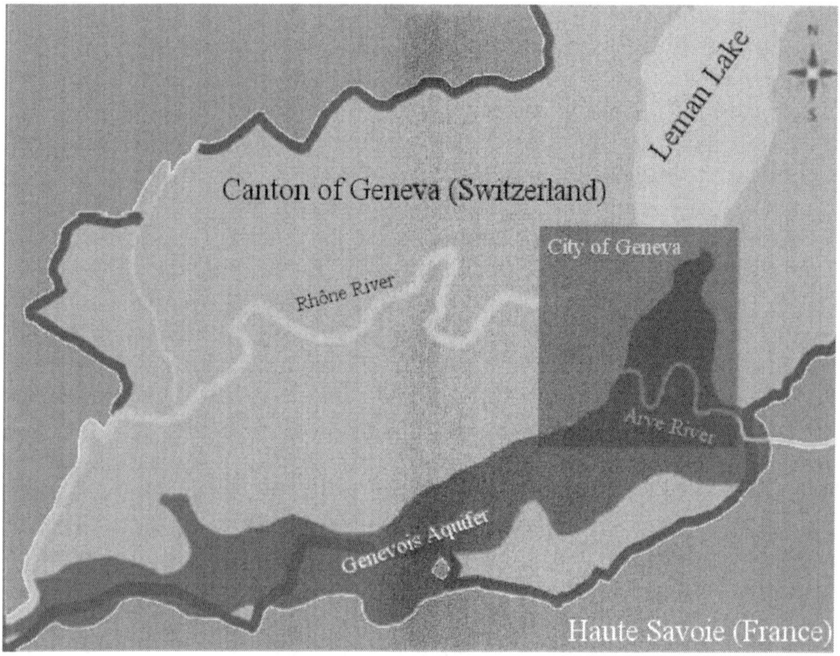

Figure 14.1 The Franco-Genevese Region and the Génévois Aquifer.

Source: Author based on outline available at http://etat.geneve.ch/dt/geologie/eaux_souterraines-270.html (accessed January 2010).

of decision-making scenarios: the original scientific assessments and local joint problem-framing that effectively reshape interests and challenge the legitimacy of established institutional frameworks.

The examination of knowledge and ideas in political processes is a classic theme in the social sciences. The literature has theorized about the different effects that new knowledge and ideas have on the emergence of new policy initiatives and institutional frameworks. Scientific knowledge and ideas are constitutive elements of this political process. They are resources of the actors in processes of political bargaining and operate as the cognitive framework according to which political preferences are shaped.

In order to equate knowledge and ideas with power resources, scholars have often pointed to the actors' awareness of the practical connection between the definition of a policy problem and the promotion of their interests (Bardwell 1991). The political agenda is indeed not determined by the issues *per se*; rather, it is determined by their social definition (Elder and Cobb 1983). Hence, political actors push particular issue-definitions into the political agenda, "policy images," that challenge existing institutional arrangements and political agreements and trigger policy change (Baumgartner 1993; Haas 1991). In fact, political struggles tend to revolve more around the political formulation of the issue than with the selection of the "right" solution (Bardwell 1991; Stone 1989). The process of problem-definition itself revolves around the specification of the politically relevant features of a phenomenon: it entails the composition of "causal stories" that attribute responsibilities and require political resolution (Stone 1989). This process is rooted in the use of particular language and symbols, and it tends to empower the actors that can legitimately mobilize information (Litfin 1994; Adler and Haas 1992). The corollary is that an idea's cogency—its intrinsic force, logic, and viability—is at least as important as its mobilization by political actors (Jacobsen 1995). Indeed, unless ideas effectively participate in the political process, meaning that they are mobilized along the specific interests of actors involved in the political process, they will have no effect on policy.

Alternative analyses of the roles of new information and knowledge in policy highlight their effects on the actors' preferred outcomes of policy (Goldstein et al. 2000). Weaver and Rockman (1993) argue that worldviews (paradigms) and culture are the factors that ultimately drive processes of institutional reform. Blyth (1999) is more blunt and claims that ideas may actually act as weapons against particular institutional systems. Indeed, new knowledge reshapes preferences in processes of policy bargaining because it is embedded with suggestions about better institutions and changing the terms of actors' interactions (John 2003). It structures paradigms and hermeneutic frameworks embedded with theoretical and ontological propositions that legitimize programmatic strategies and processes of policy change and reform (Conca 2006; Jasanoff 2004). In this sense, Hall (1989; 1993) notes that new ideas have an impact on the "interpretative frameworks" of policy: the shared understanding of the problems that policy seeks to redress, its means and its goals.

The aim of this chapter is to emphasize that water management regimes inherently reflect new knowledge and are ultimately vulnerable to it. New knowledge and information about the Génévois Aquifer influences the cognitive frameworks in which political actors formulate their policy preferences and shapes the bargaining process leading to the co-management of the shared resources.

Knowledge at the Service of Water Demand in the Génévois Region

Interventions for the progressive control of the major alpine rivers and lacustrian systems in the Franco-Genevese region began in the 19th century and continued through the 20th century. These operations played a critical role in the development of the region, which grew concomitant with the increased control of the natural resources. The control of the regional waters helped tackle historical problems such as recurrent seasonal floods or insufficient water supply in some parts of the region. Arguably for the first time, the extensive control of the water resources became an attainable political goal: water management initiatives echoed the availability of new institutional and technical tools for the control of the territory (Chappuis 1886; Vischer 2003). Through extensive operations, the regional waters became at once a source of freshwater for human consumption and a means both to improve urban sanitation and to power industry. Water management was not a politically agnostic phenomenon: it reflected ideological struggles, the changing definition of social priorities, and available technical and financial resources. The increased control and exploitation of the regional water resources is inextricable from the growth of the city of Geneva and its politics.

The Genevese expansion of the water supply systems and the infrastructure interventions, however, also created tensions among the users of the water resources. This is because Geneva's water works triggered changes in the water resources, leading, for example, to the recurrent overflowing of the Leman Lake in 1857 and to two major floods in 1877 and 1879 (Vischer 2003). Riparian communes led by the Canton of Vaud reacted by hiring experts[2] to devise alternative approaches to water management and challenged Geneva in the Federal court in 1878 (Chappuis 1886). They demanded the establishment of water relief mechanisms to regularize the region's water levels. In November 1884, the Cantons of Vaud and Geneva signed an *Inter-cantonal Convention for the Correction and Regularization of the Leman Lake Flow*.[3] The Convention, approved by the Confederation in 1885, established mechanisms for the control of the Leman Lake levels: the dredging of the Rhône River, the dismantling of constructions that affected the river's flow, and the construction of new flow-relief mechanisms. The agreement offered simultaneously a new opportunity to replace failing water supply equipment and to expand control over the water resources. In order to implement the inter-cantonal agreement, Geneva created a new intervention on the Rhône River between the Leman Lake and the Junction Bridge. The city's water utility began the construction of a new water pumping installation, the *Bâtiment de Forces Motrices de la Coulouvrenière*, in

[2]The Canton of Vaud contracted Karl Pestalozzi and Gottlieb Heinrich Legler in 1872. These experts produced a report in 1874 suggesting changes that would enhance the flow of the Rhône River, thus reducing summer floods. Their suggestion to implement a relief mechanism on the Rhône River was implemented as part of the Coulouvrenière facility (Vischer 2003).

[3]The agreement was later co-signed by the Canton of Valais.

1883.[4] Partially operational by May 1886, and fully functional by 1892, the new facility was able to supply the city as well as neighboring communes located up to 10 kilometers away. The new station also powered the flourishing city's industry, regularized Leman Lake levels, and provided the long sought flood control mechanism (Vischer 2003; Zanasco 2006).[5] Significantly, the inter-cantonal agreement proved a successful framework for the resolution of water management problems exceeding the canton's jurisdiction.

The success of the successive water management operations—the sophistication of the water supply and the flood protection mechanisms—confirmed Geneva both as the major economic and political center in the region and the most important regional center for scientific research associated with water management. Water works fostered the integration of scientific and technical expertise into the policy-making process and the development of institutional and legal instruments dedicated to the issues of water. In fact, the advent and popularization of modern science and technology were inextricable from the liberalization and radicalization of academic and political institutions (Rossier 1953). Sciences progressively gained legitimacy within academic institutions and in political decision-making. In fact, scientists actively participated in the public arena as they fought for the recognition of their respective disciplines in the academic realm. Many were involved in business and projects for public infrastructure: they designed and built the hydraulic machines, dams and new machines for industry required and fuelled by the economic development of the region. The scientific knowledge and new technologies that they developed helped control and exploit the region's waters, which was critical for the development and modernization of Geneva (Paquier and Pflieger 2008; Paquier 2007). In the process, technical expertise gained political legitimacy in decision-making about water management interventions.

The Growing Role of Groundwater in the Water Supply

The management of natural resources contributed to the continued demographic and economic growth of the region. Growth, however, was accompanied by increased water demand, and the supply of the communities located far from the city of Geneva's main water supply

[4]The original facility still stands and was reconverted into a museum and theater during the 1990s.

[5]In 1897, 13 new water-pumping turbines were added to the facility. Geneva's water jet originally resulted from the occasional release of excess pressure in the water supply system (Giacasso 1987).

systems became increasingly expensive and complicated. Historically, the communities located far from surface water resources had resorted to small surface streams and shallow groundwater wells for their water supply. As these resources became increasingly insufficient to cope with demand and other water resources remained inaccessible, waters from the regional aquifers became increasingly important for the regional water supply (Pazziani 1954).

The first users of the Génévois Aquifer knew relatively little about its extension and hydrogeologic properties. In 1865, a small private company started supplying the communes located east of Geneva with freshwater. The Canton granted a certain M. Schmidt a concession to build a dam on the Arve River and use hydraulic motors for the supply of the inhabitants located nearby. In 1866, the State authorized the transfer of M. Schmidt's concession to a newly founded company, the *Société des Eaux de l'Arve* (SEA) (Journal de Genève, August 26th 1950). The waters delivered by the new company were not those of the Arve River, but instead were abstracted from wells located on the river's margin. Groundwater was preferred because the terrain naturally filtered the river waters, making them potable at no cost. In 1902, the company began abstracting water from new deeper wells (54 meters) in the same location and would continue to do so throughout the 20th century. By 1970, the company supplied the communes of Veyrier, Trois-Chênes, Vandoeuvres, Choulex, Collonge-Bellerive, and Champel, totaling around 60,000 inhabitants (Stolz 1999).

In the west and southwest areas of the Canton of Geneva water supply was problematic. These communes were located too far away from the city center to be supplied from the city's water-pumping stations or from the SEA's network. As these communities began struggling to cope with demand, Geneva's water services—a component of the larger public utilities conglomerate, the *Services Industriels de Genève*[6] (SIG)—began exploring alternative means to supply them. The public utility was particularly interested in exploiting the aquifer that was being utilized by SEA (Journal de Genève, February 20th 1930) and decided to hire a local scientist to explore the possibility. The consultant, Etienne Joukowsky, conducted extensive investigations in the region and suggested locations for the installation of new wells (Anonymous 1948; Joukowsky 1942). Geneva's water utility installed groundwater pumping stations in Soral in 1932, Saconnex d'Arve in 1935, and Fontenex in 1941 (Calame 1943; Pazziani 1954). SIG began intensively exploiting the aquifer alongside SEA by the 1940s.

The French communes located between Gaillard and Annemasse, St. Julien-en-Génévois, and Viry along the Franco-Swiss border began

[6]The Industrial Services of Geneva are responsible for the supply of water, gas and electricity in the Canton.

exploiting the aquifer in the late 1950s. Groundwater quickly became a critical component of the supply of the fast-growing communes, as they transformed into suburban extensions of the city of Geneva (da Cunha Rebouças et al. 2005). These communes exploited shallow aquifers and small surface streams, but the quality of these waters deteriorated quickly as the communes urbanized. Alternative sources of water became too limited or too polluted to respond to demand and groundwater turned into the principal source of water (Bonnard & Gardel Ingénieurs-Conseils 2006).

The Scientific Assessment of the Génévois Aquifer

The expanded use of regional groundwater resources was supported by scientific research that developed during the 19th century. The first geologic characterization of the water-bearing formations in the Franco-Genevese area was authored by Alphonse Favre in the late 1860s. It was part of a larger geological description of the Canton of Geneva in which the geologist depicted the quaternary formations that underlie the canton (Favre 1873; Favre 1867). Although his research focused on the general tectonic features of the region, his work laid the foundations of future research on the region's hydrogeology (Amberger et al. 1988). Further investigations by other scientists filled gaps in the hydrogeological characterization of the region. In fact, by the 1940s, regional geologists and hydrogeologists had accumulated significant knowledge about the region's hydrogeology (Joukowsky 1941), and it was publicly known that the groundwater resources being exploited across the Canton of Geneva and in the French border communities belonged to a single aquifer (Journal de Genève, May 5th 1942: 4).

Geneva's Cantonal Service of Geology was founded in 1964 for the monitoring and assessment of regional hydro-geologic resources. Staffed with scientific experts, the Service surveyed regional groundwater demand and the effects of exploitation on the aquifers; it gathered information from the different users and constructed a database that evidenced the effects of the anarchic abstraction of groundwater on the regional aquifers. Furthermore, through the centralization of scattered information about the aquifers, the Service highlighted the users' interdependence on the shared resources, as well as existence of common problems. In the process, the work of the Service also led to changes in the denomination of the regional aquifers and to the formulation of the Génévois Aquifer. During an interview, Gabriel de los Cobos, the civil servant in charge of overseeing the technical aspects of the management of the Aquifer at Geneva´s GESDEC ("*Service de Géologie, Sols et Déchets*"), explained:

"We used «Arve aquifer» because the aquifer is strongly influenced by the [Arve] River, and because historically its waters were drawn exclusively by the Société des Eaux de l'Arve. *It is still part of common-speak because the Industrial Services of Geneva had a part of their water supply network based on the Arve Aquifer too. We preferred using the name "Génévois Aquifer" because it is more telling of the aquifer's transboundary nature. The 'Génévois' is a global name for the Franco-Swiss region. (...) we considered the name was more adapted than 'Arve Aquifer' because some people will still tell you that the aquifers located far from the river have different waters. (...) Here, it is often hard because people have trouble understanding the connection with the Arve River... Some still doubt the connection between them." (Gabriel de los Cobos, Interview, January 2009)*

The Génévois Aquifer was historically known as the Arve River Aquifer, due to the location of the first water pumps and the name of its historically largest user. The new denomination reflected not just the better knowledge of the regional resources more accurately, it was also a means to highlight that users located in different parts of the region were drawing water from the same aquifer. Although the aquifer's original nomenclature long persisted in many publications and technical reports, the new name helped reinforce the idea of a common transboundary resource.

The volumes of groundwater extracted from the Génévois Aquifer increased with the construction of new wells across the Canton of Geneva and in Haute Savoie. Water abstraction from the aquifer increased progressively but at a relatively slow pace until the 40s. The majority of the extraction was performed by SEA until the installation of new wells by SIG, which triggered an important increase in the extraction rates. Until 1957, the two water utilities in the Canton of Geneva were the sole users of the aquifer, at rates of approximately 8Mm^3/year. The volumes of groundwater being extracted were still below the natural recharge threshold. The development of new water pumping stations in France during the 50s tipped the balance: average water levels within the aquifer started dropping quickly by the end of the decade. The average levels within the aquifer, which had historically hovered around 377 meters,[7] began decreasing by almost two meters per year. The increases in demand were largely driven by the regional demographic and economic growth: SEA, for example, had to double its extraction rates in 1962, to 26,000 L/min to cope with new users (Journal de Genève, January 7, 1963). By 1970, the volume of water extracted from the aquifer reached around 12 and 14 million cubic meters per year, which was between 3 and 5 Mm^3/year more than average natural

[7]Average aquifer levels are measured using sea level as zero.

recharge, established at 9Mm³/year (Baroni 1979).[8] Monitoring systems, installed as the new wells became operational, evidenced the correlation between groundwater extraction and the aquifer's levels (Baroni 1979; Amberger et al. 1981) (see Fig. 14.2).

In view of the aquifer's deterioration and having connected groundwater-dependent regions to the general water supply network, SIG voluntarily decided to progressively reduce its reliance on groundwater in the early 60s (Pazziani 1954). The initiative only slowed the deterioration of the aquifer, as the other users continued increasing waters abstractions. As new French pumps became operational in the early 60s, water levels dropped even faster. The Génévois Aquifer was being exploited by wells constructed and managed by Geneva's two water utilities, the Services Industriels de Genève, and the Société des Eaux de l'Arve, and the French communes of Annemasse, Gaillard and St. Julien-en-Génévois.[9]

Data on the aquifer's deterioration contrasted starkly with the results of a survey conducted during the 70s about future regional groundwater needs (Baroni 1979). Results suggested the water supply would be insufficient to cope with regional demand unless groundwater levels were maintained above minimum levels. Were the aquifer's levels to fall below particular

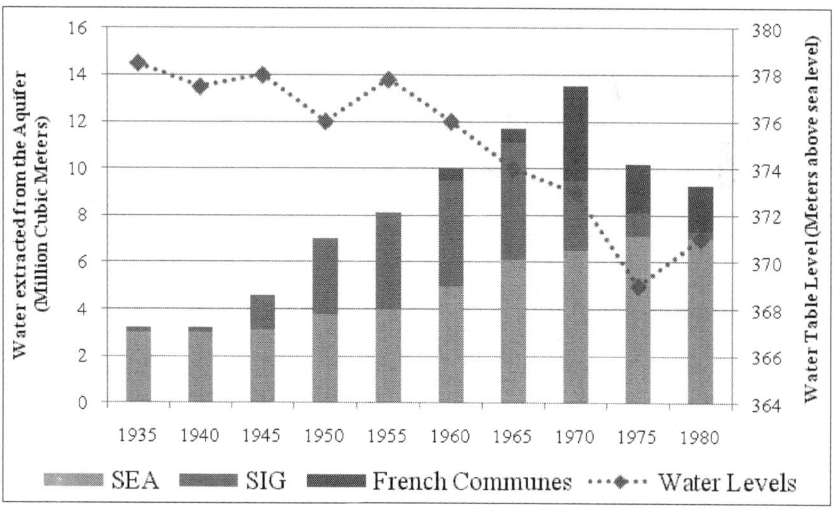

Figure 14.2 Water Abstraction and Water Levels in the Génévois Aquifer (1932–1980).
Source: Author from Data available in Baroni (1979)

[8]In addition to growing pressures from demand, drought limited the natural volumes of recharge, which further affected the aquifer's levels.

[9]In France, the water wells were managed by two multi-purpose inter-communal syndicates—public utilities serving more than one commune—in the cases of Annemasse and St. Julien, and the commune of Gaillard.

thresholds,[10] groundwater pumping would no longer be possible and the water supply would be insufficient, which posed a significant threat to the operations of users on both sides of the border.

Political Options to Deal with the Problems of the Génévois Aquifer

The over-exploitation of the aquifer created a specific set of problems. The general availability of water *per se* was not one of them, since waters could theoretically be drawn from Leman Lake and other regional surface water sources. They were instead associated with the costs of switching water supply systems from groundwater to surface resources and to the security of the water supply, two factors that could hinder the development of the communities relying on the Génévois's groundwaters. Drying wells needed to be replaced with new infrastructure and implied the loss of investments associated with their exploitation. The depletion of the aquifer also meant that groundwater would no longer constitute an alternative source of freshwater. Without groundwater, the region would be entirely dependent on the availability of surface water. Multiplying the sources of water reduced the vulnerability of the water supply. This matter was all the more relevant at the time because it coincided with the first signs of qualitative deterioration of the Leman Lake (Jouve 1995).

As the problems associated with groundwater became more apparent, scientific investigations focused on the identification of the zones of recharge, hydraulic linkages and the specific hydrogeologic features of the terrain. This information helped specify the aquifer's features and confirmed that all the users were effectively drawing water from the same source. Furthermore, it helped identify the factors causing the drop of the water level and explained the effects of these changes. It fostered the parties' agreement on both the nature of the problem and its causes.

There were two alternatives for solving the problems of the Génévois Aquifer. One of the solutions was to extract more waters from regional surface waters. This alternative required, besides the construction of new water-taking facilities, the integration of the regional water supply infrastructure. Although it was not a technologically complicated solution, it was costly. A new water treatment facility for extractions from Leman Lake was estimated at 250 million Swiss Francs (Baroni 1970; de los Cobos 2002), not including the investments necessary for the transportation of the waters across the region. Further reliance on surface waters also meant that existing groundwater exploitation infrastructures, which were significant

[10]Minimum levels varied for different pumping stations.

capital investments, would be abandoned. Moreover, this approach would also limit Geneva's diversity of supply sources, which meant that if the surface water resources became polluted, the city would be left without any sources of freshwater at all (de los Cobos 2002). The alternative option, suggested by the Director of Geneva's Geologic Service,[11] was the artificial recharge of the Génévois Aquifer. The injection of surface waters into the aquifer could potentially restore water levels and compensate for over-draft practices. The appeal of such an approach was rooted in its maintenance of the status quo: users would not be required to drastically modify their practices. In addition, the restoration of the aquifer's levels would protect existing investments, allowing water-taking infrastructure to remain operational, as well as the alternative source of freshwater in the Canton, which was useful in case of environmental emergencies affecting the surface waters.

Unlike the surface water option, the artificial recharge of aquifers presented significant technical, economic and political challenges. The technical conundrums were associated with both the aquifer's features and the means preferable to effectively recharge it. The solution needed to be able to restore water levels across the entire aquifer, and this depended on the aquifer's geomorphology and transmissivity rates. For example, if due to the properties of the aquifer, injected waters were unable to reach intended destinations in time, the artificial recharge would not be very useful; waters needed to spread across the water-bearing formation at rates that offset the user's abstraction. Also, the artificial injection facility needed to be cost-effective and politically workable. If either financial means or political support were lacking, it could not come to fruition, as it would necessarily demand the construction of new infrastructure and cooperation between the users.

The artificial recharge of the Génévois Aquifer was not an entirely original idea. Many other aquifers had already been restored using this method. It had been used since the 19th century in Glasgow (1810), Toulouse (1821) and Lyon (1854), and implemented in Germany, Sweden, Hungary, and several U.S. states during the first half of the 20th century (Baroni 1970). Probably more significantly, two Swiss aquifers were already being recharged in Basel and Neufchatel by the end of the 1960s: these experiences demonstrated that the deterioration of aquifers was potentially reversible (Baroni 1970). However, whether the artificial recharge of the Génévois Aquifer was feasible would depend on its particular hydrogeological features.

[11]The Service of Geology was in charge of the original studies on the aquifer. The first Director of the Service was Gad-François Amberger, assisted by Daniel Baroni. Both civil servants were prominent scientists in the field of geology and hydrogeology.

In order to explore the feasibility of such a solution, Geneva's Service of Geology, in association with SIG's Water Service, conducted a series of tests between 1967 and 1974. These tests filled knowledge gaps and assessed the consequences of injecting surface waters into the Génévois Aquifer. They evaluated the speed at which waters, which were radioactively marked and monitored, infiltrated and spread within the aquifer. In fact, the first experiments involved the irrigation of a non-confined area of the aquifer with surface waters of different quality.

Gabriel de los Cobos detailed the process in layman's terms:

> *"They looked at the possibility of what was called the artificial enrichment of the Génévois Aquifer. They tested it with the lake's water but results were negative because they were quite a few biological problems. It is not the same water. They then thought to use the waters that naturally infiltrate the aquifer, the Arve River. After a series of tests, and stages… because it was necessary to find the right location for the re-infiltration… several infiltration methods were tested: infiltration basins and wells. (…) They decided to infiltrate directly through the alluviums. It is important to note they were quite lucky on the location where they did the tests…. The location had great geologic qualities and is now considered one of the best places to do it. This is because the Arve alluviums are deposited directly over those that make the aquifer. The infiltration of those alluviums permitted the immediate recharge of the aquifer. They also installed filtration mechanisms to enhance the quality of the water." (Gabriel de los Cobos, Interview, January 2009)*

By 1971, the success of the initial recharge experiments led to the installation of a field laboratory dedicated to furthering the investigations. It was located close to the SEA's Arve River facility in Vessy. The new tests focused on the physicochemical properties of the groundwaters and the Arve River (Baroni 1979). They confirmed that the aquifer's natural recharge stemmed largely from the Arve River, as the aquifer's levels changed concomitant with the river's hydraulic regime and the waters' shared chemical properties.[12] The laboratory monitored the aquifer levels' seasonal fluctuations and the presence of pollutants in the groundwaters. Moreover, it was a testing ground for different methods of water injection: it evaluated alternative water-filtering techniques, as well as the effects of alternative approaches to the imperviousness of the terrain and the quality

[12]Originally, scientists believed that the aquifer received recharge from the Leman Lake. Infiltration tests and the chemical comparison of the waters demonstrated the aquifer received little recharge from the lake. The waters had different temperatures and chemical composition.

of the groundwater.[13] Investigations suggested that surface waters would require minimal treatment previous to their injection, limited to sand removal, flocculation and filtration. The tests concluded that the recharge of the aquifer was feasible.

In terms of its economic costs, the artificial recharge of the aquifer was more appealing than switching to a supply exclusively based on surface waters. In fact, besides protecting existing investments in infrastructure and avoiding the construction of elaborate water transport systems, the cost of an artificial recharge facility was calculated at about 17 million Swiss Francs (de los Cobos 2009; Baroni 1979). This was significantly less than the cost estimated for the construction of new water pumping and treatment infrastructure for the Leman Lake, which was approximately 250 million Francs. Given the economic cost and the technical viability of the recharge solution, the major factor potentially precluding the installation of the artificial recharge facility was political. In fact, the construction and long-term operation of the water injection facility required an important economic investment. Consequently, the initiative was thus ultimately dependent on the willingness of the users to finance its construction and to undertake the necessary regulations to control the exploitation of the aquifer.

The Asymmetric Impact of the Aquifer's Deterioration

De los Cobos argues that "the political will to develop a cross-border project emerged naturally in parallel with the studies and tests that were carried out on the experimental plant" (de los Cobos 2010). He suggests that the political will to support the artificial recharge was rooted in the bilateral interest to protect the resource. From his perspective, shared interests "emerged naturally" from the increased understanding of the Génévois Aquifer properties and from the assessment of alternative courses of action by its users. It reflected a process in which users progressively constructed their respective preferences.

In the mid-70s, the Canton of Geneva contacted the bordering French communes to inquire about their interest in participating in the management of the aquifer (Journal de Genève, April 13, 1976). The French communes were asked to partially fund the construction and the operation of an artificial recharge facility and to regulate their consumption of groundwater. The users of the aquifer held several meetings in which they discussed the approach and other means to deal with groundwater

[13]The Arve River is fed by the Glaciers of the Mont Blanc. During summer, which is the season of high runoff, the waters carry sediments that make them turbid. The solids in the water could potentially modify the imperviousness of the areas of recharge and the pre-injection filters, thus making recharge operations ineffective.

problems. They debated the reduction of water abstraction rates and cost-sharing mechanisms for the construction of the facility. Simultaneously, they discussed their respective reliance on different water resources, as well as the hydrogeological features of the region, and evaluated the costs associated with different sources for water supply. These assessments led some users to reduce their reliance on groundwaters. Users relying exclusively on the groundwaters, including some but not all users in Haute-Savoie and SEA in Geneva, however, were unable and unwilling to follow suit. Inability resulted from both the lack of alternative sources of freshwater and the costs associated with the supply of alternative sources of water; unwillingness reflected the strategic opportunity to free-ride unilateral actions taken by other users to restore the aquifer. Indeed, the aquifer's deterioration did not affect all users of the aquifer equally.

In the Canton of Geneva, the principal user of the aquifer, SEA, relied exclusively on the groundwater resources: the firm would not survive a transition to surface waters. SIG, the other major user of the Génévois Aquifer in Switzerland, voluntarily began reducing its reliance on the resources as early as 1965. In France, French users faced different situations depending on their respective geographical location:

> *"There were three or four different zones on the French side. This section, between Gaillard and Annemasse.... They had several wells and were interested but were seeking for water in the Arthaz Aquifer. They apparently had an alternative. [The commune of] Saint-Julien had wells here, and was considering the construction of further water wells. They were interested in maintaining water levels in the aquifer. And then, the commune of Viry didn't really have a choice: they needed minimum levels to abstract groundwater." (Gabriel de los Cobos, Interview, January 2009)*

The interest of the different communes in the restoration of the aquifer was shaped both by their respective reliance on its resources and their capacity to switch to alternative sources of water. Each commune had different stakes in the resolution of the groundwater problems. For some, groundwater deterioration posed threats analogous to those faced by the Swiss public utility: they were mainly associated with the financial cost of transitioning to alternative sources of freshwater. For others, however, such as the communes Saint-Julien-en-Génévois and Viry, the depletion of the aquifer directly threatened the freshwater supply. These communes had limited access to other sources of water. Worse, located on the margins of the aquifer, these users would not be able to free-ride a potential Swiss restoration of the aquifer. Geneva could potentially recharge the aquifer at levels that would keep their wells dry.[14]

The Alignment of Interests in Favor of the Co-Management of the Génévois Aquifer

As dependency on the groundwater resources became apparent, interest in the restoration of the water levels in the aquifer aligned. Initially at odds, the preferences of the French and the Genevese users eventually pointed to the cost-effectiveness of the joint management of the resources.

The Swiss would have likely proceeded with the recharge of the aquifer independently of French preferences. In fact, the Canton of Geneva passed a law on June 24, 1976 permitting the appropriation of the funds necessary for the construction of the recharge facility. It authorized a 15. 5 million Swiss Franc loan and a special new tax on all groundwater abstraction.[15] Geneva's Council determined that the facility would be installed on the testing grounds used to study the aquifer's reaction to artificial recharge, which it had acquired in 1970 (Journal de Genève, February 14, 1976). It would be located upstream of SEA's Arve River dam to benefit from the higher Arve River levels during the dry season.

In France, "*there were multiple turnarounds. People would agree, and then show disinterest. In fact, the French had expectations on their aquifers... they found water, but the volumes found were not as large as they expected them to be. They then expressed interest [in the restoration of the aquifer] but not in directly financing [a joint management initiative]*" (Gabriel de los Cobos, Interview, January 2009). The French communes were initially reluctant to participate in a negotiation over the management of the shared resources. However, given the increasing information about the relative costs of exploiting alternative sources of water, they eventually accepted participation in negotiations. They did so because they were aware of the costs of transitioning to a water supply based on alternative sources of freshwater and especially because free-riding the Swiss intervention would not possible for all French communes. In France, this information was used by the more vulnerable communes to influence those less reliant on the aquifer to participate in talks with Geneva.

Hence, although at first the situation seemingly fitted well with the tragic fate of common pool resources described by Hardin (1968), the strategic mobilization of knowledge about the resources and alternative management

[14]The margins of the aquifer are thinner and thus more sensible to changes in the water levels, which made wells located there more vulnerable to groundwater depletion.

[15]The tax charged 0. 13cts/m^3 of groundwater pumped and was adjusted according to the effective costs of recharging the aquifer (Baroni 1979).

approaches reshaped the incentives of the users of the Génévois Aquifer to cooperate. As knowledge about the hydrologic conditions progressed, it became increasingly clear that cooperation was strategically preferable to the depletion of the aquifer.

The 1978 Arrangement: the Pragmatic Answer to the Problems of the Génévois Aquifer

The formalization of a formula for the joint management of the transboundary aquifer was riddled with legal and administrative difficulties. The Canton of Geneva possessed the political and administrative authority to create and implement international cooperative agreements, but the French border communities lacked the necessary authority to autonomously engage in bilateral negotiations. In fact, as stated by an expert in Franco-Genevese transboundary relations, the signature of a formal agreement required the participation of French political authorities, which were at once "unaware and uninterested in the problems of the small transboundary aquifer" (Nicolas Levrat,[16] Interview, January 2009). This was a common problem in Europe during the 1970s. It was connected to the limited capacity of the regions to autonomously resolve international albeit locally relevant problems in a context of increasing regional integration. In this case, it was compounded by the absence of legal instruments specific to the management of international aquifers from which to draw principles or guidelines. At the time in which Franco-Genevese authorities considered the best means to deal with the problems of the Génévois Aquifer, international debates only summarily included aquifers—as part of larger discussions about international surface waters—and not a single formal international agreement dealt exclusively with them (Wohlwend 2002).

Despite the challenges, the State Council of the Republic and Canton of Geneva and the Prefect of Haute-Savoie formalized a (renewable) 30-yr agreement in June 1977 for the management of the Génévois Aquifer. The "*Arrangement relatif à la protection, à l'utilisation et à la réalimentation de la nappe souterraine franco-suisse du Génévois*" entered into force in January 1978 and simplified the coordination for the protection and continued use of the shared aquifer. Through its provisions, Geneva and the French neighboring communes regulated groundwater exploitation, mandated the artificial recharge of the aquifer, and established mechanisms for the systematic exchange of information about the usage of the aquifer. The agreement's provisions made existing practices sustainable by warranting

[16]Attorney responsible for the legal formulation of the 2008 Convention signed to renew the 1978 arrangement as it expired.

the maintenance of the *status quo*, rather than drastic political or behavioral change on either side of the border. Existing users were guaranteed access to the groundwaters. In addition, the arrangement created barriers to new water extraction projects. The joint management of the shared resources would permit the continued exploitation of the aquifer at limited extra cost for the users and dispel future threats posed by changes to groundwater levels. In order to accomplish these goals, the arrangement established a bi-national commission charged with the supervision of the aquifer's exploitation and allocated the aquifer's water.

The bi-national Commission (Articles 1–6) charged with steering the management of the aquifer was composed of three representatives from each country, two of whom had to be technical experts in water management. They would meet at least twice a year to assess the aquifer's condition and supervise groundwater use. The first task of the Commission was to enforce a ban on groundwater abstractions for the years in which the average aquifer level dropped below 368. 5 meters (above sea level). Then the Commission would prepare a yearly management program based on the performance of the recharge facility, users' projected requirements, and potential threats to the quality of the groundwater. The Commission centralized information and expertise on the exploitation of the aquifer. Its function, however, remained consultative: authorizations, permits and concessions to extract groundwater remained with the respective national authorities. The bilateral review process became nonetheless a necessary step for the issuance of new municipal permits for the construction of new groundwater pumping facilities or the modification of existing infrastructure. In essence, the technical expertise of the Commission became an important component of decision-making on both sides of the border.

The arrangement allocated the aquifer's waters asymmetrically. The French communes were allocated a maximum of five million cubic meters per year[17] with extractions of more than two million cubic meters charged according to a formula specified in the agreement (Article 9). The two million free quota corresponded to the volumes of groundwater extracted by French users at the time the agreement was signed. The charges for extractions beyond the quota factored the costs of operating the artificial recharge facility, depreciation, the total volume of water extracted from the aquifer, the volume of natural recharge, and the total volume of groundwater abstracted by the French. Conversely, the water allocated to the Swiss was restricted only by the aquifer's recharge capacity.[18] In exchange, Geneva assumed the (financial) responsibility for the construction and operation of the recharge

[17]This figure could in theory be reviewed by the Commission in case of necessity.

[18]The aquifer's total estimated maximum recharge (natural and artificial) was estimated at about 17 million cubic meters per year.

facility (Article 8). This approach accommodated existing users of the aquifer and created a clear economic deterrent to increased abstractions from the aquifer. It also defrayed costs associated with its restoration, which was politically instrumental to the French communal authorities. They could in fact reap the political benefits stemming from the rapid restoration of the aquifer and defer the economic costs of groundwater regulation. In other terms, the French would effectively free-ride the recharge of the aquifer, as long as demand remained constant, or slowly adjust to the progressively increasing costs of exploiting groundwater. Conversely, the Swiss obtained guarantees that the French would regulate their groundwater exploitation, as well as potential pay for groundwater use, which was necessary for the effectiveness of the artificial recharge approach.

The arrangement was negotiated by the authorities of the French border communes and signed by the Prefect of Haute-Savoie. From a legal standpoint, neither possessed the formal authority to engage in international negotiations nor in the development of joint policy for environmental management. Environmental management decisions and international relations were prerogatives of the national-level authorities, not the local representatives. In other words, local authorities were—in theory—formally restricted by national administrative law and country-level political concerns from negotiating and signing the 1978 arrangement. The centralized nature of the French political system at that time limited the legal validity of the agreement, and the agreement would have likely been dismissed had any of the parties ever challenged it:

> *"Many things were going on at the time, although they were extremely informal. The 1978 accord is a typical example of that time. I mean.... The actors directly concerned sat around a table, and after realizing that no predefined juridical agreement existed, they agreed as with a contract of private law. (...) They bargained and put down a solution on paper that was generally well respected, but lacked any legal validity. We would have faced problems had it ever been challenged.... But that never happened, which is typical of transboundary relations* [in Western Europe]." *(Nicolas Levrat, Interview, January 2009)*

The 1978 agreement was nonetheless a "pragmatic" instrument of political cooperation (Yamada 2004; Scheumann and Herrfahrdt-Pähle 2008), through which the signatories sidestepped discussions about sovereignty, equitable allocation, and administrative asymmetry (Wohlwend 2002). Although each state maintained its sovereign prerogatives over the aquifer, the agreement framed the transboundary aquifer as a shared resource to be managed for the benefit of all users—irrespective of their geographical localization. The agreement thus eluded the typical complexities of international (surface) water law (Benvenisti 1996; Eckstein 1995). Its scope was narrowed to the financial and technical aspects necessary for the joint management of the Génévois Aquifer. Compliance was independent of the agreement's formal validity because the signatories had *de facto* authority and interest necessary to autonomously enforce it. The Canton of Geneva had the necessary financial resources and technical expertise to build and operate the recharge facility, while the French communes had the power to regulate groundwater exploitation (and pay for the abstractions over the 2Mm3/y quota). The effectiveness of the agreement was rooted in the interest of the parties to abide by it, not in its legality nor in their formal authority to sign it.

Conclusion: Knowledge in the Successful Co-Management of the Génévois Aquifer

The recharge of the Génévois Aquifer began in 1980 and led to the rapid restoration of the aquifer's water levels. In the years that followed, the resolution of the aquifer's principal problems turned the Génévois Aquifer into a political non-issue. Its management was routinized and contentions over the shared resources largely disappeared from the political agenda. Furthermore, with the increasing decentralization of European policy-making, the approach chosen to manage the aquifer was progressively legitimized, as administrative reforms in France and European framework agreements validated the 1978 arrangement.

In the years that followed the 1978 agreement, European countries undertook multiple reforms aimed at the decentralization of administrative structures (Kissling-Näf and Kuks 2004). They were instrumental to goals of regional integration and the empowerment of the local authorities (Barraqué 1995). The French politico-administrative system underwent major reforms between 1982 and 1987; the central government partially transferred power to subsidiary authorities. Today, French water policy simultaneously reflects European directives (to which national legislation must adapt), national-level laws, the decisions of basin-level authorities, and those emanating from the local level. The situation in Switzerland is significantly different, as most public policies have historically been determined by the local and regional authorities. Cantons are responsible for decisions regarding the water supply, sanitation and infrastructure, according to their respective constitutions and political institutions (Mauch et al. 2004). Indeed, although the revision of certain articles of the Federal Constitution has conferred progressively more power on the Federal agencies, the Swiss cantons remain the principal wielders of political authority. In 2007, as the 1978 arrangement expired, legal frameworks supported regional-level agreements and the subsidiary governance of natural resources. Consequently, and in view of successful management of the Génévois Aquifer, the renewal of the agreement revolved around legal technicalities, rather than political or operational dysfunctions. The central provisions of the original agreement that specified the mechanisms used to manage the aquifer were maintained. Changes concerned the explicit acknowledgement of the legal instruments that legitimized the arrangement and the direct inclusion of the now formally empowered French communes. The arrangement was renewed for 30 yr in December 2007. The renewed agreement, now deemed "Convention," was signed by the Communes of the "Annemassienne" region, the Commune of the "Génévois" Rural Districts, and the Rural District of Viry and the State Council of the Republic and Canton of Geneva.

The successful co-management of the Génévois Aquifer was based on three elements: 1. the existence and recognition of problems associated with the aquifer; 2. the alignment of political interests in favor of the cooperative resolution of the problems; and 3. the effectiveness of the measures devised for the resolution of the problems of the aquifer. Indeed, before the aquifer's existence was socially acknowledged, water management in the Franco-Genevese region dealt exclusively with surface water resources. In the 19th century, as scientific discoveries and technological developments gave way to new means to control and exploit natural resources, regional water infrastructure projects thrived. The successive interventions on the waterscape led to the progressive extension of the water supply and control infrastructures, and to economic and demographic growth. The region's expansion also led to increasing demands for water and to the increasing exploitation of the Génévois Aquifer. As the usage of the resources increased, so did the scientific knowledge associated with them. Regional hydrogeological assessments permitted the social recognition of environmental problems in terms useful to policy-makers: it pushed groundwater issues on to the political agenda. Simultaneously, the hydrogeological assessments of the Génévois Aquifer demonstrated the particular interdependence existing among the users of the aquifer. The recognition of the asymmetric impact of the aquifer's deterioration and the lack of economically competitive alternative approaches to resolve the water management problems aligned French interests with those of Geneva and supported the joint management approach. The artificial recharge effectively mitigated the deterioration of the shared aquifer, which reconfirmed the legitimacy of the co-management approach and guaranteed the enforcement of the political arrangement.

In the different stages of the policy-making process, knowledge of the Génévois Aquifer operated as both the cognitive framework around which preferences were formulated and a power resource that leveraged the preferences of particular actors in the bargaining process. It empowered new actors in the local political arena and supported the pragmatic self-enforcement of the co-management approach.

Acknowledgements

The author would like to acknowledge the Northwestern University Graduate School and the Institute of the Americas (IHEAL) for their support, and commentaries by Hendrik Spruyt, Gabriel de los Cobos, and Diane Eikenberry.

References

Adler, E. and P.M. Haas. 1992. Conclusion: epistemic communities, world order, and the creation of a reflective research program. International Organization 46(1): 367.

Amberger, G., U. Siegenthaler and P. Verstraete. 1981. Etude en cours de la nappe souterraine de l'Arve. Eclogae Geologicae Helvetiae 74(1): 225–232.

Amberger, G., A.V. Carozzi and J. Charollais. Histoire Géologique du Salève: d'Horace-Bénédict de Saussure à nos Jours. *In*: Le Salève. 1988. Proceedings of the Colloquium organized by the Société de Physique et d'Histoire Naturelle de Genève-Société de Physique et d'Histoire Naturelle. Geneva.

Anonymous. 1948. Nécrologie: E. Joukowsky. Bulletin technique de la Suisse romande 7(74): 16–17.

Bardwell, L.V. 1991. Problem-framing: A perspective on environmental problem-solving. Environmental Management 15: 603–612.

Baroni, D. 1970. Station expérimentale de réalimentation d'une nappe souterraine à Vessy (Genève). Bulletin technique de la Suisse romande 96: 185–195.

Baroni, D. 1979. Réalimentation artificielle de la nappe souterraine franco-suisse de l'Arve à Vessy-Genève. Ingénieurs et architectes suisses 105: 2230–2315.

Barraqué, B. 1995. Les Politiques de l'Eau en Europe. Revue Française de Science Politique 45(3): 420–453.

Baumgartner, F.R. 1993. Agendas and Instability in American Politics. University of Chicago Press, Chicago.

Benvenisti, E. 1996. Collective action in the utilization of shared freshwater: The challenges of international water resources law. American Journal of International Law 90: 384.

Blyth, M. 1999. Great transformation: economic ideas and political change in the Twentieth Century. Ph.D. Dissertation, Columbia University, New York.

Bonnard & Gardel Ingénieurs-Conseils. 2006. Réalisation du Schéma Directeur d'Alimentation en Eau Potable du Canton de St. Julien-en-Génévois. Rapport de Phase 1; Inventaire et Diagnostic. B&G, St. Julien-en-Génévois.

Burchi, S., K. Mechlem, UNESCO and FAO 2005. Groundwater in International Law Compilation of Treaties and other Legal Instruments. FAO legislative study 86, Rome.

Calame, J. 1943. La 'Grande Genève' du Point de Vue Technique: Problèmes Résolus-Problèmes qui se Posent. Schweizerische Bauzeitung 121/122.

Chappuis, J. 1886. Régularisation du Niveau du lac Léman et Utilisation des Forces Motrices du Rhône à Genève. Bulletin de la Société Vaudoise des Ingénieurs et des Architectes 13: 13–24.

de los Cobos, G. The aquifer recharge system of Geneva (Switzerland): a 20 year successful experience. *In*: P.J. Dillon [ed.]. 2002. 4th International Symposium on Artificial Recharge of Groundwater: Management of Aquifer Recharge for Sustainability. Balkema Publishers-AIH. Adelaide, South Australia.

de los Cobos, G. 2009. La Recarga Artificial de Acuífero como Ayuda a la Gestión de los Recursos Hídricos: el Ejemplo del Sistema de Ginebra (Suiza). Boletin Geológico y Minero 120: 305–310.
de los Cobos, G. 2010. The Transboundary Aquifer of the Geneva Region (Switzerland and France): Successfully Managed for 30 years by the State of Geneva and France. *In:* UNESCO-IHP [ed.] ISARM 2010 International Conference on Transboundary Aquifers: Challenges and new directions. Paris.
Conca, K. 2006. Governing water . Contentious Transnational Politics and Global Institution Building. MIT Press. Cambridge.
da Cunha Rebouças, A., P. Knoepfel, J.P. Leresche and S. Nahrath. 2005. Enjeux du Développement Urbain Durable: Transformations Urbaines, Gestion des Ressources et Gouvernance. PPUR Presses Polytechniques. Paris.
Delli Priscoli, J. and A.T. Wolf. 2009. Managing and transforming water conflicts. Cambridge University Press, Cambridge.
Eckstein, G.E. 1995. Application of International Water Law to Transboundary Groundwater Resources, and the Slovak-Hungarian Dispute over Gabcikovo-Nagymaros. Suffolk Transnational Law Review 19: 67.
Eckstein, G.E. 2005. Protecting a Hidden Treasure: the UN International Law Commission and the International Law of Transboundary Ground Water Resources. Columbia Journal of Environment 30(3): 525–564.
Eckstein, G.E. and Y. Eckstein. 2003. Hydrogeological Approach to Transboundary Ground Water Resources and International Law. American University International Law Review 19: 201.
Elder, C.D. and R. Cobb. 1983. The political uses of symbols. Longman, New York.
Favre, A. 1867. Recherches géologiques dans les parties de la Savoie, du Piémont et de la Suisse Voisines du Mont-Blanc. Masson, Geneva.
Favre, A. 1873. Description géologique du canton de Genève. Masson, Geneva.
Giacasso, P. 1987. L'usine des Forces motrices de la Coulouvrenière à 100 Ans. Gas, Wasser, Abwasser 67 (1): 31–34.
Goldstein, J., M. Kahler, R.O. Keohane and A.-M. Slaughter. 2000. Introduction: Legalization and World Politics. International Organization 54: 385.
Haas, E.B. 1991. When Knowledge is Power: Three Models of Change in International Organizations. University of California Press, Berkley.
Haas, P.M. 1992. Introduction: epistemic communities and international policy coordination. International Organization 46(1): 1.
Hall, P.A. 1989. Conclusion: The political power of economic ideas. pp. 361–391. *In*: The Political Power of Economic Ideas: Keynesianism across Nations. Princeton University Press, Princeton.
Hall, P.A. 1993. Policy Paradigms, Social Learning, and the State: the Case of Economic Policymaking in Britain. Comparative Politics 275–296.
Hardin, G. 1968. The Tragedy of the Commons. Science 162(3859): 1243–1248.
Henry, P. and U. Mocafico. 1991. De la Roue à Eau à la Turbine Hydraulique. Miroir des techniques XIIIe-XXe siècle. Ecole Polytechnique Fédérale de Lausanne. Lausanne.
Hochstetler, K. 2002. After the Boomerang: Environmental Movements and Politics in the La Plata River Basin. Global Environmental Politics 2(4): 35–57.
Jacobsen, J.K. 1995. Much Ado About Ideas: the Cognitive Factor in Economic Policy. World politics 47(1): 283–310.
Jarvis, T., M. Giordano, S. Puri, K. Matsumoto and A.T. Wolf. 2005. International borders, ground water flow, and hydroschizophrenia. Ground Water 43: 764–770.
Jasanoff, S. 2004. States of knowledge : the Co-production of Science and Social order. Routledge. New York.
John, P. 2003. Is there life after Policy Streams, Advocacy Coalitions, and Punctuations: Using Evolutionary Theory to Explain Policy Change? Policy Studies Journal 31(4): 481–498.

Joukowsky, E. 1941. Géologie et eaux souterraines du pays de Genève. Imprimerie Kundig, Geneva.
Joukowsky, E. 1942. Esquisse de la Géologie du Plateau Genevois. Bulletin technique de la Suisse romande 68.
Jouve, B. 1995. Collectivités Locales et Relations Internationales: une Emancipation Délicate. Swiss Political Science Review 1: 137–158.
Kissling-Näf, I. and S. Kuks. 2004. The evolution of national water regimes in Europe: transitions in water rights and water policies. Springer. London.
Litfin, K. 1994. Ozone discourses: science and politics in global environmental cooperation. New Directions in World Politics. Columbia University Press, New York.
Llamas, M.R. and P. Martinez-Santos. 2005. Intensive Groundwater Use: Silent Revolution and Potential Source of Social Conflicts. Journal of Water Resources Planning and Management 131(5): 337–345.
Matsumoto, K. 2002. Transboundary Groundwater and International Law: Past Practices and Current Implications. M.S. Thesis in Environmental Law. Oregon State University, Oregon.
Mauch, C., I. Sangare, E. Reynard and C. Larrue. The Evolution of the Water Regime in Switzerland. pp. 293–328. *In:* S. Kuks [ed.]. 2004. The Evolution of National Water Regimes in Europe. Transitions in Water Rights and Water Policies. Kluwer Academic Publishers, Dordrecht, Boston, London.
Moench, M. Groundwater and Poverty: Exploring the Connections. pp. 441–456. *In:* M.R. Llamas and E. Custodio [eds.]. 2003. Intensive Use of Groundwater: Challenges and Opportunities, Balkema, Rotterdam.
Morris, B.L., A.R.L. Lawrence, P.J.C. Chilton and B. Adams. 2003. Perspectives on Groundwater Management.pp.105-126. *In:* Groundwater and its Susceptibility to Degradation: a Global Assessment. UNEP, Nairobi.
NATO (Advanced Study Institute on Overexploitation and Contamination of Shared Groundwater Resources), and C. J. G. Darnault. 2008. Overexploitation and Contamination of Shared Groundwater Resources: management, (bio) technological, and political approaches to avoid conflicts. Springer, London.
Paquier, S. 2007. L'eau à Genève et dans la région Rhône-Alpes, XIXe-XXe siècles. Harmattan, Paris.
Paquier, S. and G. Pflieger. 2008. L'Eau et les Services Industriels de Genève: aux Sources du Modèle Suisse des Services Urbains. Entreprises et histoire 36–51.
Pazziani, P. 1954. Le service de distribution d'eau de Genève. Schweizerische Bauzeitung 72.
Phillips, D., M. Daoudy, S. McCaffrey, J. Öjendal and A. Turton. 2006. Trans-boundary Water Cooperation as a Tool for Conflict Prevention and for Broader Benefit-sharing. Swedish Ministry for Foreign Affairs, Stockholm.
Puri, S. and A. Aureli. 2005. Transboundary aquifers: A Global Program to Assess, Evaluate, and Develop Policy. Ground Water 43: 661–668.
Risse-Kappen, T. 1995. Bringing Transnational Relations back. *In:* Non-State Actors, Domestic Structures and International Institutions. Cambridge Studies in International Relations. Cambridge University Press, New York.
Rossier, P. 1953. Coup d'œil sur l'Histoire des Sciences Exactes à Genève. Revue d'Histoire des Sciences et de leurs Applications 231–249.
Salman, S.M.A. 1999. Groundwater. Legal and Policy Perspectives. World Bank technical paper. World Bank, Washington D.C.
Scheumann, W. and E. Herrfahrdt-Pähle. 2008. Conceptualizing cooperation on Africa's transboundary aquifer systems. Deutsches Institut für Entwicklungspolitik, Bonn.
Schnitter, N. 1992. Die Geschichte des Wasserbaus in der Schweiz. Oberbözberg, Olynthus.
Scott, C.A. and A. Milman. 2010. Beneath the Surface . Intra-National Institutions and Management of the United States-Mexico Transboundary Santa Cruz aquifer. Environment and Planning: Government and Policy 28(3): 528–551.

Stolz, P. 1999. Le Patrimoine Industriel Genevois: Infrastructures. Dossier et Contenus. Ecole Polytechnique Fédérale de Lausanne, Geneva.

Stone, D.A. 1989. Causal Stories and the Formation of Policy Agendas. Political Science Quarterly 104(2): 281–300.

UNECE. 2007. Status of Transboundary Rivers, Lakes and Groundwaters in the UNECE region Purpose of the Assessment and Policy, Status and Management Responses. United Nations Press, Geneva.

Vischer, D.L. 2003. Histoire de la Protection contre les Crues en Suisse: des Origines jusqu'au 19e siècle. Rapports de l'OFEG, Série Eaux. Office Fédéral des Eaux et de la Géologie, Berne.

Weaver, R.K. and B.A. Rockman. 1993. Do institutions matter? Government capabilities in the United States and Abroad. The Brookings Institution, Washington D.C.

Wohlwend, B.J. An Overview of Groundwater in International Law. A Case Study The Franco-Swiss Genevese Aquifer. *In:* ESCWA and The German Technical Cooperation Agency [eds.]. 2002. Workshop On Harmonization of Diverging Interests in the Use of Shared Water Resources. Beirut.

Yamada, C. 2004. Second Report on Shared Natural Resources: Transboundary Groundwaters. Special Rapporteur on Shared Natural Resources UN International Law Commission. United Nations Press, New York.

Zanasco, J.-L. 2006. Les Cahiers de l'Energie. Services Industriels de Genève, Geneva.

Zeitoun, M. and J. Warner. 2006. Hydro-hegemony—a framework for analysis of trans-boundary water conflicts. Water Policy 8: 435–465.

Zeitoun, M. and J.A. Allan. 2008. Applying hegemony and power theory to transboundary water analysis. Water Policy 10: 3–12.

Zekster, I.S. and L.G. Everett. 2004. Groundwater Resources of the World and their Use. IHP Series on Groundwater 6. UNESCO, Paris.

15

Guarani Aquifer Co-Management and the Brazilian Experience

Pilar Carolina Villar

Introduction

The increased use of groundwater in several parts of the world draws attention to the degradation risk of the aquifers and the lack of experience in their management. If the challenge of the last century was to regulate the use of surface waters, in the present one it is to consolidate this regulation to avoid a water crisis and include the slower and more hidden hydrologic-cycle dimension: groundwater and aquifers.

Groundwater management requires the participation of several agencies and players. In the case of transboundary aquifers, such as the Guarani Aquifer that extends through Argentina, Brazil, Paraguay and Uruguay, the institutional design is complex, since its management depends on the cooperation of numerous players: international, national, regional and local. Specific groundwater policies are just starting to be developed in the four countries and face difficulties to be implemented. In this context, co-management offers some interesting perspectives to groundwater management. This chapter aims at analyzing the existing institutional framework to promote the co-management of the Guarani Aquifer, focusing

Environmental Sciences Post-Graduate Program, University of São Paulo.
Email: pcvillar@gmail.com

on the Brazilian case. Brazil was chosen for three main reasons: a) it has more advanced water policies in comparison to the other countries; b) the biggest part of the aquifer is located in the Brazilian territory; c) and this country is its main user, since of the 1,04 km^3/year extracted, 94 percent comes from Brazilian wells (GW MATE 2009).

The research was performed through the qualitative analyses of primary and secondary sources, including the results of the Guarani Aquifer Project, the Agreement on the Guarani Aquifer, the MERCOSUR documents, international rules related to water resources, Brazilian legislation, and the existing literature on the subject.

This chapter is divided in five parts. Section 2 analyzes the co-management as a transboundary-aquifer management strategy. Section 3 characterizes the Guarani Aquifer and the international efforts to promote its management. Section 4 analyzes the Brazilian groundwater policy and the co-management perspectives, and Section 5 presents final considerations.

Co-Management and Transboundary Aquifers

The intensive use of groundwater is a phenomenon that started from the second half of the 20th century and is capable of significantly changing the hydrologic cycle[1] (Fornés et al. 2005). The increased extraction is a result of several causes: the degradation of surface waters, the perception of the superior quality of groundwater, its availability in arid and semi-arid areas, its use for irrigation,[2] advances in drilling techniques and the low price of energy to pump these resources. The groundwater exploitation allowed for social-economic shifts (Jarvis 2010), but it also led to aquifer over-exploitation, contamination and salinization risks. The exploitation and degradation of these resources tends to get worse due to the climate-change phenomenon[3] (Loáiciga 2003).

[1]According to Fornés et al. (2005), the intensive use of groundwater can affect springs and river base-flow, water table depth, piezometric levels, groundwater storage, groundwater-dependent wetlands, groundwater quality, river-aquifer relations and generate land surface subsidence.

[2]According to Llamas and Martınez-Santos (2005) the use of groundwater generated a silent revolution in the rural areas. This phenomenon means the action of millions of independent farmers in arid and semi-arid countries, implemented the necessary means to irrigate their land with groundwater without governmental participation on these groundwater developments.

[3]Aquifers have a high storage capacity and are less sensitive to climate change than surface water bodies on short term. However the climate change impacts may aggravate the pressure on groundwater resources by diminishing recharge capacities and intensifying the need of groundwater to fill gaps in surface water availability due to increased variability of precipitation.

Several authors explain the increase in groundwater exploitation considering that is a common-pool-resource (Feitelson 2006; Gunn 2009; Jarvis 2010), which refers to "a natural or man-made resource system that is sufficiently large as to make it costly (but not impossible) to exclude potential beneficiaries from obtaining benefits from its use" (Ostrom 1990).

In the case of groundwater, anyone who has the financial and technical means (Moech 2004) can easily exploit them, as long as, "the marginal value product of the water is greater or equal to the marginal pumping costs" (Feitelson 2006). This characteristic produces externalities because the water pumped is no longer available to others and decreases the aquifer levels for everyone. Also, locally-based contamination impairs all groundwater users (Palma 2003). The hidden character of these waters and its intrinsic relation with water, added to the right to ownership, make their extraction exclusion and control more difficult.

Groundwater has been subjected to serious degradation through overuse or contamination all over the world. This is a classic example of what Hardin (1968) labeled the tragedy of the commons, which results from a conflict between collective and individual interests. Generally, the possible solutions for this dilemma can be found in the following strategies: a) privatizing the commons (individual management); b) regulation by an external government agency (state-level management); and c) collective agreements among local resource users (local level management) (Laver 1984; Ostrom 1990). Several institutional arrangements may be built based on these strategies and used for the management of the common-pool resources (Sick 2002).

Empirical studies show that the most successful management experiences were those where the resource was restricted to the national limits and managed by small to relatively large groups counting on the support of nested institutions at varying levels (Ostrom et al. 1999). Environmental goods require collective action and cooperation. This also applies to transboundary water resources, although the process is more complex. Managing the hydrological cycle in an integrated perspective demands the co-ordination of a range of existing social, administrative, economic, and political boundaries (Sick 2002).

Within this context, co-management offers some interesting prospects for dealing with groundwater exploitation. The definition and nature of co-management varies considerably. It can be interpreted as "the sharing of power and responsibility between government and local user" (Berkes et al. 1991). The World Bank defines it as "the sharing of responsibilities, rights and duties between the primary stakeholders, in particular, local communities and the nation state" (World Bank 1999). Borrini-Feyerabend et al. (2000) described co-management as "a situation in which two or more social actors negotiate, define and guarantee amongst themselves a fair sharing of the management functions, entitlements and responsibilities for a given territory, area or set of natural resources". Also, it can be interpreted as "a collaborative and participatory process of regulatory decision-making between representatives of user-groups, government agencies, research institutions, and other stakeholders" (Jentoft 2003).

The fundamental premise in co-management is the construction of alliances among stakeholders, who become partners in the management of the water resources. Since the Guarani Aquifer is a transboundary aquifer, it is necessary to include the international level, which is not usually contemplated in co-management experiences. International institutions, projects or agreements may influence the countries positively in managing the natural resources (Bernauer 1995). Governments can provide administrative, regulatory and infrastructural framework, while the local level contributes with knowledge, presence in the resource setting and support mobilization (Sick 2002).

International cooperation, allied with the decision-power division between the governmental agencies and community where these resources are located, allows for a more democratic management process and mutual governance (Pinkerton 1993). The state has a complex position with regards to the environment, since it is expected to promote economic growth, maintain that the systems are productive and simultaneously ensure the balance of the ecological systems (Walker 1989). At the same time, if the individuals do not contribute to the maintenance of the resources, they may threaten them and, consequently, harm their business and the business of other stakeholders as well. Unfortunately, aquifer problems are often socially invisible. Since society fails to perceive the problem, public policies neglect it (Walker 1989).

Water-resources managers base water quality and quantity coordination on the watershed[4] level, which is considered the planning unit by excellence. The idea is that a state agency with jurisdiction over the watershed should develop a "partnership with other relevant stakeholders (primarily including local residents and resources users) that specifies and guarantees their respective functions, rights and responsibilities with regard to the protected area" (Borrini-Feyerabend et al. 1996).

The watershed can be divided into different physical units (e.g., upper catchments, river valleys, coastal estuaries) that allow for a better understanding of the water resources and the characteristics of the areas

[4]The watershed can be defined as a "geographical entity" which "possesses a basic structure and a topographic pattern that, together with the law of gravity and the flow of water, shape its biotic and abiotic characteristics and processes with considerable regularity" (Lovelace and Rambo 1991).

that influence water management (e.g., socio-economical activities, identification of main users and possible sources of contamination). Since it is not a political boundary, the watershed is usually subject to the influence of several federative bodies (Porto and Porto 2008). With regards to groundwater, this unit presents further challenges, since the aquifers do not respect the limits of the watershed and they are not always connected to it.

The adoption of the co-management strategy to manage the watersheds, especially the aquifers, makes us consider the notion of "scaling up" environmental management systems (Sneddon 2002), which evokes "the application of propositions or models about micro-scale systems to meso-scale and macro-scale phenomena" (Young 1994). The benefits of co-management can be transferred to larger geographic scales or can be directly influenced by the decisions made on them. As Swyngedouw (1997) affirms: "Scale, both in its metaphorical use and material construction, is highly fluid and dynamic, and both processes and effects can easily move from scale to scale and affect different people in different ways, depending in the scale at which the process operates".

The Guarani Aquifer management demonstrates the need to build institutional arrangements integrated between the different levels. The next section describes aquifer's characteristics, the joint projects established to stimulate co-management among and within the countries, and analyze the International Agreement on the Guarani Aquifer.

Building International Joint Management: From Technical Cooperation to the Agreement on the Guarani Aquifer

The Guarani Aquifer

The Guarani Aquifer System is a transboundary aquifer with an area of 1.100.000-km^2 spanning four countries: Paraguay, Uruguay, Argentina and Brazil (GW MATE 2009). It is located in east-central South America, between 12° and 35° south latitude and 47° and 65° west longitude, in the Paraná Sedimentary Geological area. The average thickness of the aquifer is 250 meters and the volume of water is estimated to be 30,000 km^3, which is equivalent to 100 yr of cumulative flow in the Paraná River (GW MATE 2009). The quality of water is good, with low mineralization rates in most places (OAS 2009; GW MATE 2009).

The studies performed show that the aquifer is a very heterogeneous structure, which requires different management approaches according to the aquifer characteristics. Figure 15.1 shows GW MATE's (2009) suggestion of zoning the aquifer into management areas, according to their hydrogeological characteristics: I—non confined recharge and discharge zone;

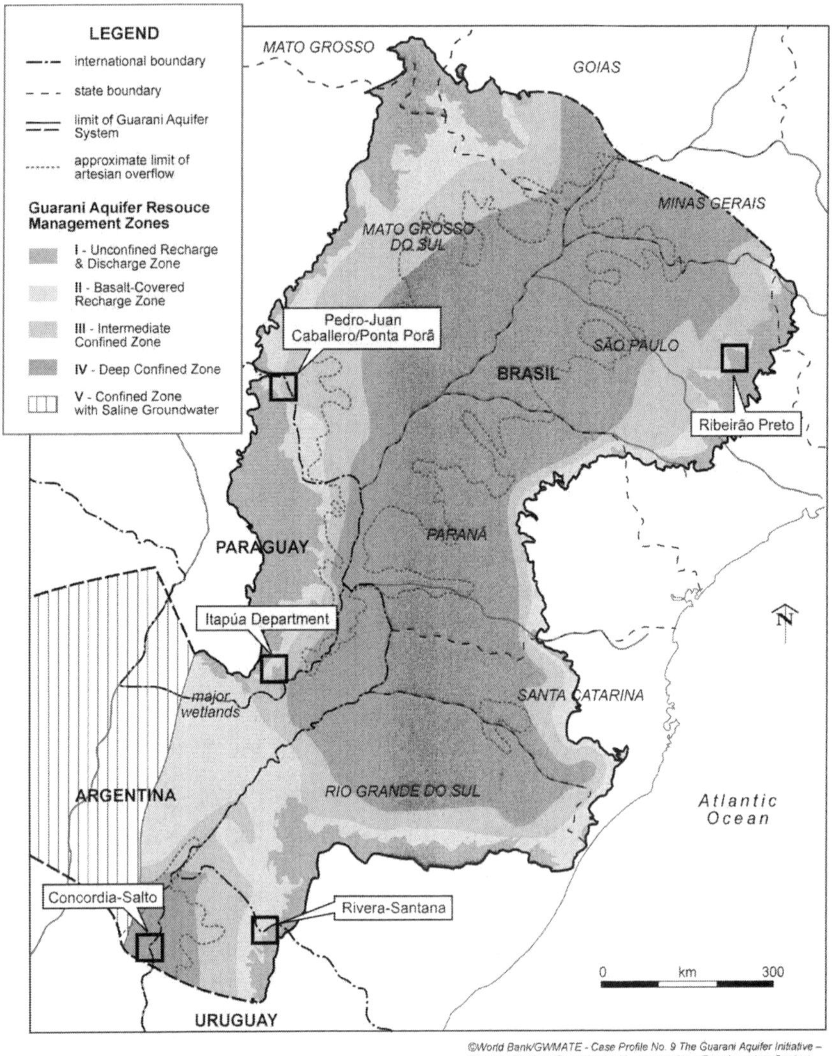

Figure 15.1 Guarani Aquifer System—Management Zones.

Color image of this figure appears in the color plate section at the end of the book.

II—basalt-covered recharge zone; III—intermediate non-confined zone; IV—deep confined zone; and V—confined zone with saline groundwater.

According this management zones division, the areas most vulnerable to contamination and that allow for the recharge of the aquifer are the non-confined recharge and discharge zone (zone I) and the basalt-covered recharge zone[5] (zone II) (GW MATE 2009). The confined zones (intermediate, deep, and with high salinity level) present low vulnerability to pollution. However, there is no significant recharge and the extraction of groundwater results in mining the aquifer because there is no water replacement (OAS 2009; GW MATE 2009).

The largest part of the aquifer is located in Brazilian territory where it extends across eight states: Rio Grande do Sul, Santa Catarina, Paraná, São Paulo, Mato Grosso do Sul, Minas Gerais, Mato Grosso and Goiás. The state of São Paulo is the main user, corresponding to 80 percent of the national extraction of waters from this aquifer (GW MATE 2009). The use in this region is intense. In the municipality of Ribeirão Preto (SP) the aquifer shows signs of over-exploitation and measures for the restrictions to the drilling of new wells have been adopted (Villar and Ribeiro 2009).

The aquifer is located in a region with high water availability. On the surface is the La Plata river basin, considered the second largest watershed in extension occupying 3,100,000 km^2 (Morong 1892). Despite having registered several conflicts over the use of the La Plata basin surface waters, there are no registered conflicts over the use of the Guarani Aquifer. The cooperation experiences concerning groundwater were mainly motivated by the epistemic community and international organizations within a prevention and precaution context.

Technical Cooperation: A Stimulus to Management

International cooperation with regards to groundwater in Latin America was induced by the epistemic communities from the South-Cone universities. At the Congress of the Latin American Underground Hydrogeology for Development Association (ALHSD) in 1992, the idea for the first cooperation project for a Latin-American aquifer appeared (Borghetti et al. 2004). The Proyecto Sostenible del Acuífero Botucatu[6] (1995), financed by the International Development Research Centre (IDRC—Canada), integrated public and private institutions aiming at the establishment of legal mechanisms and joint measures to manage this aquifer (Borghetti et al. 2004)[7]. From that point, the first technical-scientific base on the Aquifer

[5]This zone is overlain by very thick and fractured basalt, which allows for the recharge, but at a lower capacity than zone I.

[6]The Botucatu Aquifer is one of the Guarani Aquifer System geological formations. The denomination Guarani Aquifer was only approved by Argentina, Brazil, Paraguay, and Uruguay in May 1996, and intended to include all the different geological formations that are part of the system as well as honor the Guarani indigenous people who live in the area (Borghetti et al. 2004).

[7]This initiative allowed for the Technical-Scientific Meet on the Botucatu International Aquifer (1995) and the International Workshop on the Mercosur Gigantic Aquifer.

was put together and a joint project was established between universities in Argentina and Uruguay[8] (Borghetti et al. 2004).

The availability of this preliminary technical base, the articulation of the researchers and their dialogue with the Global Environment Facility (GEF) resulted in the Environmental Protection and Sustainable Development of the Guarani Aquifer System Project (also known as the Guarani Aquifer System Project), which represents a cooperation mark in groundwater matters in these countries. This project lasted six years (2003–2009) and involved the four countries and several international agencies, such as: Global Environment Facility (GEF), the World Bank (WB), the Organization of American States (OAS), the Dutch and German Governments, and the International Atomic Energy Agency (OAS 2005, 2009).

Financing from international institutions resulted in highly ambitious cooperation goals, turning this project into an important tool to stimulate joint management at different levels. The objective was to support the countries *"to elaborate and implement a shared institutional, legal and technical framework to preserve and manage the Guarani Aquifer System (GAS) for the current and future generations"* (OAS 2005). The information generated allowed for a better understanding of the aquifer's dynamics. Conclusions of the project pointed to an aquifer without conflicts or degradation, a recharge flow limited to the borders of the Guarani aquifer system and the local nature of the problems related to the use of the aquifer (OAS 2009; GW MATE 2009). Despite the transboundary nature of the Guarani Aquifer System, the project emphasized that the appropriate scale to manage the Guarani Aquifer System is local:

> *"current and potential transboundary effects of the GAS are restricted to a narrow strip of territory of no more than a few dozen kilometers depending upon local specific hydrodynamic conditions" (OAS 2009).*

> *"Confirmation that groundwater management would be carried out at the local level and that it would involve local stakeholders and water users was, unquestionably, one of the strong points of the Project" (OAS 2009).*

The project used a "bottom-up" approach. The peculiarities in the Guarani Aquifer System make the local level an important player in the management of these waters, especially in face of the absence of significant transboundary conflicts, the number of stakeholders involved and its

[8]The co-operation included Universidad Nacional del Litoral and Universidad de Buenos Aires (both in Argentina) and Universidad de la República (Uruguay). The results of the technical cooperation were published in Montano et al. 1998.

geological characteristics.[9] Such conclusions were based on the four pilot projects.[10] These pilot projects had a positive influence in the Guarani aquifer management. For example, in the Ribeirão Preto municipality (SP) the first well-drilling restriction zone of the Guarani Aquifer was established. However, despite some improvements in the groundwater management, the water levels continue to decrease in the Ribeirão Preto area (Villar and Ribeiro 2009).

The management of a transboundary aquifer requires the establishment of joint cooperation schemes with a multi-sectorial, multidisciplinary and multi-player approach. The Guarani Aquifer requires cooperative management between the several levels of governments and various stakeholders.

The convergence of national and international participants and funds stimulated scientific production and focused social attention to the management challenge. The social, academic and international pressure created a favorable environment for the execution of an international agreement between the Guarani Aquifer countries and the establishment of specific policies in these countries.

The Agreement on the Guarani Aquifer

In contrast to surface waters, the international law of transboundary aquifers is still in an undeveloped stage and the experiences for aquifers joint management are scarce[11] (Feitelson 2005). The main international instrument for the management of transboundary waters, the United Nations Convention on the Law of the Non-Navigational Uses of International Watercourses, still hasnot been ratified and its approach

[9]There are highly vulnerable zones (recharge areas), low-vulnerability zones (confined zones) and zones whose waters are not renewable. Within this context, the next section will analyze how the management of groundwater is performed and the movements of the several Brazilian players.

[10]The pilot projects were: a) Concordia (Argentina)/Salto (Uruguay), b) Rivera (Uruguay) / Santana do Livramento (Brazil), c) Encarnación—Cuidad del Este—Caaguazú (Paraguay) and d) Ribeirão Preto (Brazil). The first two sought a better understanding of issues linked to the management of transboundary groundwater and the others sought the development of management strategies within a groundwater vulnerability context (presence of recharge areas, contamination or overexploitation risk).

[11]The agreements that focus on transboundary aquifers are: (a) the Convention relative à la protection, à l'utilization, à la réalimentation et au suivi de la nappe souterraine Franco-Swiss du Genevois, signed in 2008 by France and Switzerland (replacing the previous agreement of 1977); (b) two technical cooperation agreements for monitoring and exchanging data related to the Development of a Regional Strategy for Utilization of the Nubian Sandstone Aquifer System Program; (c) and a technical cooperation agreement for the establishment of a consultative mechanism for the northwestern Sahara Aquifer System (Burchi and Mechlem 2004).

towards groundwater is highly limited[12] (Eckstein 2005). The international legal loopholes on the theme motivated the International Law Commission of the United Nations to prepare a project, which was approved on December 11, 2008, by the U.N. General Assembly, with the title of Resolution 63/124 —The law of transboundary aquifers.[13]

In August 2010, Argentina, Brazil, Paraguay and Uruguay signed the Agreement on the Guarani Aquifer, which incorporates the guidelines of Resolution 63/124, honoring the principles therein: sovereignty, equitable and reasonable use of water resources, the obligation not to cause harm, cooperation, and exchange of data and information. Its signing affirms the intent of a joint international-management process. This is the first international agreement for a transboundary aquifer executed in Latin America and the only one within the prevention perspective; it was not driven by conflicts for the use of the water (Villar and Ribeiro 2011).

One of the most controversial points of the Agreement is the emphasis on the States' sovereignty over their territorial portion of the Guarani Aquifer. Articles 1 and 2 highlight the sovereign territorial domain of the Guarani Aquifer, whilst Article 3 sets the sovereign right over natural resources. The extent of the aquifer's joint management is limited, since each country will be solely responsible for "promoting the management, monitoring, and sustainable use of water resources of the Guarani Aquifer System" (Article 3). On the other hand, the States' freedom to act is limited by the "criteria of rational and sustainable use, and respecting the obligation not to cause appreciable harm to the other Parties or the environment" (Article 3). If the sovereignty is an important principle in the agreement, so are: the principle of equitable use of water resources (Articles 3 and 4), the obligation not to cause harm and cooperation initiatives (Articles 8, 9, 10, 12, 13, and 14). The cooperation actions are concentrated on the following points: the need for the exchange of information on water resource utilizations, the duty of prior notification, the right to seek additional information, the establishment and development of joint projects and cooperation programs for technical, scientific, and management aspects (Villar and Ribeiro 2011).

The great promise of the agreement for joint management lies in Article 15, which establishes the creation of a Commission composed of the states involved, which will coordinate the cooperation process. If implemented, this Commission will be the first in Latin America to deal with this subject.

[12]To better understand the international-rights deficiencies of the aquifers, please see: Eckstein 2005, Mechlem 2009, McCaffrey 2009.

[13]The U.N. Resolution 63/124 is available in the Official Documents System of the United Nations website at: http://daccess-dds-ny.un.org/doc/UNDOC/GEN/N08/478/23/PDF/N0847823.pdf?Open Element.

At the moment, it is not possible to determine what its future will be, since its bylaws, powers, members and budgets have not been defined.

Relevant themes, such as the protection of the recharge areas[14] or the extraction of non-renewable resources from the Guarani Aquifer were excluded from the agreement. Moreover, the dispute settlement system will be defined later in an Additional Protocol (Article 17). The countries refused Mercosur's (Olivos Protocol) controversial settlement process and opted for the creation of a new system, whose mechanics are still not established (Villar and Ribeiro 2011).

The analysis of the agreement reveals its fragility, since the cooperation mechanisms are limited and require regulation in the national and international spheres. Its main message is that despite a few joint actions, each country will be responsible for managing and exploiting the Aquifer in its territory. The countries adopted an innovative position when they signed an agreement within a precaution/prevention context; nevertheless, they simultaneously maintain a conservative position with regards to the agreement content. In face of the absence of conflicts over and degradation of the aquifer, it should be straight forward to accept the philosophy defined by the UNDP (2006), "cooperation [over transboundary waters] need not always be deep .... Indeed, given the different strategic, political and economic contexts in international basins, it makes sense to promote and support cooperation of any sort, no matter how slight".

The negotiation of an international agreement happens in two complementary spheres: the international and the national. In the international sphere, the States are the main players; nevertheless, the results of the negotiations must be incorporated and accepted by the national sphere (Feitelson 2006). The emphasis on the national sovereignty over the aquifer and the exclusion of the controversial points facilitate the social consent of the nationals, which is essential for the ratification of the agreement and implementation of joint projects. Cooperation is a process, the countries' conservative position may be justified as a way to allow for a faster acceptance of the agreement and gradually expand it to include more ambitious cooperation goals.

The application of national and international standards depends on the countries' institutional capacity to mobilize the several players involved in the policy established (Feitelson 2006). In the national jurisdiction, the management of the aquifer is influenced by several players and legal

[14]The recharge areas could be included in Article 14, which makes a vague allusion to the possibility of "identifying critical areas, particularly in border areas that require specific treatment measures". Nevertheless, it seems like the focus of Article 14 is the application of restriction and control measures, and not the design of a management based on precaution and prevention.

standards, since the aquifer goes beyond the administrative limits of several states and provinces, including several municipalities and a great number of users (irrigation, public supply, thermal tourism, the industry). The next section will analyze how the groundwater management has been dealt with in Brazil.

The Brazilian Waters Policy and the Management of Aquifers

The Brazilian re-democratization process driven by the 1988 Federal Constitution transformed the water management in the country, thanks to the inclusion of social participation and tax decentralization for states and municipalities. The Constitution promoted a new water-resource management system, regulated by Law 9.433/97, that relies on a decentralized management model based on the watershed and participative management performed by means of collegiate structures, composed of the public power, users and civil society (Porto 1998).

The constitutional arrangement established divides the powers between the Union, member States, Federal District and Municipalities. The standards establish the ownership and decision rights with regards to the resources, establishing the degree of freedom of each one of the entities and shaping how co-management will be executed. Water is defined as common-use public asset and authority over water is divided between the Union and the States. In contrast to surface water, whose location will determine if it belongs to Union[15] or states, groundwater will always be considered state assets (Article 26, III). Thus, the Guarani Aquifer domain is divided between the eight Brazilian states through which it extends. The Union has private legislative competency in the waters theme[16] (Article 22, item IV), nevertheless, that does not imply that the other entities cannot establish administrative rules over the waters that are under their respective domains. Union, States and municipalities have common administrative responsibility to protect the environment (Article 23) and states can legislate concurrently with the Union topics such as forests, hunting, fishing, fauna, environmental protection, soil preservation and pollution control (Article 24). The municipalities do not have authority over water, but are the ones responsible for legislating on matters of local interest, supplementing Federal and State legislation when necessary (Article 30, items I, II) and promoting urban planning(Article 30, VIII).

[15]Brazilian Constitution established the property of the Union over lakes, rivers, and any watercourse it owns; interstate waters, waters that serve as borders with other countries and water that extend into or come from a foreign territory (Article 20).

[16]The Union's private legislative competency means it is responsible for editing legal norms, rules and determining principles to discipline the political and administrative activities referring to this matter.

The division of constitutional competencies in environmental matters forces the formulation of federative agreements, since the domain of the waters and the capacity to legislate environmental protection are shared. The lack of interaction between the entities has generated inefficient, disarticulated and/or conflicting environmental policies. To get around this probiem, Federal Law 9.433/97 instituted the National Water Resources Management System (*Sistema Nacional de Gerenciamento de Recursos Hídricos*—SINGREH), shown in Fig. 15.2. This institutional organization intends to integrate water policies among the Union, member states and municipalities and allow for the participation of non-governmental players in the management decisions. The SINGREH is composed at the Federal level by the Ministry of the Environment, represented by the Water Resource and Urban Environment Secretary (Secretaria *de Recursos Hídricos e Ambiente Urbano*—SRHU), the National Water Agency[17] (*Agencia Nacional de Água*—ANA) and the National Water Resource Council (*Conselho Nacional de Recursos Hídricos*—CNRH). At a State level, the SINGREH is formed by the state secretaries responsible for water management, the state water resource council and technical bodies. Moreover, the Union, state and municipal bodies related to water issues are part of the system. The base of the system is the watershed committee[18] and water agencies.[19] The presence of collegiate

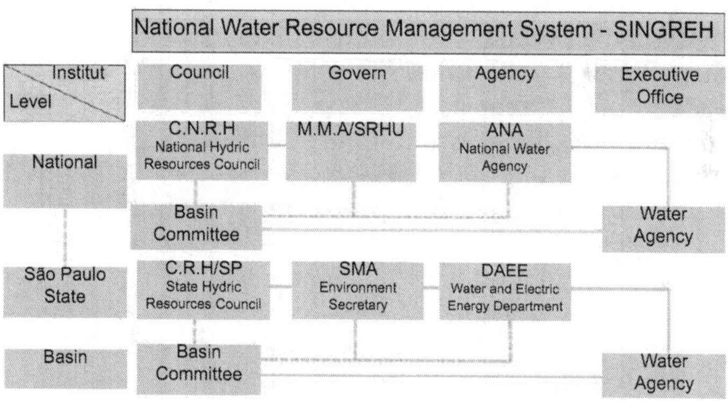

Figure 15.2 The National Water Resource Management System (SINGREH).
Source: ministério do Meio Ambiente, 2001.

[17]ANA is a Federal agency created to implement the National Water Resource Policy and coordinate the national water resources management system.

[18]The watershed committees are composed of representatives of the public administration (Federal Government, the states or federal district and municipalities in which the committees are locate), water users and civil organizations related to water.

[19]Water agencies serve as executive secretariat of each committee and are responsible for the maintenance of a cadastre of water users, collection for water use, organization of studies for the water management and the assistance the watershed committed. The creation of these entities is request by the watershed committee and needs the authorization by the National or State Council on Water Resources.

structures, represented by the CNRH, the state water resource council and the watershed committee, minimizes the centralized-government character of water management and allows for the participation of social and private players.

The extent of system management is limited, since it excluded several sectors or functions related to the water, hindering the way it works (Saleth and Dinar 2000). For example, mineral and bottled waters are classified as mineral resources, being completely excluded from the water-resource management system. They belong to the Union, are ruled by mineral-resource regulations and are under the wing of the National Mineral Production Department (*Departamento Nacional de Produção Mineral*–DNPM). The SINGREH bodies do not have information about the amount of water exploited by the mineral-water sector, which reports only to the DNPM.[20] The system also does not integrate either the bodies responsible for environmental or soil management or the users related to this theme (Villar 2010).

Despite the difficulties, the water policy includes elements of integrated water resources management. The previous centralized model that focused on the production of hydraulic energy was replaced with an integrated, decentralized and participative management model, which includes a series of instruments capable of promoting water management. The water is classified as a limited natural resource having an economic value and subject to multiple uses (Porto 1998).

The adoption of the watershed as a management unit aims at the decentralization of the decision power and facilitates the approach of cause/effect relations that occur in the watershed. Federal Law 9.433/97 established a new territoriality that is not tied in with the classic administrative divisions: municipality, State and Union. This new way of managing water faces difficulties in having the recommendations established in the watershed level incorporated in the public policies of the traditional administrative divisions, which hold the power to implement environmental and urban policies (Porto and Porto 2008; Villar 2010).

The possibility of social participation in the watershed committees allowed for the politicization of the management[21] (Guivant and Jacobi 2003). The watershed committees are collegiate structures that ensure the plurality of interests in defining water policies and allow for greater supervision of the management actions. The strengthening of the deliberative spaces is essential for the consolidation of a democratic water management. These

[20]To overcome the problem, the CNRH edited Ruling n° 76/2007 determining the integration of mineral- and bottled-water management, but how such integration will be done is ambiguous.

[21]For more information about the Brazilian participation, see Brannstrom et al. 2004.

committees must seek the increase of its representation capacity to make sure its format and results are effectively public. Conflicts, contradictions and tensions are natural in the democratic processes and must be seen as advancement, since they allow the conflict to be discussed, negotiated and voted. The plurality of stakelolders and the stimulus to their participation in the decision processes tend to contribute to the construction of a more democratic water management, although there are important power asymmetries between the stakeholders (Guivant and Jacobi 2003).

Groundwater has been gradually inserted in the management instruments foreseen by Federal Law 9.433/97: the watershed plans, water quality monitoring framework, water-resource information system, water pricing and water permits. These instruments have different and complementary purposes. The watershed plans and the water quality monitoring framework aim at the construction of consensus and joint planning negotiated between public administration, civil society and economic agents (Porto and Porto 2008). The content of the watershed plans have a technical and political nature, but do not impose binding obligations upon municipalities or the other Federation entities in the use of the soil (Villar 2010). Thus, the bodies from SINGREH have the duty of stimulating municipalities to adopt the measures established in the watershed plans in their Municipality Plans. The SINGREH cannot impose the content of water plans on municipalities because soil-use management is a municipal attribution.

The CONAMA's Ruling n° 396/2008 extended the water-quality monitoring framework to aquifers. This instrument determines the maximum polluting loads that can be launched according to the classification of the aquifers. Unfortunately, in face of the difficulties to apply this instrument to surface water, the lack of detailed information on the aquifers and the economic and social impacts generated by the regulation, are an impediment for watershed committees to apply the instrument to aquifers (Villar 2010).

The permit and the charging for groundwater intend to control and stimulate rational use and exploitation. The permit is the instrument through which the Government attributes to the public or private interested party the right to exclusively use the water resource. This is not the transfer of water ownership, but the concession of usage rights, which may be paid or for free. There are an increasing number of permit requests, but it is still far from supplying reliable groundwater exploitation standards. One major problem concerning this instrument is the lack of a specific consolidated-analysis methodology for all state permit authorities. Moreover, it is necessary to determine the recharge volume, as well as define criteria that consider the relation of groundwater with surface waters. The interaction and articulation between State water-use permit authorities and the National

Mineral Production Department—DNPM must also be built to conciliate groundwater and mineral extraction.

The charge falls within the category of payment for the use of the water. This instrument is a valuable source of resources to invest in water projects, but faces restrictions from water users, especially in agriculture. Only a few watershed committees have established prices for the extraction of groundwater.

The water-resource information system is a communication instrument under National Water Agency's responsibility. Nevertheless, most of the information on groundwater is in the Groundwater Information System (SIAGAS) coordinated by the Brazilian Geological Service (CPRM). Both systems operated in an autonomous and independent fashion, without proper communication. The edition of the CNRH resolution n° 107/2010 raises the opportunity to join the systems, since a groundwater monitoring network will be planned and coordinated by the National Water Agency. Meanwhile, the implementation and operation of the system will be done by CPRM. Groundwater management in Brazil is still undergoing a structuring process. There are a series of challenges to be overcome in face of its newness, and the difficulty of managing a hidden resource. The Guarani Aquifer project was essential to demonstrate the need for groundwater to be included in the country and states' water policies, as well as the role of the municipalities in their implementation. The Union has had very modest participation in groundwater management, probably because the resource is situated under the state's authority. The States addressed groundwater only indirectly or incidentally as part of surface water, or were simply not addressed at all by state policies. At the local level, the aquifers were not considered a priority in most watershed committees or in municipal laws.

Perspectives to Stimulate Groundwater Co-Management

The concern with groundwater management is recent, especially in the case of transboundary aquifers. In Brazil, the lack of mobilization by the stakeholders may be justified because the Guarani Aquifer is still underused, so aquifer water is not seen as a scarce resource subject to competition over its use. The lack of information on the aquifers stimulates this perception, since there is no clear measure of its actual status. The extraction intensification trend of these waters will probably change this scenario and stimulate management and social mobilization.

The regulatory system that allows for the co-management of the Brazilian water resources is recent, and despite the difficulties, it already registers successful experiences for surface water. This is the case of the Piracicaba, Capivari and Jundiaí Committee and Cotia-Guarapiranga and Billings-Tamanduateí Subcommittees, which despite the limitations have achieved participative management. The experience acquired in the management and social mobilization of surface water will contribute to the co-management of groundwater resources, even if adaptations are necessary. The core stakeholder in groundwater are slightly different from those in surface waters, since aquifers are not appropriate for certain uses (hydro power generation, navigation or fishing), which occupy a prominent position in the discussions of public water policies. The stakeholders associated with soil and the mineral exploitation of the waters will have to be incorporated.

The nature of co-management is based on a continuous problem-solution process (Carlsoon and Berkes 2005). A heightened social perception of risks can stimulate the strengthening of the co-management of the aquifers. In the case of the State of São Paulo, the main user of the aquifer (OAS 2009), a greater mobilization of the public administration, users (especially in the supply companies) and the emergence of a social movement can already be seen. Public institutions started investing in programs to characterize the geology of aquifers and developing strategies to integrate the several watershed committees actions, since the territorial aquifer limits are different from the watershed ones.[22] The creation of the Aquifer program[23] is a São Paulo state's effort to promote the management of the aquifers in this state. This Project aims at (1) identifying the critical areas, (2) creating prevention and control mechanisms and (3) defining water-use guidelines. The most ambitious goal of the process is transforming, through law, the recharge area of the Guarani Aquifer of the state into an environmental-protection area, which would require the inclusion of seven watersheds. These efforts are still concentrated in the epistemic-community sphere connected to the governmental bodies, and their proposals must be submitted to the political process (watershed committees and legislative assembly).

Another stimulus that can help the mobilization of the social players is the payment for environmental services. Although this instrument has not been foreseen in the water law, it has proven to be highly efficient in protecting the resource. The trend is that this instrument will be increasingly used by the government, corporate-responsibility projects and in the negotiation of public-supply service contracts. The state of São Paulo already authorizes the payment for environmental services to the farmers and the municipality of Botucatu (SP), located in the Guarani recharge area, obligating the company that wins the concession bid for basic

[22]For example, the recharge areas of Guarani Aquifer located in the State of São Paulo covers an area of approximately 23,500 km^2, which spreads for seven watersheds and under the management of seven different watershed committees.

[23]http://homologa.ambiente.sp.gov.br/aquiferos/15052009_cap_dos_agentes_publicos.asp.

sanitation services to apply 1 percent of its net revenue to the payment of environmental services. These experiences are focused on surface waters, but nothing prohibits its applicability in the aquifer recharge areas. The possibility of financially benefitting those that protect such areas brings multiple advantages, since it allows for the stimulus to the protection of the aquifer, social mobilization and environmental education.

Final Considerations

The co-management of the Guarani Aquifer at all levels is, at best, in its infancy. The knowledge generated by international projects raises the importance of the aquifers, although groundwater considerations in the public water policies are very modest. The different management levels and users face difficulties to generate integrated policies. Neither is there true social mobilization to promote groundwater management, which is still heavily managed by technical bodies, although these decisions are submitted to the approval of the Watershed Committees. This reality does not interrupt the co-management discussion. To the contrary, among the causes that stimulate the precariousness and difficulty of managing the resource is the lack of it. The interaction between the location and the other scales, as well as social mobilization and that of the users, is essential for the management of these waters to be highlighted.

References

Berkes, F., P. George and R. Preston. 1991. Co-management: the evolution of the theory and practice of joint administration of living resources. Alternatives 18(2): 12–18.

Bernauer, T. 1995. The effect of international environmental institutions: how we might learn more. International Organization 49: 351–377.

Borghetti, N., J.R. Borghetti and E.F. Rosa Filho. 2004. Aquífero Guarani: A verdadeira integração dos países do Mercosul. GIA, Curitiba. BR.

Borrini-Feyerabend, G. 1996. Collaborative management of protected areas: Tailoring the approach to the context. Issues in Social Policy. IUCN, Gland. Switzerland.

Borrini-Feyerabend, G., M.T. Farvar, J.C. Nguinguiri and V.A. Ndangang. 2000. Co-management of natural resources: Organising, Negotiating and Learning-by-Doing. GTZ and IUCN, Kasparek Verlag, Heidelberg. GE.

Brannstrom, C., J. Clarke and M. Newport. 2004. Civil society participation in the decentralisation of Brazil's water resources: assessing participation in three states. Singapore Journal of Tropical Geography 25(3): 304–321.

Burchi, S. and K. Mechlem. 2004. Groundwater in international law Compilation of treaties and other legal instruments. D L Service UNESCO/FAO, Rome. Italy.

Carlsoon, L. and F. Berkes 2005. Co-management: concepts and methodological implications. Journal of Environmental Management 75(2005) 65–76.

Eckstein, G.E. 2005. Protecting a hidden treasure: The U.N. International Law Comission and International Law of Transboundary Ground Water Resources. Sustainable Development Law & Policy 5(1): 5–12.

Feitelson, E. 2005. Political economy of groundwater exploitation: The Israeli case. Water Resources Development 21(3): 413–423.
Feitelson, E. 2006. Impediments to the management of shared aquifers: A political economy perspective. Hydrogeology Journal 14(3): 319–329.
Fornés, J.M., A. Hera and M.R. Llamas. 2005. The silent revolution in groundwater intensive use and its influence in Spain. Water Policy 7(3): 253–268.
Guivant, J.S. and P. Jacobi. 2003. Da hidro-técnica à hidro-política: novos rumos para a regulação e gestão dos riscos ambientais no Brasil. Cadernos de Pesquisa Interdisciplinar em Ciências Humanas. 4(43): 2–26. Available online: http://www.periodicos.ufsc.br/index.php/cadernosdepesquisa/issue/view/492. [Accessed 7 July 2010].
Gunn, E.L. 2009. Governing shared groundwater: the controversy over private regulation. The Geographical Journal 15(1): 39–51.
GW MATE (Groundwater Management Advisory Team), 2009. The Guarani Aquifer Initiative —Towards Realistic Groundwater Management in a Transboundary Context. The World Bank, GW MATE and Water Partnership Program. Case number 9. Available from: http://siteresources.worldbank.org/INTWAT/Resources/GWMATE_English_CP_09.pdf [Accessed 4 August 2010].
Jarvis, W.T. 2010. Water Wars, War of the Well, and Guerilla Well-fare. Ground Water. 48 (3): 346–350.
Jentoft, S. Co-management—the way forward. pp. 1–14. *In*: D.C. Wilson, J.R. Nielsen and P. Degnbol [eds.]. 2003. The Fisheries Co-management Experience: Accomplishments, Challenges and Prospects. Kluwer Academic Publishers, Dordrecht, NE.
Haas, P.M. 1992. Introduction: epistemic communities and international policy coordination. International Organization 46: 1–35.
Hardin, G. 1968. The Tragedy of the commons. Sciences, new series 162(3859): 1243–1248.
Laver, M. 1984. The Politics of Inner Space: Tragedies of Three Commons. European Journal of Political Research 12: 59–71.
Llamas, M.R. and P. Martínez-Santos. 2005. Intensive groundwater use: Silent revolution and potential source of social conflicts. American Society of Civil Engineers Journal of Water Resources Planning and Management 131(4): 337–341.
Loáiciga, H.A. 2003. Climate Change and Ground Water. Annals of the Association of American Geographers 93: 30–41.
Lovelace, G.W. and A.T. Rambo. Behavioral and social dimensions. pp. 81–90. *In:* K.W. Easter, J. Dixon and M. Hufschmidt [eds.]. 1991. Watershed resource management: Studies from Asia and the Pacific. East-West Center, Honolulu.
McCaffrey, S.C. 2009. The International Law Commission Adopts Draft Articles on Transboundary Aquifers. The American Journal of International Law 3(2): 272–293.
Mechlem, K. 2009. Moving Ahead in Protecting Freshwater Resources: The International Law Commission's Draft Articles on Transboudary Aquifers. Leiden Journal of International Law 22(4): 801–821.
Moench, M. Groundwater: The challenge of monitoring and management. pp. 79–100 *In:* P. Gleick [ed.]. 2004. The World's Water 2004–2005. Island Press,Washington, USA.
Montaño, J., O. Tujchneider and M. Auge et al. 1998. Acuíferos regionales en América Latina sistema acuífero Guaraní. Universidad Nacional del Litoral, Santa Fé. AR. Uruguay.
Morong, T. 1892. The Rio de la Plata. Its Basin, Geography and Inhabitants. Journal of the American Geographical Society of New York 24: 479–509
OAS—Organization of American States. 2005. Office for Sustainable Development & Environment. OAS, Water Project Series, number 7. Washington. Available from: http://www.oas.org/dsd/Events/english/Documents/OSDE_7Guarani.pdf [Accessed 7 July 2010].
OAS—Organization of American States. 2009. Guarani Aquifer: strategic action program=Acuífero Guaraní: programa estratégico de acción—Bilingual edition.

Organization of American States (OAS), Brazil. Available from: http://iwlearn.net/iw-projects/Fsp_112799467571/reports/strategic-action-program/view [Accessed 7 July 2010].

Ostrom, E. 1990. Governing the Commons: The Evolution of Institutions for Collective Action. Cambridge University Press, New York, USA.

Ostrom, E., J. Burger and C.B. Field. 1999. Revisiting the Commons: Local Lessons, Global Challenges. Science 284(5412): 278–282.

Palma, C.R. 2003. Joint Quantity/Quality Management of Groundwater. Environmental and Resource Economics 26(1): 89–106.

Pinkerton, E. 1992. Translating Legal Rights into Management Practice: Overcoming Barriers to the Exercise of Co-Management. Human Organization 51(4): 330–341.

Pinkerton, E.W. 1993. Co-management efforts as social movements: The Tin Wis coalition and the drive for forest practices legislation in British Columbia. Alternatives 19(3): 33–38.

Pomeroy, R.S., B.M. Katon and I. Harkes. 2001. Conditions affecting the success of fisheries comanagement: lessons from Asia. Marine Policy 25: 197–208.

Porto, M. 1998. The Brazilian Water Law: A New Level of Participation and Decision Making. International Journal of Water Resources Development 14(2): 175–182.

Porto, M.F.A. and R.L.L. Porto. 2008. Gestão de bacias hidrográficas. Estudos avançados. 22(63): 43–60.

Saleth, M.R. and A. Dinar. 2000. Institutional changes in global water sector: trends, patterns, and implications. Water Policy 2: 175–199.

Sneddon, C. 2002. Water Conflicts and River Basins: The Contradictions of Comanagement and Scale in Northeast Thailand. Society Natural Resources 15(8): 725–741.

Sick, D. 2002. Managing Environmental Processes Across Boundaries: A Review of the Literature on Institutions and Resource Management. International Development Research Centre, Ottawa. CA.

Swyngedouw, E. Neither global nor local: "Glocalization" and the politics of scale. pp. 137–166. *In:* K.H. Cox [ed.]. 1997. Spaces of globalization: Reasserting the power of the local. Guilford Press, New York, USA.

The World Bank. 1999. Report from the International Workshop on Community-Based natural Resource Management (CBNRM), Washington, DC.

UNDP. 2006. Beyond Scarcity: Power, Poverty and the Global Water Crisis. Human Development Report. United Nations Development Programme. New York, USA.

Villar, P.C. 2010. Moving toward Guarani Aquifer management: the Brazilian case. *In:* UNESCO-IAH-UNEP. Pre-proceedings—ISARM International Conference Transboundary Aquifers: Challenges and new directions [CD ROM]. UNESCO, 6–8 December 2010 UNESCO, Paris.

Villar, P.C. and W.C. Ribeiro. 2009. Sociedade e gestão do risco: o aquífero Guarani em Ribeirão Preto-SP, Brasil. Revista geografia Norte Grande 43: 51–64. Available from:http://www.scielo.cl/scielo.php?script=sci_arttext&pid=S0718-34022009000200003&lng=es&nrm=iso [Accessed 4 March 2011].

Villar P.C. and W.C Ribeiro. 2011. The Agreement on the Guarani Aquifer: a new paradigm for transboundary groundwater management? Water International. Groundwater Special Edition 36(5): 646–660.

Walker, K.J. 1989. The State in Environmental Management: The Ecological Dimension Political Studies. XXXVII, 25–38.

Young, O. 1994. The problem of scale in human=environment relationships. Journal of Theoretical Politics 6(4): 429–447.

16

Institutional Dynamics of Khorezmian Water Management on the Farmer Level in Uzbekistan: From Ancient Practices to the Present Situation

Darya Hirsch

Introduction

The Republic of Uzbekistan covers a territory of 447,000 km^2. It is situated in Central Asia between the Amudarya and Syrdarya Rivers (Fig. 16.1) and borders on Kazakhstan, Kyrgyzstan, Tajikistan, Afghanistan and Turkmenistan.

The Republic of Uzbekistan like other Central Asian countries depends on irrigation for productive agriculture. In Uzbekistan irrigation was introduced many centuries ago. In the past, in the majority of cases the governance of irrigation water was a responsibility of the communities or of people selected from the society, based on public esteem. In Soviet times,

Independent Environmental Services Professional, Email: daryahirsch@email.de

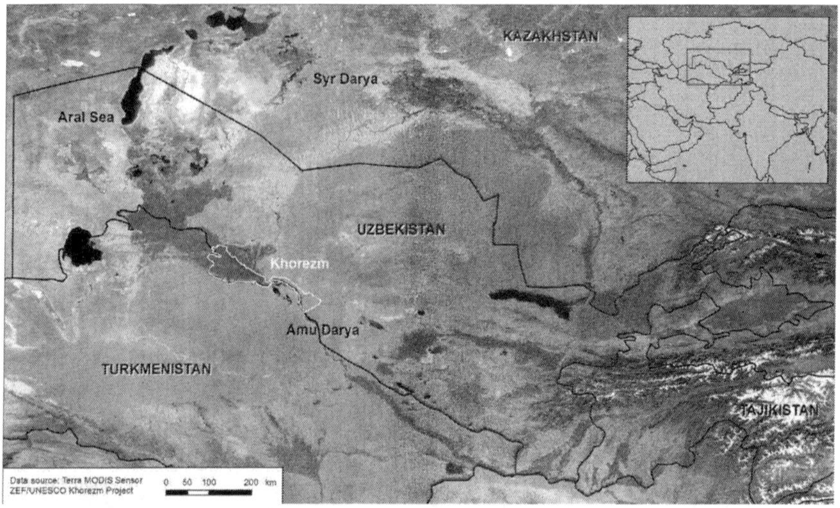

Figure 16.1 Map of Uzbekistan.
source: ZEF/UNESCO project in the Aral Sea Region.

the governance of water resources was done under strong regulation of the state. Only recently, after Uzbek independence (1990), the introduction of farmer-oriented management has become possible. Existed and existing traditions, culture, and social norms on the governance of water resources for irrigation play important role in the evolvement of this type of water co-management.

The described area, Khorezm, was an ancient and medieval state of central Asia, located in and around the basin of the lower Amudarya River. It is now a part of a region, namely the North West of Uzbekistan. Khorezm is one of the oldest centres of civilization in Central Asia and is located on in the catchment area of the Aral Sea.

Khorezm belongs to the semi-arid regions where distribution of precipitation is irregular. Most rain falls from winter to spring. The mean annual rainfall does not exceed 80–90 mm (Forkutsa 2005). In some years only 40 mm rainfall occurs. Given the evaporation rate of 1600–2000 mm measured in Khorezm one can assume that crop production heavily depends on irrigation (Atashev et al. 1966).

The Khorezm oasis is one of the most ancient regions of irrigation in the world (Tolstov 2005; Hillel 1992). According to Soviet orientalist scholars, historians and archaeologists, such as Vasili Bartold, Alexander Yakubosvky, Sergey Tolstov, Yahya Gulyamov, and Boris Andrianov, the construction of canals in Khorezm started in the middle of the second century B.C. Originally, farming was based on the natural flood of the Amudarya. Later it was based on natural delta channels where water was taken from into the irrigation canals. From the social and institutional point of view, ancient *khorezmian* water management, mostly for irrigation needs, involved co-management arrangements and structures already in the early years. In spite of strong power of feudal tribal chiefs within the *Khan* system of that time, the community was involved in decision-making and power and decision-making was shared in some situations with the community. Beside this, some sharing of responsibility for a resource between the resource manager and the community as well as platform for consensus existed. The Soviet period was characterized by technical progress, but at the same time neglected well-organized institutions of water co-management of ancient Khorezm. Elements of these institutions and also written sources about them come to the fore with the independency of Uzbekistan from the Soviet Union.

In this chapter, elaborative discussion and description of local water management (on the level of water users associations) aims to identify the dynamics between the practices at present and the practices set forth by tribal heads before the Soviets and later under the Soviets in 1926. Moreover, the explanation in the present situation of the water management in this chapter gives a basis for the analyses of how vulnerable the present water management system is in regard to its special ecological, social and technical dimensions.

The chapter presents the essential historical phases as the follows:

1. The pre Soviet period was labour intensive, small scale lift with the description of technical and management and socio-economic characteristics.
2. The Soviet period is characterized by technical methods, mechanization and collective farming.
3. Post- independence presented beginnings of transition.

The information contained in the chapter could be a useful resource of information that will lead others to consider co-management as an option for managing resources in their communities.

Essential Historical Phases of Khorezmian Water Management: From Ancient Practices to the Present Situation

The pre-Soviet Period: Organization of Irrigation System Management and Water Use in the Period from the 17th to the Middle of the 20th Century

Khorezmian management of irrigation systems carried out in the17th century and later based on the so-called "*khorezmian* model", originates from *mahalla* (neighbourhood) rules and conditions (Kadirov 1998) that were in place long before the Soviets introduced their irrigation system. The "*khorezmian* model" of irrigation therefore differs from the approach used in other parts of Uzbekistan since its administration is based more on customary local rules. Some names of the current khorezmian canals and locations still refer to the "module".

The head part of large canals was called "*Sokka*", the main canal was named "*Arna*", the distributive canal was "*Yap, yab*", a network or branch of a canal was called "*bedaklar*", and the canals providing water to the fields were named "*solmalar*". The first three terms such as Sokka, Arna and Yab are still in use, and are found in today's names of canals such as Tashsokka, Daryalik-Arna and Shikhyab.

In the 17th century, organization and maintenance of this highly sophisticated irrigation system needed skilled and disciplined professionals and users. The local officials became responsible for the organization and supervision of irrigation works. In the *Khan* period, the officials were elected by the community for organization and control of irrigation works. For example, in the epoch of Feruz Khan (fourth quarter of XIX—beginning of XX centuries) the secretary of the *Khan*'s court, the poet Ogakhi, was responsible for the irrigation management. Ogakhi was named the "main Mirob (responsible for maintenance of canals) of the State".

In Kadirov's view, the application of self-financing and self-governance principles in organization of irrigation works such as water use, maintenance, establishment of water or irrigation districts for improvement of water use as well as the transfer to a democratically selected board of representatives, are all practices, which were inherited from the far or near past.

The term of Water Users Association (WUA) might be new for Uzbekistan, but not the principle of the common use, governance and management of water resources. In the 19th as well as in the beginning of the 20th centuries, such institutions like *arykaksakals, mirobs* and *tuganchi* had been widespread in Central Asian countries and especially in the territory of Uzbekistan (Kadirov 1998). *Arykaksakals* dealt with water supply from the head to the tail-end of the canal ("*aryk*"). They knew exactly what user, when and how much water was necessary to be supplied. *Arykaksakal* was responsible for the main (large) "*aryk*", which irrigated the land of a few villages.

Mirob is, according to ICARDA (2002) as well as O'Hara and Hannan (1999), an official in charge of water distribution for irrigation and is a key figure in the operations and maintenance of the irrigation system and water distribution according to the agreed arrangement. The inspection of the system and pertinent structures is carried out periodically by peasants and the *mirob*. Although regular maintenance is scheduled to be carried out periodically when sufficient self-help labour is available, major repairs cannot be carried out without governmental support.

Some *mirobs,* under the direction of *arykaksakal,* were responsible for their canals. The promotion to the post of *arykaksakals* occurred among *mirobs;* the most enterprising and competent *mirob* was selected as an *arykaksakal* (Kadirov 1998). *Arykaksakals* and *mirobs* were on one hand the title of a post and on the other hand it was a status of knowledge and experience of the person. With time, the posts of *arykaksakal* and *mirob* were transferred from father to son by right of succession in some districts. This way, dynasties of *arykaksakals* and *mirobs* were established. However, the public control of their activities and assessment of their work persisted (Kadirov 1998).

Tuganchi literally translated as "who constructs dams or dykes" were responsibe for the construction, operation and maintenance of canals (Kadirov 1998; Bazarov et al. 2007). *Tuganchis* used their practical knowledge and experience for constructing simple nomadic water intake structures by using local materials such as wood, stones etc.

Besides the institutions of *arykaksakals, mirobs* and *tuganchi,* the small territorial village associations for the water distribution from a water source to users played an important role. The name of these associations varied from region to region. So, for example, in Khorezm they were called *"djabdi"* (equipper), in Zarafshan valley—*"kush"* (double, twin, pair of ox) and in Fergana valley—"ketmon" (kind of mattock) (Kadirov 1998).

All regional irrigation associations had a leader or *aksakal* (*djabdiboshi —main djabdi, kushboshi—main kush, ketmonboshi—main ketmon*) who collaborated with *mirobs.* These leaders or *aksakals* helped in organizing *khashars* (voluntary, collective work done for the sake of the common good) and were responsible for conducting them. However, according to Kadirov there is no information in the literature about the responsibility of *aksakals* to higher level organizations (Kadirov 1998).

The *Khashar,* which was included into duties of the central authorities, played an extremely important role in maintenance and development of the irrigated facilities. In Central Asia, the population, by way of labour duty, carried out the repairs and services, such as construction of new channels, clearing of heads of main channels, construction of dams, and coastal dams.

Works and distribution of irrigating water between water-users were carried out between small territorial rural communities of water-users as a part of *Khashar*. During the *Khashar* period, each community of water-users did not depend on one person. The collective maintenance of irrigating systems and water-intake constructions in serviceability, and also the maintenance of water administration demanded an establishment of the strict account of irrigated water and its distribution between water-users.

The *Khashar* as a social element contributes to the predictability of resources supply as well as indicates the shared norms of peasants. One can argue whether such behaviour was voluntarily or forced. The *Khashar* is a practice that is in keeping with Uzbek tradition and characterizes the Uzbek community over generations. The *Khashar* survived the Soviet period and is still an essential element, e.g., in irrigation management.

The goal of above mentioned small territorial village associations' work was to supply and distribute irrigation water together with users at self-sufficient basis (Kadirov 1998).

As in every feudal system in Khorezm, peasants had two types of duties against landowners: corvee (in form of labour) and tribute (in kind). Kadirov (1998) describes the corvee as follows:

> *Uzbek ancestry, who lived in Central-Asian region, since emancipation from slavery, learned how to implement large-scale and labour-intensive irrigation works such as digging of canals with the length over many tens and thousands kilometres, their yearly cleaning and maintenance, construction and maintenance of different devices from local materials jointly. These joint activities were called "Khashar"*

Undoubtedly, these activities could be organized only where members understood community interests well, are disciplined, and fulfil instructions of work organizers (*arykaksakals, mirobs and tuganchi*) clear and quickly. Kadirov (1998) stresses that 'since "*Khasharchi*" generally were ordinary peasants, certain human characteristics such as decency, diligence, self-discipline, call of duty and feeling of solidarity were needed'. Pokrovskii (1927) mentioned:

> *....the nations of Central Asia, before the occupation of the land by the Russian government had a formed land and water regime, which was built on customs and traditions. The field of application was restricted by tight limits of tribal and intertribal interaction as well as water-land communities. Nevertheless, we encounter a heterogeneous water regime due to the influence of heterogeneous customs (Pokrovskii 1927).*

The above-mentioned examples show that in Uzbekistan and particularly in Khorezm peasants were quite likely committed to the idea of the collective user-based water management.

This ability for collective action characterized the Pre-Soviet period of the *khorezmian* irrigation management. At that time, irrigation was labour intensive, based entirely on group effort. The irrigation method was gravity/ surface irrigation (Fig. 16.2).

Figure 16.2 Chigirs (Fedorovich 1945).

Soviet Period: Stages of Irrigation and Land-reclamation Constructions

Since little literature about irrigation development in Khorezm is available, this overview will start from 1929. This was the period of the beginning of Soviet power and was characterized by technical and managerial changes.

In 1935, the chigirs (see Fig. 16.2) started to be replaced by pump irrigation. In 1937, in the major part of the region gravity irrigation was widespread. With the introduction of pump irrigation the expenses for irrigation of Khorezm were economized (Atashev et al. 1966)

Cleaning of the canals also started to be mechanized in 1935. By 1940 the total volume of cleaning works in the canals was reduced due to the construction of control devices in the irrigation systems that improved the hydro-geological conditions of the oasis.

The realization of large reconstruction projects of the South Khorezmian irrigation system between 1932 and 1941 allowed the transition to gravity irrigation (Morgunenkov and Poslavski 1932–1941 cited in Khamidov 1993). This led to the extension of irrigation land, the increase of the land use coefficient as well as the increase of agricultural production (Khamidov 1993). With the transition to gravity irrigation, the water losses have increased due to infiltration which led to a raised groundwater table and increased soil salinity (Ibrakhimov 2005). In order to prevent the secondary salinization, the construction of a drainage system was started by the end of 1941 (Khamidov1993). The systems were equipped with modern hydro-technical structures, high-powered mechanisms; the structure and staff of operation services were re-organized. Most of the Uzbek vast network of irrigation and drainage (I&D) systems were built during the 1960s–1980s.

To link these activities with the present, the large-scale irrigation system of Uzbekistan was neither extended nor rehabilitated during the last 30 yr. This lack of maintenance of the irrigation networks hampers current irrigation management on the different levels starting from national through regional to local (Fig. 16.3),

Beside the technical changes, new elements of participation in water management by the local level were (re)introduced in the Soviet "format" of semi-voluntary associations.

Figure 16.3 Condition of irrigation network in Khorezm, 2003 (Own presentation).

In the early 30's of the XX century, peasants (dekhkans) were urged to contribute to the reconstruction and rehabilitation of irrigation systems (Rikunov 1923). Logically, there was a need to organize peasants in collective associations of water issues. This form of organization was quite understandable to the peasants. Rikunov wrote in 1923:

> *In Turkestan the population has got used to collective labour over the centuries, because a cleaning of canals, maintenance of irrigation heads, digging of new canals is beyond a small peasant's power. This requires joint efforts of the whole village.*

In this way, land reclamation associations were established.

Land reclamation associations do not relate to cooperative organizations in the true sense of the word (Kilichevski 1927). Half-forced membership and conferment of same functions with public-legal character distinguish land—reclamation associations and cooperatives. At the same time, water and land reclamation associations in Central Asia get extraordinarily close to the patterns of life and thought of local population, meet solid, traditionally established relations (Kilichevski 1927).

A water- and land reclamation association was created when two thirds of peasants decided to establish this organization. These associations were composed of juridical persons, who could take out credit and were financed through user fees. However, during the collectivization in the late 30s, these water- and land reclamation associations collapsed.

After the breakdown of the Soviet Union, an interest in co-water management was rekindled and considered both the above mentioned practices: ancient experience, where whole water use and organization of water supply have been based on tribal and intertribal relationships as well as failed attempts to establish and to keep functioned water- and land reclamation associations.

Post-independence: Irrigation Management Transfer and Water Users' Association

Since 1990, one of the first endeavours of the government and ministry officials of the Republic of Uzbekistan has been an adjustment of water law to the political changes. By that time, the Republic of Uzbekistan did not belong to the Soviet Union and had some experience in independence. The first reforming steps have increasingly devoted themselves to the change and development of the agricultural sector. Huge agricultural enterprises-kolkhozes and sovkhozes were reorganized into three types of rural enterprises: collective farm-*shirkat*, semi-private entity-farmer (*fermer*, in local language) and peasant farm-*dekhkon*. The principal difference is shown in the Fig. 16.4.

While until 1998 the reforms were aimed at land issues (privatization), in 2003 the changes in agricultural sector got a new focus on water resources. Principles such as irrigation management transfer, and integrated water management have been given great importance.

In the "Program of development of land-reclamation of the Republic of Uzbekistan till 2010" it is stressed that WUAs are the most widespread and effective form of association of land users in different countries (MAWR 2001). These users exploit and maintain irrigation systems collectively.

The establishment of WUAs and its development in the Republic of Uzbekistan is one of the main issues for the conservation of existing irrigation infrastructure, involving farmers in the process of rational use of irrigation water.

Initiators of the establishment of WUAs in the Republic of Uzbekistan have been the Ministry of Agriculture and Water Resources (MAWR) of Uzbekistan as well as international organizations such as International

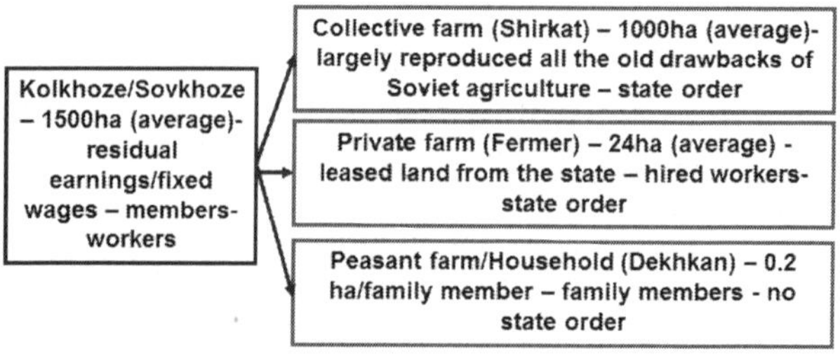

Figure 16.4 Enterprise restructuring in Uzbekistan in 1990.

Water Management Institute (IWMI), World Bank (WB), Asian Development Bank (ADB), and United States Agency for International Development (UsAID). The WUAs which have been established under the aegis of the MAWR have been based on state agricultural enterprises. Because of unprofitability, these enterprises have been eliminated by the state. By the establishment of the majority of WUAs, the government of the Republic of Uzbekistan passed special resolutions, which have broken up agricultural enterprises and to the creation on their territory farms as well as different subsidiary sub-units including WUAs. A draft of a WUA by-law was suggested by the Ministry of Agriculture and Water Resources. The "Program of development of land-reclamation of the Republic of Uzbekistan till 2010" emphasized that "however, the main questions, which referred to financial and economic activity of WUA, have not been elaborated and determined till now" (MAWR 2001).

The Khorezm region is a pioneer in the field of the complex formation of the new form of water governance-WUAs, since first Uzbek WUAs were organized in Khorezm in 2000 on the basis of the liquidated and unprofitable shirkats. The organizational setup of WUAs on the basis of abrogated shirkats is referred to as an "administrative-territorial" form of WUA.

According to Bocharin and Ergashev (2004) "the establishment of WUAs in Uzbekistan before 2000 was not expanded further due to the specific conditions of development of *fermers*" (Bocharin and Ergashev 2004). The role of *fermers* was marginal at that time, in terms of either their number or their land area in each *shirkat*. Being only the secondary water users after *shirkats*, *fermers* were already provided with maintenance of irrigation and drainage at the level of on-farm system, because the operation and maintenance was carried out by the *shirkat* as the primary water user. It was also difficult to organize this for the *fermers*, because their territories were located as enclaves dispersed through the *shirkats* area, and they did not have a common goal such as managing any united water thoroughfare.

The resolution of Cabinet of Ministers in 1999 concerning the dismantlement of non-profitable *shirkats* and distribution of their land to the *fermers* initiated scaled liquidation of unprofitable *shirkats*. WUAs became the dominant form of operation of on-farm irrigation systems.

A WUA was conceived as a *fermer*-run organization. However, for the beginning the state as an initiator took the full responsibility for the establishment and first run of the WUA.

In order to establish WUAs, the local administration convokes a general assembly of potential WUA members, where they explain the goals, objectives and tasks of WUAs. In accordance with a decision of the general assembly, the commission of WUA establishment (later, commission) is created. This commission has different tasks such as carrying out the explanatory work among potential WUA members, organization of the inventory of means of dismantled *shirkat*, drawing up a transfer balance, and preparation of constituent documents. The water users are obligated to pay user fees for services provided on water distribution. However, WUAs suffer chronically from non-payments which are essential for a WUA as non-commercial organization. There are also weak contractual patterns: neither WUA administration nor *fermers* as WUA members abide with the contract. The third problem is acceptance of WUA by members. The members in Uzbekistan equate WUAs with fees or taxes collectors that do not improve water management structure.

As a result, the established WUAs do not perform their functions in full measure.

New Wave of the Reforms in Irrigation Management in 2003

A new wave of the reforms in agrarian sector of Uzbekistan started in March of 2003. The beginning of the reforms was initiated by the presidential decree from 24.03.03 № УП-3226 "On the most important extension directions of reforms in agriculture".

In the decree № УП-3226 the main emphasis was placed on a development of fermers who should become the main producers of agricultural production in the future (Cabinet of Uzbekistan 2003a). The roles of newly established players on the national, provincial and district levels were defined and specified.

In the decree № УП-3226, the focus laid on fundamental review of management systems of agricultural production keeping in view the liberalization of MAWR from distributive functions and refusal from administrative-command governance methods of agriculture.

According to Taksanov "such forms of 'governance' remained from the Soviet time and carried negative impacts on management of agricultural systems, often breaking whole reform processes of agrarian sector" (Taksanov 2003).

In the decree № УП-3226, the main goals of MAWR were determined. One of the goals was formulated as follows: water resources management providing transition from administrative-territorial to basin principle of irrigation systems, and also introduction to all levels of market principles in irrigation water use (Cabinet of Uzbekistan 2003a).

In pursuance of the presidential decree № УП-3226 and in an effort to radically improve the management of systems of agricultural production in accordance with the market economy requirements, the organizational structure of MAWR was revised (Cabinet of Uzbekistan 2003b).

The reforms gathered up speed and one month later the next regulation was passed. In pursuance of decree № УП-3226 and with the view to transfer from administrative-territorial to basin principle of management of irrigation systems, the resolution of Ministry of the Republic of Uzbekistan from July 21, 2003 No. 320 "The improvement of water management organisation" was passed (Cabinet of Uzbekistan 2003c). In the new structure basin management boards of irrigation systems (BUIS), management boards of main canals as well as the management boards of irrigation systems (UIS) were presented for the first time. Ten BUISes were established based on existing structures of water management.

The establishment of two-level management system for the national water resources through BUISes and WUAs became a most important component of water reform in Uzbekistan.

The Government of Uzbekistan has decided to reorganize water management by the hydrological principle. A new type of WUAs was established. The hydrographic WUA is the unification of *fermers* considering the location of irrigated area and the canals; users obtain water from the same canal. In the period from 2005 to 2006 all remaining *shirkats* throughout the country were dismantled and WUAs established in their place. The WUAs were again established with administrative boundaries, i.e., largely with same boundaries as the former *shirkats* (Veldwisch 2008). Issues of concern with respect of WUA functioning such as problems with fee collection, participation of members in the management of a WUA, whether within administrative-territorial or hydrological boundaries, remained.

After all these reorganizations, the foci of reforming become land reclamation for water saving purposes. Today, for the successful implementation of the reclamation and improvement of efficient use of irrigated land in the country, there are a number of ordinances and decrees of the President of the Republic of Uzbekistan, as well as documents of the Government regulating this activity. In particular, the decree of the President "On measures to comprehensively update the land reclamation system" is an important guide for action in this direction, improvement of irrigated lands.

In accordance with the decree, in October, 2007 a Fund for Ameliorative Improvement of the Irrigated Land was established. The Fund's main tasks are improvement and cleaning of drainage system, land planning as well purchasing of equipment needed for above mentioned activities. Therefore the State allocates annually US$ 100 million . Construction works of new irrigation canals is not planned and not conducted. The Fund for Ameliorative Improvement of the Irrigated Land rehabilitates and reconstructs existing systems, which lead to water saving.

Conclusions

There were governmental and donors attempts to embody the idea of self-management practices used in ancient Uzbekistan into the present water user associations. The initiators of WUAs establishment in Uzbekistan followed a goal to find such prototype of farmer-driven self-sufficient organization in the local literature.

The central theme of the historical background of *khorezmian* irrigation management is, besides the topic of self-management characteristics, the transition to a new type of irrigation (surface, mechanized maintenance, centralized management, the USSR concept) from the previous labour intensive system. The Soviets introduced technical and managerial changes. These days a new transition has begun institutional changes on the management side.

In the literature on irrigation management in Central Asia during the Soviet time, it is hard to find descriptions about the role of water users in the irrigation management process. However, those historical sources showed that many traditions have disappeared during the Soviet period. Nevertheless, there is a potential to reanimate or at least refresh and introduce old elements of self-organized water administration. The reanimation or, in other words, reviving of knowledge of old traditional basis of irrigation management might be helpful in the establishment of new structures of irrigation management in Uzbekistan. There is a need for anthropological research in irrigation management.

Besides this, the role of water users increases with the occurrence of WUAs. The transformation of irrigation management from administrative to hydrographical approach has advantages and obstacles. The advantages of the hydrographical principle are the management according to irrigation canals on the one hand, and strengthening of public activeness and mobilization on the other hand. The administrative-territorial principle had already established social networks that were created over many generations of peasants. The blind establishment of new water management structures such as WUAs destroyed these social networks physically, but they remain advisory and trustworthy structures for the former members of the community. The reforms in agriculture and water management should be carried out with intensive capacity building of involved stakeholders using, for example, regular meetings, and a signed contract between WUA and its members. While stakeholder capacity building is important, it is probably not wise to neglect the technical side of transformation such as infrastructure rehabilitation as well as introduction of monitoring infrastructure (e.g., discharge measurement devices).

Acknowledgements

This research was conducted under the ZEF/UNESCO Khorezm project "Economic and Ecological Restructuring of Land- and Water Use in the Region Khorezm (Uzbekistan): A Pilot Project in Development Research" funded by the Federal Ministry of Education and Research (BMBF), Germany (Project Number 339970D) and supported by the European Commission (Contract No. 511179—NEWATER).

The author would like to thank colleagues from the ZEF/UNESCO project for comments and discussions and Uzbek colleagues, Bakhtiyor Babajanov, Nodira Khodjaeva for their research support.

References

Atashev, A., A. Rachinsky and G. Horst. 1966. Vodnoe hozyastvo i meliorativnoe stroitel'stvo v Khorezmskoi oblasti. Fan, Tashkent, Uzbekistan (in Russian).

Bazarov, D., Ju. Kazbekov and Sh. Rakhmatullaev. 2007. History and present condition of urban water supply system of Tashkent city, Uzbekistan. *In:* P. Juuti, T. Katko, H. Vuorinen (eds.). Environmental History of Water- Global views on community water supply and sanitation. IWA Publishing, Alliance House, London, UK, pp. 213–221.

Bocharin A. and I. Ergashev. 2004. Problema integrirovaniya upravleniya, razional'nogo ispol'zovaniya i ohrani vodnyh resursov v basseine Aral'skogo morya Zadanie 02 "Razrabotat' meropriyatiya po sovershenstvu upravleniya i kompleksnomu ispolzovaniyu vodnih resursov na transgranichnyh potokah" Tema 02.07 "Razrabotat novie i usovershenstvovat sushestvuyushie dokumenti vodnogo prava gosudarstv CA, vklyuchaya status mezhgosudarstvennih i nazional'nih ob'ektov" (dlya Respubliki Uzbekistan). Annual report 9029. SANIIRI, Tashkent, Uzbekistan.

Cabinet of Uzbekistan. 2003a. O vajneishikh napravleniyah uglubleniya reform v selskom khozyaistve. Presidential decree № УП-3226 (in Russian).

Cabinet of Uzbekistan. 2003b. O Sovershenstvovaii Organisazii Deyatelnosti Ministerstva Selskogo I Vodnogo Hozyaistva Respubliki Uzbekistan. No. 290 from 28.06.2003 (in Russian).

Cabinet of Uzbekistan. 2003c. O Sovershenstvovanii Organisazii Upravleniya Vodnym Khozyaystvom. No. 320 from 21.07.2003 (in Russian).

Fedorovich, B. 1945. Lik pustini. Molodaya Gvardia, Tashkent, Uzbekistan (in Russian).

Forkutsa, I. 2005. Modelling water and salt dynamics under irrigated cotton with shallow groundwater in the Khorezm region of Uzbekistan. PhD thesis. University of Bonn, Germany.

Global Views on Community Water Supply and Sanitation. IWA Publishing.

Hillel, D. 1992. Out of the Earth. Civilization and the Life of the Soil. University of California Press, Berkley, Los Angeles, USA.

Ibrakhimov, M. 2005. Spatial and temporal dynamics of groundwater table and salinity in Khorezm (Aral Sea Basin), Uzbekistan, PhD thesis. University of Bonn, Bonn, Germany.

[ICARDA] International Center for Agricultural Research in the Dry Areas. 2002. Needs assessment on soil and water in Afghanistan, Aleppo, Syria.

Kadirov, A. 1998. Ozbekison Irrigaciyasi tarihidan lavhalar (XIX asr ortalaridan 1920 yillargacha) Adbulla Kodirii Nomidagi Halk Merosi Nashrieti ,Tashkent, Uzbekistan (in Uzbek).

Khamidov, M. 1993. Nauchnie osnovi sovershenstvovaniya vodoispol'zovaniya na oroshaemih zemlyah Khorezmskogo oasisa, Ph.D. thesis. Uzbekskaya Akademiya Sel'skohozyaistvennih nauk, Tashkent, Uzbekistan (in Russian).

Kilichevski, V. 1927. Rayoni vodno-meliorativnikh tovarishestv' Vestnik Irrigazii 1:15 (in Russian).

[MAWR] Ministry of Agriculture and Water Resources of the Republic of Uzbekistan. 2001. Program of development of land-reclamation of the Republic of Uzbekistan till 2010, part 7: Restructuring of agricultural enterprises and establishment of WUA on irrigation canal systems, (Governmental paper) (in Russian).

O'Hara, S.L. and T. Hannan. 1999. Irrigation and Water Management in Turkmenistan: Past Systems, Present Problems and Future Scenarios. Europe-Asia Studies 51(1): 24.

Pokrovski, E. 1927. Razvitie vodnogo prava v Srednei Asii.Vestnik Irrigazii 1:10 (in Russian).

Rikunov, M. 1923. Organisuyte vodnye tovarishestva! Vestnik Irrigazii 1:3 (in Russian).

Taksanov, A. 2006. Farmer movement in Uzbekistan. Association of Middle East Studies (AMES) 1–5.

Tolstov, S. 2005. Following the Tracks of Ancient Khorezmian Civilization. UNESCO, Tashkent, Uzbekistan.

Veldwisch, G.-J. 2008. Cotton, Rice & Water: Transformation of Agrarian Relations, Irrigation Technology and Water Distribution in Khorezm, Uzbekistan. PhD thesis. University of Bonn, Germany.

Index

About the Editors

Velma I. Grover is an Adjunct at the Arcelor Mittal DOFASCO School of Engineering and Public Policy at McMaster University in Hamilton, Canada, and an Adjunct Professor at the Faculty of Environmental Studies at York University in Toronto. Velma has worked with nonprofit organizations, consulting companies and international organizations mainly in the field of water and waste management. Some of the research appointments include: visiting fellow at Fudan University, China; visiting fellow at Smith School of Enterprise and the Environment, Oxford University and Kalmar University, Sweden.

She has a Masters in Environmental Management and a Ph.D. from University of London, UK. Velma has published on various environmental management issues, including seven edited volumes.

Gail Krantzberg is Professor and Director of the Dofasco Centre for Engineering and Public Policy in the School of Engineering at McMaster University offering Canada's first Master's Degree in Engineering and Public Policy. Gail completed her M.Sc. and Ph.D. at the University of Toronto in environmental science and freshwaters. She worked for the Ontario Ministry of Environment from 1988 to 2001, as Coordinator of Remedial Action Plan and Great Lakes Programs, and Senior Policy Advisor on Great Lakes. She is a past president of the International Association of Great Lakes Research. Dr. Krantzberg was the Director of the Great Lakes Regional Office of the International Joint Commission from 2001 to 2005. She has authored more than 100 scientific and policy articles on issues pertaining to ecosystem quality and sustainability and is a frequent speaker to media and the public.

Color Plate Section

Chapter 3

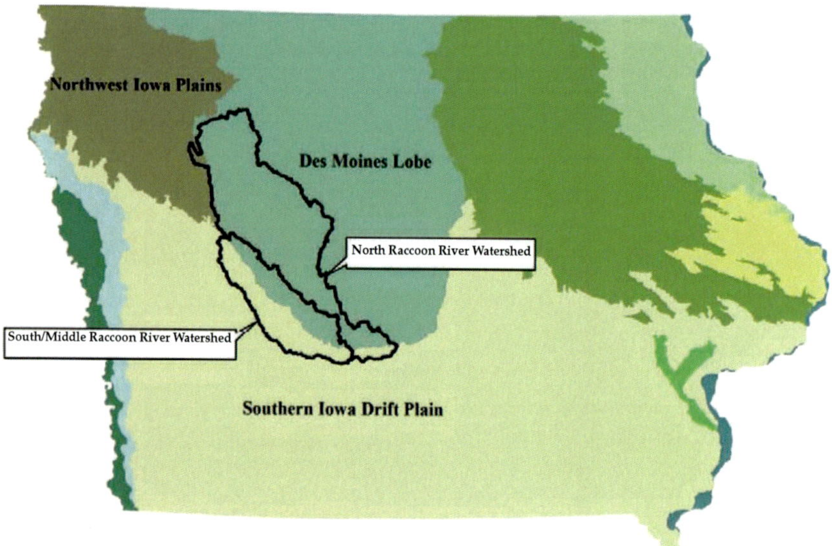

Figure 3.1 Location within Iowa and land forms.
Source: Raccoon River Masterplan 2011

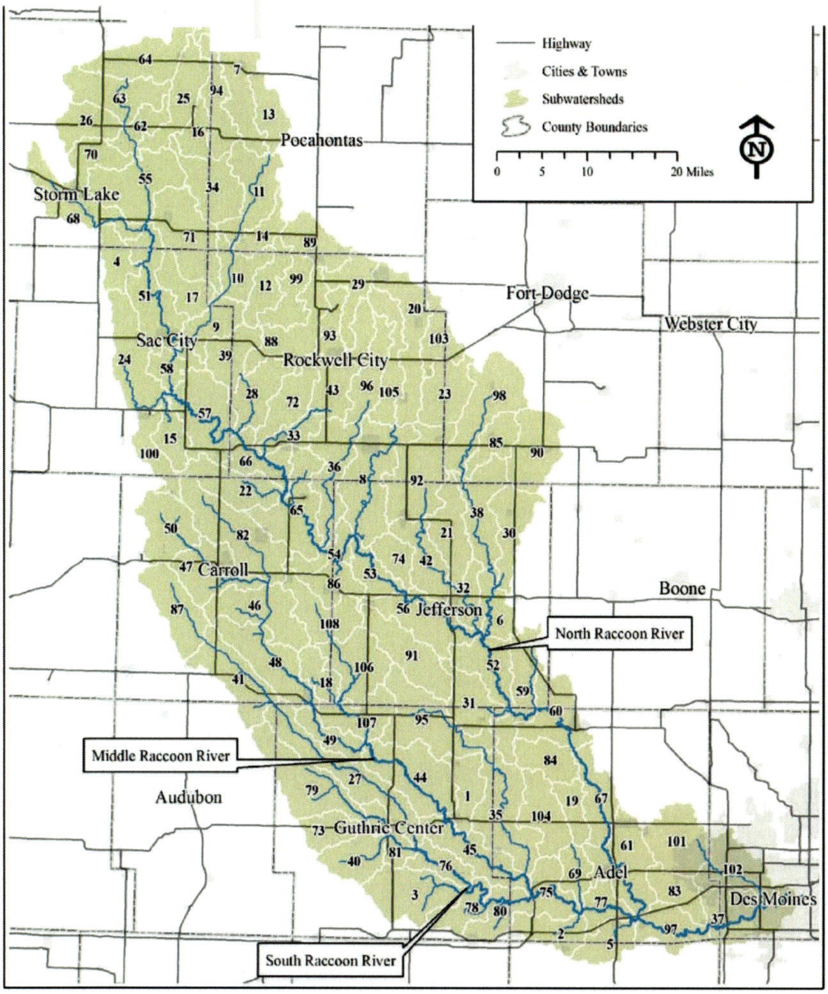

Figure 3.2 Raccoon River Watershed.

Source: Raccoon River Masterplan 2011

Chapter 11

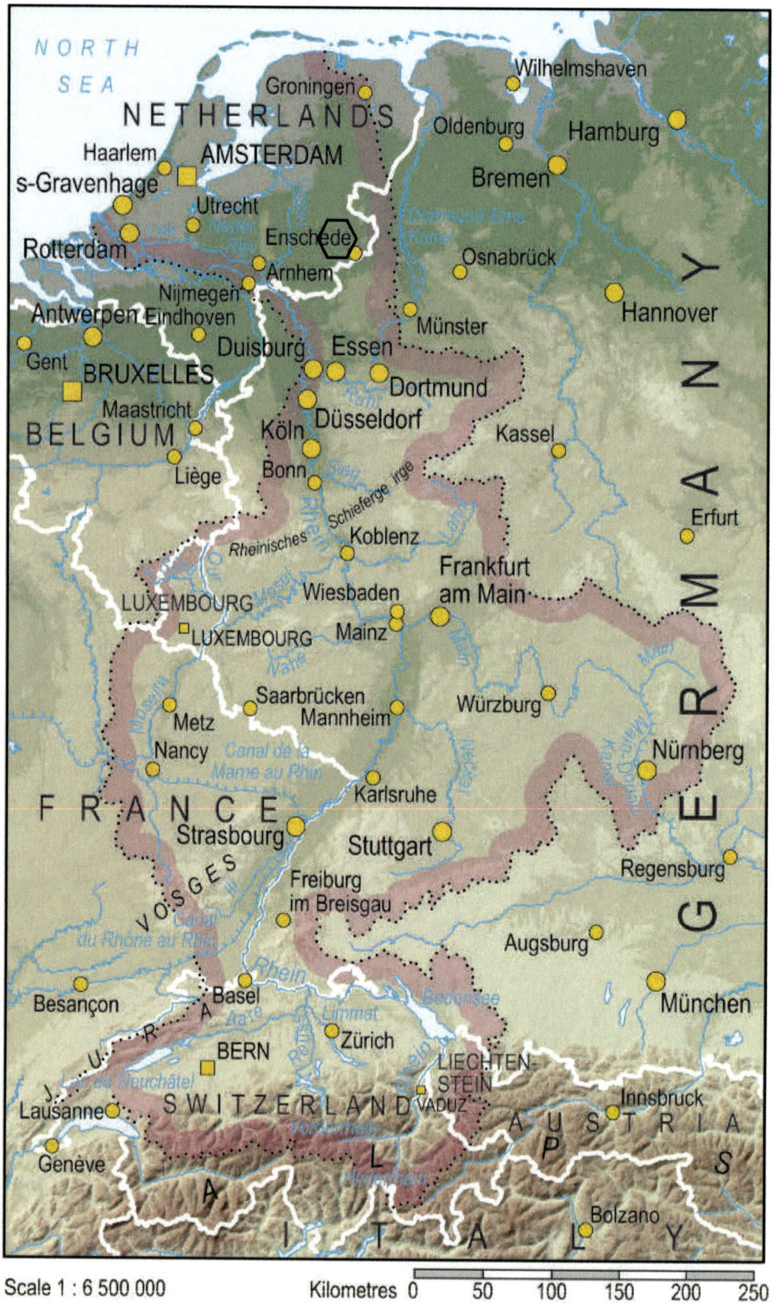

Figure 11.1 Rhine basin (Source: UNEP DEWA/GRID)—case study area indicated by hexagon.

Chapter 12

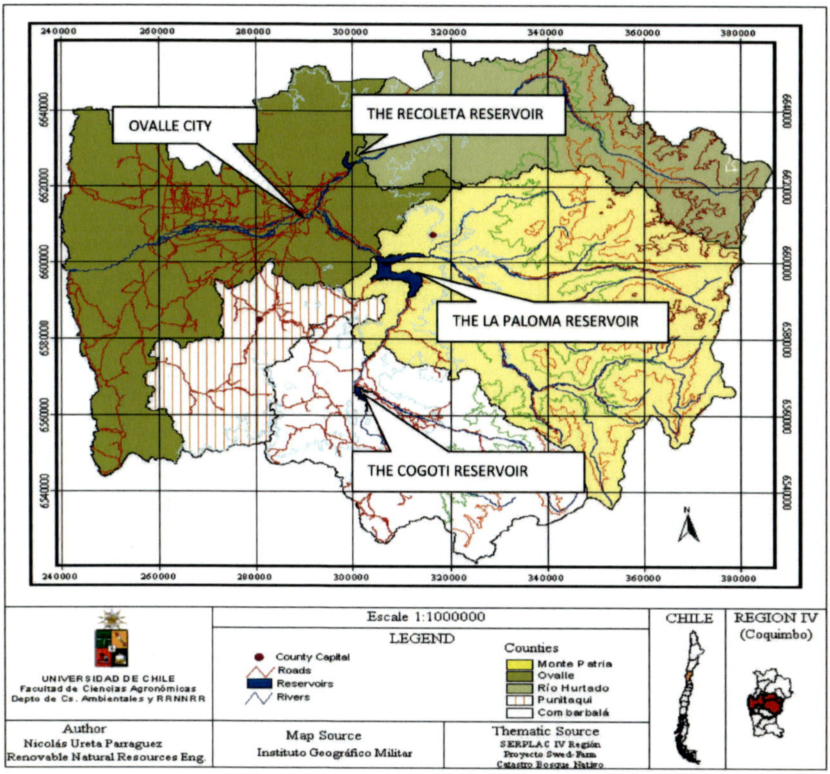

Figure 12.1 The Limarí Province and the Limarí River basin (adapted from Fuster 2006).

Chapter 15

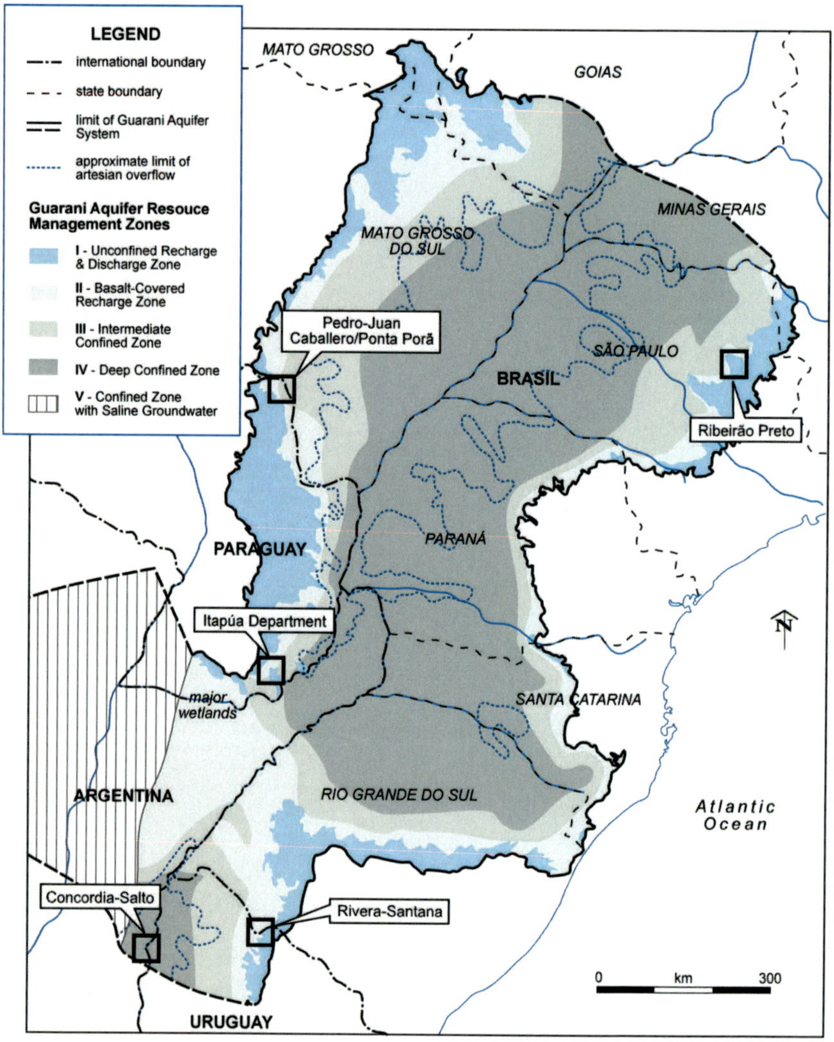

©World Bank/GWMATE - Case Profile No. 9 The Guarani Aquifer Initiative – Towards Realistic Groundwater Management in a Transboundary Context

Figure 15.1 Guarani Aquifer System—Management Zones.